'Cultural profiles play a crucial and often underestimated role not only in the science, but also in the practice of risk and crisis management. This discussion can't be limited to the large scale patterns between societies and continents, but should include the diversities within a society or geographical region, creating awareness on the range of perspectives, attitudes and priorities within different societal sectors. It is one of the major strengths of this very timely book that it addresses the many ways and levels in which cultures can influence disaster risk.'

Professor Jakob Rhyner, UNU-EHS Bonn

'At last . . . the long-awaited, definitive book that explores the complexities of culture and disaster risk. *Cultures and Disasters* is a highly ambitious work that interprets the subtle dimensions of the management and reduction of risk. The authors consider diverse questions that include the philosophy of risk, politics and power relationships, local knowledge and learning, social divisions, religion, celebrity culture, coping strategies and adaptive behaviour. Researchers, students and practitioners will all benefit richly from this splendid work produced by five eminent editors and over twenty international experts in the field.'

Ian Davis, Visiting Professor in Disaster Risk Management in Copenhagen, Lund, Kyoto and Oxford Brookes Universities

CULTURES AND DISASTERS

Why did the people of the Zambezi Delta, affected by severe flooding, return early to their homes or even choose not to evacuate? How is the forced resettlement of small-scale farmers living along the foothills of an active volcano in the Philippines impacting on their day-to-day livelihood routines? Making sense of such questions and observations is only possible by understanding how the decision making of societies at risk is embedded in culture, and how intervention measures acknowledge, or neglect, cultural settings. The social construction of risk is being given increasing priority in the understanding of how people experience and prioritise hazards in their own lives and how vulnerability can be reduced, and resilience increased, at a local level.

Cultures and Disasters adopts an interdisciplinary approach to explore the cultural dimension of disaster with contributions from leading international experts in the field. Part I provides a discussion of research in both theoretical and practical considerations to better understand the importance of culture in hazards and disaster management. Culture can be interpreted widely with many different perspectives; this enables us to critically consider the culturally-bound nature of research itself as well as the complexities of incorporating various interpretations into disaster risk reduction (DRR). If culture is omitted from consideration, related issues of adaptation, coping, intervention, knowledge and power relations cannot be fully grasped. Part II explores what aspects of culture shape resilience and how people use culture in everyday life to establish DRR practice. It looks at what constitutes a resilient culture and what role culture plays in a society's decision making. It is natural for people to seek refuge in tried and trusted methods of disaster mitigation; however, cultural and belief systems are constantly evolving. How these coping strategies can be introduced into DRR therefore poses a challenging question. Finally, Part III examines the effectiveness of key scientific frameworks for understanding the role of culture in DRR and management. DRR includes a range of norms, and breaking these through improved cultural understanding will challenge established theoretical and empirical frameworks.

Fred Krüger is Professor of Geography at the University of Erlangen-Nürnberg, Germany.

Greg Bankoff is Professor of Modern History at the University of Hull, UK.

Terry Cannon is a Research Fellow at the Institute of Development Studies, UK.

Benedikt Orlowski is a Research Fellow at the Institute of Geography, University of Erlangen-Nürnberg, Germany.

E. Lisa F. Schipper is a Research Associate at the Overseas Development Institute, London, UK, and Associate at the Stockholm Environment Institute, Sweden.

ROUTLEDGE STUDIES IN HAZARDS, DISASTER RISK AND CLIMATE CHANGE

Series Editor: Ilan Kelman, Reader in Risk, Resilience and Global Health at the Institute for Risk and Disaster Reduction (IRDR) and the Institute for Global Health (IGH), University College London (UCL).

This series provides a forum for original and vibrant research. It offers contributions from each of these communities as well as innovative titles that examine the links between hazards, disasters and climate change to bring these schools of thought closer together. This series promotes interdisciplinary scholarly work that is empirically and theoretically informed, with titles reflecting the wealth of research being undertaken in these diverse and exciting fields.

Published:

Cultures and Disasters
Understanding cultural framings in disaster risk reduction
Edited by Fred Krüger, Greg Bankoff, Terry Cannon, Benedikt Orlowski and E. Lisa F. Schipper

Forthcoming:

Recovery from Disasters
Ian Davis and David Alexander

CULTURES AND DISASTERS

Understanding cultural framings in disaster risk reduction

Edited by
Fred Krüger, Greg Bankoff,
Terry Cannon, Benedikt Orlowski
and E. Lisa F. Schipper

Routledge
Taylor & Francis Group

LONDON AND NEW YORK

First published 2015
by Routledge
2 Park Square, Milton Park, Abingdon, Oxon OX14 4RN

and by Routledge
711 Third Avenue, New York, NY 10017

Routledge is an imprint of the Taylor & Francis Group, an informa business

British Library Cataloguing in Publication Data
A catalogue record for this book is available from the British Library

Library of Congress Cataloging in Publication Data
A catalog record for this book has been requested

ISBN: 978-0-415-74558-1 (hbk)
ISBN: 978-0-415-74560-4 (pbk)
ISBN: 978-1-315-79780-9 (ebk)

Typeset in Bembo
by Florence Production Ltd, Stoodleigh, Devon, UK

'Aftershock' by Haitian artist S. Aubane is a striking painting depicting the disastrous effects of the 2010 Haiti earthquake. This catastrophic 7.0 magnitude event killed tens of thousands of people, displaced over a million and caused major damage to buildings and infrastructure. It has also left countless victims traumatised and with little means to survive without assistance by 'outside' relief agencies – these personal aftershocks and their cultural embeddedness are often overlooked in technocratic disaster recovery efforts.

CONTENTS

FIGURES

TABLES

CONTRIBUTORS

David Alexander is a Professor at the Institute for Risk and Disaster Reduction at University College London, UK. He works in emergency planning and management, and also conducts research on earthquake epidemiology and theoretical models of disaster risk reduction.

Greg Bankoff is Professor of Modern History at the University of Hull, UK. His research to date has focused on disaster risk adaptation and reduction, societal vulnerability and resilience, and resource management. With the editors of this book, he has also co-edited the IFRC World Disasters Report 2014 on Culture and Risk.

Jörn Birkmann is Professor of Spatial Planning and the Environment, and Director of the Institute of Regional Development Planning at the University of Stuttgart (Germany). He researches natural hazards, integrated adaptation strategies, global disaster response and reconstruction, vulnerability assessment in coastal and flood-prone communities, and linking disaster risk reduction and adaptation to climate change.

Terry Cannon is a Research Fellow at the Institute of Development Studies, UK. His research interests include vulnerability analysis and methodologies, problems with the resilience approach, adaption at the local level and problems of the concept 'community', rural livelihoods, and disaster vulnerability and climate change adaption. He is co-author with Wisner, Blaikie and Davis of *At Risk: Natural Hazards, People's Vulnerability and Disasters*. With the editors of this book, he has also co-edited the IFRC World Disasters Report 2014 on Culture and Risk.

Brian R. Cook is an Assistant Professor in the School of Geography at the University of Melbourne, Australia. His work has looked at risk and human vulnerability, human-environment relations, science and technology studies, and water management, typically exploring expert knowledge and decision making through analyses of scientific and local knowledge production.

Andrew Crabtree is an Adjunct Associate Professor at Copenhagen Business School, Denmark. His research has covered climate change, flooding vulnerability and psychosocial issues, sustainable development indicators and ethics.

Kate Crowley works as a risk engineer at NIWA in New Zealand. She advises on the impact of natural hazards on society by exploring aspects of disaster management such as social vulnerability and risk and how this can be addressed through educational outreach and science communication programmes.

Georg Fiedler is a PhD student at the Institute for Biology and Environmental Sciences at Carl von Ossietzky University, Germany. He is a member of a working group for Applied Geography and Environmental Planning.

Maureen Fordham is Professor of Gender and Disaster Resilience in the Department of Geography at Northumbria University, UK. Her research focuses on sustainable hazard and disaster management, the impact of gender and age group on disaster risk, vulnerability and capacity analysis, community based disaster risk reduction, flood warning systems, environmental perception and management, and policy analysis.

Leberecht Funk is a PhD student and Research Associate at the Institute of Social and Cultural Anthropology, Freie Universität Berlin, Germany. His research focuses on emotion, childhood, socialisation, personhood and animism.

JC Gaillard is Associate Professor at the School of Environment, the University of Auckland, New Zealand. His research looks at disaster risk reduction with a particular focus on ethnicity, gender minorities, prisoners and homeless people. JC also develops participatory tools for DRR. More details available from: http://web.env.auckland.ac.nz/people_profiles/gaillard_j/

Klaus Geiselhart is a Research Fellow at the Institute of Geography, Friedrich-Alexander University Erlangen-Nürnberg, Germany. His work focuses on pragmatism and practice theory, stigma and discrimination, and geography of health.

Kenneth Hewitt is Professor Emeritus and Research Associate in the Department of Geography and Environmental Studies at Wilfrid Laurier University, Canada. His past work has looked at environmental risks and disasters, high mountain environments (geomorphology, glacial hydrology, natural hazards and disasters), peace research, and environmental and civil hazards of war in the twentieth century.

Ilan Kelman is a Reader at the Institute for Risk and Disaster Reduction and the Institute for Global Health, University College London and a Senior Research Fellow at the Norwegian Institute of International Affairs (NUPI), Oslo, Norway. His current research focuses on how and why disaster related activities affect conflict and health, as well as health and disaster topics for islands.

Fred Krüger is Professor of Geography at the University of Erlangen–Nürnberg, Germany. His research interests are in Developmental Geography: vulnerability and resilience, risk concepts, man-environment links, social geography, and cross disciplinary social and cultural science. With the editors of this book, he has also co-edited the IFRC World Disasters Report 2014 on Culture and Risk.

James Lewis of Datum International, UK works and writes on causes of vulnerability to natural hazards, corruption and poverty, climate change, island vulnerability, behavioural responses, and interconnections with socioeconomic capacity and development.

Sarah Marsh is specialised in disaster response and recovery management. Her work focuses on immediate response to crisis situations while integrating sustainable recovery and rehabilitation in both conflict and natural disaster/chronic environmental emergency settings.

Jessica Mercer works as an independent consultant with Secure Futures in Winchester, UK. Her work focuses largely on disaster risk reduction and climate change adaption.

Julie Morin is affiliated with Laboratoire Géosciences Réunion – IPGP, Université de la Réunion, France. Her research focuses on risk assessment and mitigation in volcanic and tsunami hazards, people's vulnerability, risk perception, evacuation proceedings and modelling, sensitisation tools and crisis management.

Anthony Oliver-Smith is an Emeritus Professor at the University of Florida, USA. His past publications have looked at natural hazards and disasters, vulnerability analysis, recovery/reconstruction, disaster risk reduction and climate change adaption, and socio-ecological systems.

Benedikt Orlowski is a Research Fellow at the Institute of Geography, Friedrich-Alexander University Erlangen-Nürnberg, Germany. His research focus is on human geography, particularly the geography of risk and cross-disciplinary approaches to volcanic risk, and risk and the media.

Kristinne Sanz is affiliated with the School of Geography, Politics and Sociology at Newcastle University, UK. Her research focus is on gender relations, non-heterosexual identities, disaster development nexus, anthropology and women's networks.

Gerrit Jasper Schenk is a Professor in the Institute of History, Darmstadt Technische Universität, Germany. His research has centred on historical disaster research, visual culture of disaster, relation between society and environment, and urban history.

E. Lisa F. Schipper is Research Associate at the Overseas Development Institute, London, UK and an Associate with the Stockholm Environment Institute, Stockholm, Sweden. Her research looks at adaption to climate change, community based adaption, sociocultural vulnerability, religion and risks, and links between adaption and DRR. With the editors of this book, she has also co-edited the IFRC World Disasters Report 2014 on Culture and Risk.

Fabian Schlatter is a Research Fellow at the Institute of Geography, Friedrich-Alexander University Erlangen-Nürnberg, Germany. His research focuses on health geography and public health, in particular the adherence to antiretroviral therapy in Botswana, culture and health, risk and resilience and social determinants of health.

Neysa Setiadi is a Research Associate for the United Nations University, Institute for Environment and Human Security (UNU-EHS), Bonn. Her work focuses on vulnerability, disaster risk reduction, early warning and urban planning.

Martin Voss is a Professor in the Department of Political and Social Sciences, Freie Universität Berlin and Head of the Disaster Research Unit (DRU). His work focuses on sociology and politics of disaster and catastrophe, global environmental and social change, vulnerability and resilience.

INTRODUCTION

Exploring the links between cultures and disasters

*Greg Bankoff, Terry Cannon, Fred Krüger
and E. Lisa F. Schipper*

Why are people still so vulnerable to natural hazards, even after decades of scholarship, information raising and capacity building on Disaster Risk Reduction (DRR)? The answer is, in part, that some crucial aspects have been missing from these efforts. One aspect is culture. It has only been in recent years that 'practical' DRR has gained more from an approach that sees hazards, vulnerability and resilience as social constructions. The shift to giving greater priority to the social and, as such, cultural embeddedness of risk has led to a widespread acceptance of the need to focus more on people's interpretations, negotiations, experiences and creative adaptations to hazards when it comes to analysing, or intervening in, disasters. In short, there is now a greater appreciation of linkages between disasters and culture(s) as important elements in the social dynamics of disaster related preparedness (including mitigation) and response. But questions need to be asked. What is the nature of these linkages? Or more to the point, what role does culture have in DRR and why is the cultural dimension often undervalued in DRR and Disaster and Risk Studies? How can this neglect of culture in disaster preparedness and management be overcome and which methodologies and conceptual frameworks do justice to the significance of culture(s)? Acknowledging people's cultural production of risk, and their responses to it – how they perceive, experience and respond to disasters – can help us to better understand why people are affected by hazards and why they do or do not take action to minimise them. We are convinced that this insight will facilitate more meaningful DRR interventions.

Much academic work on risk and disasters is still just that: academic. It is entrenched in a specific realm of disaster studies, with the debate largely confined to a community of scholars (and sometimes practitioners) that is defined by the topic of DRR alone. However, if risk is seen as embedded in culture, then the study of how risk and disasters are produced, interpreted and acted upon should be carried into a much broader multi- and transdisciplinary field, not disparate from, for instance, development studies or environmental sciences but embracing or including them (cf. Varley 1994, Wisner *et al.* 2004, Casimir 2008). This book therefore attempts to open the debate on the significance of culture(s) for dealing with disaster management, including preparedness and mitigation, and encompasses in the debate a wide range of disciplines, approaches and concepts. Such a comprehensive approach will hopefully facilitate the mutual understanding of diverging viewpoints about the linkages between risks, disasters and cultures, and help to overcome the neglect of culture in much of disaster-related research and action. The premise from which we start is that the significance of 'culture' must be understood and incorporated into any attempt to deal with natural hazards, rather than being viewed as largely irrelevant.

The significance of 'culture' in creating risk

Calling something 'culture' is a way to frame complex behaviours and render them understandable. Linking culture to risks and disasters in DRR-related studies and practice is attractive because a) culture is a framing that can offer explanations for dealings and doings in the disaster context that might, if culture was left out, be considered as 'confused', 'irrational', 'unwise', 'weird' or 'irrelevant'; b) culture provides a self-reflexive base to think about the role and nature of DRR; c) culture has a (often hidden) power that enables, or hinders, people and populations to deal with hazards appropriately and enables, or hinders, researchers to contribute to improving DRR.

People supposedly deal with the dangers they face from natural hazards in ways that are in accordance with their worldviews. Sometimes this includes ways that are emotionally appropriate and that do not challenge their relations with their peer group. Beliefs and perceptions of risks can be shared within a particular hazard location (but not necessarily so) and are often connected with the livelihood activities people engage in and the environments they construct. On provisional terms – and we will broaden our perspective considerably below – we can call these behaviours 'culture', which may include ways of producing, perceiving and dealing with risk that may remain hidden to outsiders, or, if detected, sometimes not seem appropriate but traditional or unnecessary. These cultural, behavioural and design arrangements enable people to live with extreme risks, and can include religious beliefs and related activities on the one hand and social organisation and adaptation in the built environment on the other.

Ways of dealing with risk may include adherence to group attitudes that cannot easily be avoided: they are about 'belonging' and being part of a shared experience of life and the spiritual forces that affect it. Breaking with the beliefs that form part of peer group behaviour means running the risk of losing acceptance to that group, which can be crucial to all other aspects of life (often referred to as 'social capital'). Such peer-affected behaviour can include gendered attitudes to risk, such as *machismo* (where men behave as if taking a risk seriously is a sign of weakness), and explains different attitudes to risk by men and women. It is also demonstrated when people from one household are unwilling to take up a disaster risk-avoidance action because they do not want to appear different or 'weak' in comparison with the rest of the group. Obviously, the opposite may also sometimes be the case. Risks and disasters that have not been experienced before and/or create unusual, exceptional or unfamiliar situations might be dealt with by actions that are not deemed 'appropriate' by peers or institutional authorities, or are not in line with 'established' codes of behaviour. Some of these behaviours are interpreted through the disciplines of psychology and behavioural economics (Ariely 2009), although these have not been widely applied to disaster contexts – especially not those in the so-called Global South. These interpretations suggest that many people ignore risks, or consider that their ability to influence those risks is minimal (because they are natural or god-given). This therefore also reduces the success of DRR policies and projects.

As the peer-group example shows, risk is a result of societal perceptions, decisions and actions and, therefore, a social construct. But risk is not always linked to 'belonging'. In anthropology, sociology and more recently geography, there have been many attempts to explore the ways risk is connected to social contexts and, as such, cultures. There has been substantial academic investment in the role of culture in relation to disasters. In the social and cultural sciences, this has mainly been in anthropology and, while the topic has been addressed for several decades, it has had little impact on mainstream DRR institutions. Over thirty years ago, Douglas and Wildavsky (1982), Yates (1982) and Douglas (1986, 1992) differentiated between risk-averse and risk-taking societies, drawing on the 'willingness' of people to accept, or avoid, risks and thus highlighting how risk and culture are inseparable. In later research, risks have been seen more as a precondition for an effective response to threats than a threat in their own right. Societies that manage to actively produce risks by way of calculating them and negotiating options to deal with hazards will be more likely to avoid, or successfully respond to, damaging impacts (Beck 1992, Krüger and Macamo 2003). Hewitt (1997) describes four factors that determine risk: hazardous events, social vulnerability and adaptation to threats, structural traits of the natural and social environments ('habitat'), and coping strategies and intervention measures. Risk thus becomes a feature not only of culture and society, but also of a specific location or region; i.e., it is rendered a spatial quality – this notion has been further explored recently by a number of authors in Müller-Mahn (2013).

Other areas of academic work that are most relevant have also not yet made much impact on DRR – there is a growing literature on risk perception, psychology, culture and climate change, especially in relation to cultures of climate change 'denial' (Hulme 2009). The other main area of relevant work – which surprisingly seems to also have had almost no impact at all on DRR – is of public health and the parallel issues of risk perception and culture that very significantly affect people's behaviour in relation to disease and other public health issues, as well as poverty. The HIV/AIDS pandemic has been used widely to explore this in detail (cf. Stillwaggon 2006, Becker and Geissler 2009, Geiselhart 2009, Niehof *et al.* 2010).

Even when it has been acknowledged that culture must be taken more seriously, putting that into practice is not easy. Culture is complex and difficult to analyse, often precisely because those who are doing the analysis are usually outside of 'it' and therefore not fully able to comprehend it as a lived experience. However, this also means that there are inherent challenges in trying to analyse or be objective about our own 'culture', and especially having self-awareness of the beliefs and assumptions, the framings and logics of our own behaviour as 'outsiders' who want to understand 'other' experiences or beliefs, or to bring DRR to people. Understanding our own institutional culture is therefore also vital. It is not the intention here to explore the debates regarding the methodological validity of who is best placed to study culture – an insider or outsider; an anthropologist, geographer or sociologist; an academic or a practitioner who has worked for decades 'in the field' – but it is important to note that such debates exist for DRR professionals to explore. What can be recognised here is that culture forms an important and intractable part of any context where risk- and disaster-related research is carried out and DRR may be applied.

But what is (are) culture(s)?

When addressing linkages between cultures and disasters (and then even arguing that these linkages must not be ignored in DRR), a first and most obvious step would be to define exactly what we mean by culture. This, however, is where the challenge begins. Culture cannot be interpreted as a given set of social factors, but as a constantly changing and shifting configuration of social practices, or as outcomes of experiences, social arrangements and situations that are inscribed into society. It is manifest through subcultures and an array of discourses on different scales from the individual to the global, which are permanently being negotiated on these different scales. Different cultures are lived and lived-in at the same time, along parallel trajectories and based on parallel histories – often interwoven, sometimes detached. All this is further complicated, in the context of risks and disasters, by environmental degradation and climate change, and how people respond to it. Culture is 'messy' and thus adds to the 'messiness' already inherent in the area of disasters and DRR. The problem is that if culture is left out of the analysis of disaster and risk (risk already being a social construction and therefore always 'cultural') then the extent and importance of hazards, DRR and related issues of adaptation, coping, intervention, knowledge and power relations cannot be fully grasped. But because culture is so complex, it eludes a clear and simple definition.

Whatever its manifestation, culture in one sense is also an adaptation to the risks inherent in the local environment, whether those threats are real or imagined. In particular, people adapt the built environment (the totality of created, modified or constructed spaces and places) to accommodate the risk of living for many generations in places where they are regularly exposed to hazards. These patterns then become embedded in cultures over time (Moore 1964, p. 195). This cultural adaptation, however, is shown to depend on three crucial factors: that the hazard is repetitive, that it is of a nature to allow forewarning, and that it inflicts significant damage to human and material resources (Wenger and Weller 1973, p. 9). Over the centuries, communities have adapted to risk in the shaping of their environments. Where the hazard was frequent and of a magnitude to regularly cause loss of life and property damage, people in the past developed the pragmatic and theoretical knowledge of learning to live with threat on a day-to-day basis, an accommodation that is reflected in the design of buildings, the materials used and the construction techniques. This vernacular architecture is (and, alas, was) a very visible manifestation of cultural adaptation to risk over time. It is obvious that this adaptation is not an (almost) involuntary or 'automatic' process triggered and determined by natural factors. It is an active, creative way of dealing with threats and uncertainties based on ideas and negotiations, communication, social institutions and agency.

Given the difficulty of defining culture, it comes as little surprise that, when editing this book, we were confronted by many colleagues, students and attendees of two conferences we had organised on 'Cultures and Disasters' with some tricky questions: if you do not exactly know what culture is how can you include it in the study of disasters and in DRR? How can you edit a volume on cultures and disasters and then not define what you mean exactly when you talk of 'culture'? Or 'cultures'? (With the added comment that 'you are not even specific about singular or plural'!) For the purpose of this book, we take two positions: first, we accept the way the different chapter authors define culture(s) for the sake of their arguments – some authors are quite clear about their take on culture while others are more open-ended.

Second, our own view as editors, and the underlying premise that provides the framework for this book, is that we assume a more functional and practical position in the context of disasters and implementation. We examine how aspects of risk and disasters are linked to individuals', peoples', populations' and organisations' perceptions, interpretations and activities. For this purpose, we are approaching culture as beliefs, attitudes, feelings, experiences, values and narratives, and their associated behaviours, actions and day-to-day routines that are shared by, or at least abided by, most people in respect to threats and hazards. Culture in relation to risk and disasters is 'functional' and provides the ways that people interpret, assess and adapt with their peers and other members of society to the risk environment in which they live and work. It is about ordering this risk environment through institutions (such as rules or laws) and governed by interests (and sometimes oppression and violence) and re-inventing it as new risks arise. Culture is thus situated in a process by which people shape their diverse beliefs, institutions and surroundings around disasters.

The interaction between culture and risk in this pragmatic sense relates to many aspects of human and institutional behaviour and action, including where people live and in such aspects as the design of their houses, their attitude to health, livelihood needs and their connections with others in specific hazard locations. This encompasses the ways in which societies organise themselves, including aspects of authority and power relations that are 'cultural' (such as gender, class, ethnicity, caste, or age group bonding, and the exercise of power through political or media campaigns etc.) and will inevitably have an influence on people's ability and agency to cope with, or adapt to, hazards and disaster outcomes.

Although we are employing this rather functional interpretation of culture, we are also aware of the danger that identifying something as cultural can easily slide into essentialism. We are in no way discarding other, broader conceptualisations or those intepretations that stress more the fluidity and multiplicity of cultures. These see cultures as outcomes of fields of practices, i.e., contexts where practices are institutionalised in order to reduce contingencies in everyday life (cf. Dewey 1929, 2005, Schatzki 2003, Barry 2006). Dealing with hazards and disasters becomes particularly challenging when they trigger singular situations where common assumptions fail, institutionalised practices are contested and corrupted, and the actors involved come up with unconventional suggestions, exceptional solutions and creative responses.

We also see risk of disasters as a potential catalyst for cultural change. Considering disaster as a historical process allows its impact to be evaluated as more than only a destructive event. It can be seen as simply an agent of change in its broadest perspective capable of offering opportunities to push through needed policy solutions (Birkland 1998, Johnson *et al.* 2005). As catalysts of change, disasters have long been transformative agents in their own right, causing political, economic and social adjustments, triggering needed adaptations in human behaviour and the built environment, as well as perhaps contributing to the overthrow of dynasties, economic systems and even civilisations. The rebuilding of Lisbon after the great earthquake of 1755 is just one such instance as the Portuguese state became intimately involved in the recovery and reconstruction process, innovating policies for urban planning and developing seismic-resistant building codes, many of which continued to be standard practice until the 1920s (Dynes 2000, Chester 2001). Once the historical importance of disasters as transformative agents is more fully appreciated then policy makers and planners are more likely to consider longer term developmental goals as well as satisfy more immediate needs (Bankoff 2011). The future reconstitution of the Haitian state, as well as the rebuilding of Port-au-Prince following the January 2010 earthquake, is a case in point.

This leads us to the question of whether we should talk of culture (sing.) or cultures (pl.). As is obvious from the title of this book, we opted for the plural form in order to acknowledge the multiplicity of cultural expressions and dimensions, and their emergence and coexistence on many social and spatial scales. The two conferences we organised prior to publishing this volume (one held in Bielefeld, Germany, in 2011, the other in Erlangen, Germany, in 2013[1]) had the same title as this chapter (Cultures and Disasters I/II), and for the same reasons. Therefore, it seemed only logical to stick to the plural for a book that draws extensively on the papers and discussions given in these two events. The plural denotes that there is not just one way or one set of specific practices (or values, norms, beliefs, narratives, arrangements etc.) and that we are cognisant of different viewpoints from a variety of perspectives (whether they are ethnic, disciplinary, national etc.). However, when we talk of a culture, we are aware that it can also refer to specific values, conventions and social practices; cultures would thus refer to different sets of specific values, conventions or practices – to cultures as specific entities, an essentialist notion that we are trying to avoid.

Using culture(s) as an entry point for analysis, this volume addresses what we consider to be significant problems that affect DRR, as attempted by many different types of organisations around the world. Various chapters argue that there is something significantly wrong in a lot of what happens in DRR, including ineffectiveness and waste. The problems can be characterised as gaps between what 'outsiders' consider disaster risks to be and the very different ways that risks are perceived, understood and dealt with by 'insiders', in their culture at the 'community' level. The focus of this book is mainly at this local 'community' scale, because the community level and community-based DRR have become highly significant in recent years for all types of actors. It is also at this scale that a great deal of analogous work is going on in 'community-based adaptation' (CBA).

The culture 'gap'

DRR initiatives arise from a range of organisations that act at different levels, but which share a framework of understanding based on a practical, 'rational' or 'scientific' approach, and a belief that their concerns about significant natural hazards are universal and shared by all people who have been affected, or are likely to be affected, by known and anticipated hazards. There are two key processes that significantly reduce or distort the success of disaster risk reduction measures, and although these are mostly familiar to persons involved in DRR activities, they are often ignored (Bankoff 2004). These processes are also difficult to overcome because of the ways that institutional behaviours ('cultures' of the institutions themselves) and funding arrangements make it difficult to take them into account.

The first of the two issues that concerns us is that 'outsiders' (people from organisations attempting to support DRR) often hold different conceptions of risk in comparison to the priorities of the communities they are trying to help. However, these differences are not always self-evident or, if recognised, acknowledged. For example, the foreign aid system may require interventions and outcomes to be predetermined for dealing with large scale hazards at community level, so there is little or no operational space to allow for these differences. Sometimes priorities are determined by an organisation's headquarters that is set far from the intervention site. There is, therefore, a significant concern that interventions are most likely to be unsuccessful due to poor uptake or ownership of the project by local people who do not share the priorities of outsiders. Different perspectives on risk may simply be overlooked by the DRR agencies who believe that they either know what should benefit a community and try to change people's behaviour or risk priorities, or believe that they have included, through their community-based approach, all relevant stakeholders and are thus acting according to what the local people have expressed as their priorities or most important concerns. Working with the 'community', however, will often not include everybody – those who have little say and cannot articulate themselves due to unfair power relations often remain undetected by the intervening agencies. Tensions, fractions and inequalities often appear to be overlooked, or at least are not dealt with appropriately, by outside organisations when programmes and projects are actually implemented, or they are not considered to be significant barriers. These power relations are almost always present (in a wide variety of configurations), especially on grounds of gender, class, ethnicity, caste, patron–client relations or age group bonding. They are also sometimes justified by ideology and religion, making it difficult to intervene from outside. Communities are never neatly aligned to fit with the needs and priorities of the external institution that wants to work with local people. Local people themselves are also not uniform; the idea of community implies unity, collaboration, cooperation and sharing, which is usually not the case (cf. Etzioni 1996, Oliver-Smith 2005, Twigg 2009, Jones *et al.* 2013). In fact, the 'community' remains a myth (Cannon *et al.* 2014) and sometimes a romantic idea (de Beer 2012). Therefore, even when participatory risk assessments are carried out, it is highly likely that people's priorities will be subsumed within activities that focus on major hazards and/or on what local elites have communicated as being of major concern. Evaluations that show the work has been successful normally come at the end of a project, which is not a relevant time to test whether they have been effective in disaster preparedness. A significant reduction in disaster risk may only be evident long after the project is ended and this is not captured in the evaluation of the project activities.

The mismatch of priorities explained above is reinforced by a second key factor: people often choose to live in dangerous places for good reasons (Cannon 2008). The ways that 'culture' contrasts with the prevailing DRR rationalities is most evident when people seemingly ignore risks that outsiders regard as serious. But people often consider that their ability to influence those threats is minimal. Other important factors also override conventional DRR notions of risk. Many authors have noted that people knowingly live in areas 'at risk' and exercise substantial choice in doing so. The significance of place and emotional attachments to it are well understood in the anthropology literature, and this is where our first argument (about people's different priorities and their livelihood needs) overlaps with the cultural beliefs that help people to live with risks.

Although it is true that power relations and poverty force many people to live in unsafe locations and unsafe houses, there is plenty of evidence that hundreds of millions of people choose to live on the sides of volcanoes, in earthquake fault zones, in river flood plains and on coasts that are exposed to storms and tsunamis. These are the places where they get their livelihoods, which are often better in those danger zones (valley and volcanic soils are very fertile, coasts are good for fishing and farming, fault zones in arid areas often have associated water supplies). People – more or less willingly – trade off everyday benefits against the danger of the less frequent physical hazards that can affect those locations. In effect, they 'discount' the future risk of a big event in order to reap the day-to-day benefits of their livelihood. Of course people do not – cannot – engage in any formal cost–benefit analysis to work out their 'discount rate', but in effect that is what is going on.

There are also studies of people wanting to move back to their homes and towns after they have been destroyed and where the hazard may be repeated (cf. Oliver-Smith 1986, Cannon 2008). This poses a number of problems for DRR. Most critically, it calls into question the implicit and underlying logic of DRR interventions that adopts a certain type of rationality, which dictates that, given sufficient knowledge and awareness, people would not live in 'risky' areas. The implications of this for DRR practitioners are inherently problematic. As Cannon (2008, p. 355) notes, 'improving their [vulnerable communities] security means persuading people to act against what they think are their own interests, or denying their culture or psychological preferences'.

In a similar vein, the external experts may not realise or fully appreciate the accommodation that many local communities may have reached with the risks inherent in their environment over time. Especially where hazards may be 'a frequent life experience', that is a recurring and oft experienced event in people's lives, local communities may have developed ways of 'normalising threats'. This normalisation of risk may be expressed through practices that deal with the emotional and psychological requirements of living with uncertainty and which may influence character traits, normative values and religious beliefs. They may also be more physical manifestations discernible in the built environment and the historical record, such as in the design and construction of buildings to accommodate climatic and seismic contingencies, in agricultural systems that emphasise minimising risk rather than maximising surplus, and in the constant relocation of settlements and migration to find safer locations and more sustainable livelihoods. Whether such practices are sufficient to constitute a coherent body of local knowledge, separate from and in contrast to the notion of a 'universal' and Western scientific thought, is a matter of considerable debate. These practices do, however, often represent a distinctive pattern of activity and behaviour to that premised upon the assumptions of avoidance-loss so much favoured by Western social science, and they do raise important questions about how cultural forms and traits are interpreted (Bankoff 2004).

Related to this second key factor are people's own explanations of threats and disasters that, to other groups or outside 'experts', sometimes seem irrational or even absurd, but make perfect sense in a lifeworld that has been collectively framed by people's experiences and practices for years and generations. However, one must be wary of viewing these produced lifeworlds as expressions of fixed cultural entities. A conceptualisation of culture as acting upon individuals merely as a constraint or as a facilitator implies seeing people as hostages of values and their institution-alisations and culture as robbing individuals' agency (Macamo 2013). Sometimes in DRR action, intervening organisations yearn for a culturally coherent response by those affected by threats and hazards and are convinced of finding it when working with 'communities'. What is presented to them on location, however, is more a system of authority, hegemony and power relations than a stable and easily discernible 'culture'. Culture as a constant, everyday production of 'contexts of action' (Macamo 2013) cannot easily be grasped by intervening agencies, despite their hope to be able to do so at the 'community' level. Taking culture seriously in DRR intervention, and doing credit to the importance of culture in risk and disaster contexts, means having to take the complex everyday dealings and livelihood activities of the people one is working with into account. Risk management and disaster-related intervention should thus not be a sectoral undertaking to mitigate threats, but must entail a more holistic approach that takes the broad range of livelihood and lifeworld realities into account. The same goes for the study of disasters and risks: the chapters in this book demonstrate the challenge of bringing culture into the disaster discussion, not only because culture in itself is a contested topic but also because it influences risk in so many different ways.

Culture(s) as the missing dimension in disaster and risk studies: some key suggestions

Given the complexity of culture(s), and the vast array of approaches to conceptualising the topic, it is helpful to suggest some entry points into studying the links between cultures and disasters. Most of these were identified at the two conferences mentioned above. Despite the multiplicity and varied approaches to the question of culture and risk, there were some central points that the majority of experts from different disciplines and agencies agreed upon. The discussions at the 2013 Erlangen Conference, in particular, were marked first by what appeared to reflect the 'gap' between insiders and outsiders described above, and also between practitioners working in DRR and scholars working on DRR. It became quite evident, however, that the sharing of expertise of these two groups of actors has potential to lead to a more coherent understanding of culture–risk–linkages without having to negate the complexity and multi-varied interpretations and cultural framings of risks and disasters. The board diagram (Figure 1) is a graphic recording of one of the conference discussions. It evolved as the discussion continued, serving as a 'real-time', simultaneous visual documentation of the debate, but also influencing it as discussants and audience elaborated arguments and spotted contradictions as the visual transcript emerged. Turning the spoken words into images, the professional graphic recorder, as an interpreter, summarised the topics addressed, thus highlighting central issues as the debate went on. While the difficulty in defining disasters and culture(s) could not be overcome, a multitude of other aspects were brought to light that have found entry into the chapters of this book and need to be brought to the attention in future DRR-related work. These include: the relationship of hazards, disasters and local knowledge; interpretations, values and taboos; the nature of globalised strategies of dealing with risks; the mismatch of traditions and 'modern' DRR-related technologies and policies; the arrogance of not taking note of people's experiences, priorities and

FIGURE 1 The complexity of linking culture, risk and disasters: graphic recording of a discussion held at the 2013 'Cultures and Disasters' Conference, Erlangen, Germany

Source: Graphic recording: Gabriele Schlipf.

actions in DRR; the importance of acknowledging power relations in the study of how risk is constructed in DRR; and embedding risk and disasters in the contexts of justice and human rights.

Five key suggestions and conclusions emerged from these debates. While these reflect a diagnosis of scholarly and practical approaches to risk and disasters, and are targeted mainly at disaster-related intervention, they also have implications for the study of culture and risk inter-linkages.

1   Paying more attention to people's own priorities, perceptions and belief systems: involving the people affected and their interests in the planning and implementation of DRR and aid measures may increase the acceptance and sustainability of DRR concepts within societies.

2 DRR priorities have to be set to the needs of people at risk by getting rid of a supply driven DRR and introducing a demand driven DRR.
3 Being aware of one's own 'cultural' perspective and the boundedness of DRR, and its effects on practical work and disaster risk research, is crucial and unveils the constructedness of DRR relevant concepts such as 'community', 'resilience' or 'local knowledge'.
4 Definitions of 'disaster' and 'culture' also are cultural conventions themselves. The terms therefore should be used with wariness about their social constructedness.
5 Culture can be seen as situated in a process, by which people shape their diverse beliefs, narratives and practices around disasters. Culture is never homogeneous. It can be based on a specific place, but not necessarily so, and it depends on the diversity and mobility of people. It can be both, sweet and cool as well as destructive and creative.

How this book is structured

As is easily discernible from this short, and by no means complete, list of key points, more often than not a combination of approaches will be employed to come to grips with the 'messiness' of cultures and disasters as epitomised in the graphic recording above (Figure 1). It is interesting to note that the five conclusions can be 'read' differently depending on what one is primarily concerned with: how culture shapes vulnerability (i.e., how people are exposed to hazards via culture), or how it influences people's ways of dealing with disaster risk and disasters (i.e. how culture enables or facilitates an appropriate 'handling' of threats). Most authors in this volume are in fact approaching their topic from various viewpoints that include most, if not all, of the above five findings. For ease of presentation, the chapters are grouped into three sections that allow for a step-by-step entry into the challenging analysis of the more than complex relationships between culture(s) and disasters.

The first section combines chapters that deconstruct conceptualisations of disasters and highlight different framings of risk. The chapters explore how disasters are constructed by societies and how culturally embedded perceptions and ideas determine the (potential) creation or elimination of risks. Kenneth Hewitt leads the way by exploring the reconfiguration of social relations of safety and calamity and examines the current paradigms of securitisation that dominate attempts to control risk and disasters. Next, Anthony Oliver-Smith links disasters to neoliberalism, which he interprets as a cultural frame that expresses, largely in economic terms, how risk reduction strategies are designed and implemented. He argues for DRR policies that must contend with a dominant cultural frame that undermines a successful implementation of meaningful risk management strategies. Greg Bankoff emphasises cultural adaptation by people to natural hazards and, drawing on the example of earthquakes, shows how the built environment provides a visible manifestation of how culture is ineluctably intertwined with disaster risk management, not only in terms of practical adjustments to hazards but also as a form of long term educational transmission through heritage. Gerrit Schenk analyses the chances and problems of learning from the history of disasters and discusses cognitive concepts and models of 'nature'. Terry Cannon concludes this section by linking the significance of culture to disasters and climate change adaptation, explaining how the latter is embedded in beliefs and behaviours.

The second section links cultures to issues of social vulnerability. These chapters draw on how perceptions of and attitudes towards disasters are embedded into societies as day-to-day social practices and how risk-related discourses are shaped and communicated. James Lewis initially illustrates how aspects of behaviour, often over long periods of time, may affect the safety of others and become more significant to disaster risk than technologies, or what is generally termed 'resilience'. Klaus Geiselhart, Fabian Schlatter, Benedikt Orlowski and Fred Krüger explore notions of culture as fields of (often institutionalised) practice. The HIV/AIDS pandemic serves as an example to illustrate how exceptional events and processes are leading to creative solutions in coping with risks and how these solutions (often deemed irrelevant by DRR actors) may help to improve disaster preparedness and response. Lisa Schipper then links disasters to religion and belief systems, and discusses these cultural expressions as drivers of vulnerability and as potential entry points for resilience building. Finally in this section, Andrew Crabtree, using a Bourdieuan interpretation of risk approaches, analyses structural and cultural changes since the onset of the 2008 flood in Bangladesh and how socioculturally embedded expectations affect psychosocial risk in the aftermath of the flood.

In the third section, the chapters draw on linkages between cultures and the realisation of disaster reduction programmes, stressing inequalities in DRR as a cultural institution. David Alexander introduces this final section of the book by making observations on two aspects of the changing role of human culture in DRR, namely celebrity and victimhood. He demonstrates how the vigorous promotion of commercial and entertainment values by the mass media is radically changing the public's attitudes towards disasters. Then, using the example of flood management measures, Brian Cook shows a widening gap between those who decide on these measures and those who are governed by these decisions. Ilan Kelman, JC Gaillard, Jessica Mercer, Kate Crowley, Sarah Marsh and Julie Morin examine approaches to understanding the role of culture in DRR by focusing on how different knowledge systems can be combined, drawing on case studies from Small Island Developing States (SIDS). JC Gaillard, Maureen Fordham and Kristinne Sanz then reflect upon the importance of integrating gender perspectives in DRR, demonstrating how women in post-disaster situations adapt to, accommodate and resist the socially and culturally defined gender roles and expectations to ensure their survival and contribute positively to the larger society. Jörn Birkmann, Neysa Setiadi and Georg Fiedler examine how religious symbols have been used to promote the acceptance of evacuation and temporary migration in the light of tsunami risk in Indonesia. Martin Voss and Leberecht Funk end this section and the book with a chapter that uses the example of the Tao people in

Taiwan to explain an epistemic gap between privileged 'experts' who advocate the rules of scientific discourses on disaster, and people whose livelihoods are at risk but who do not employ scientific interpretations of hazards.

Conclusion

Despite (or because of) the complexity of cultural framings of risks and disasters, we realise that this volume is only a tentative step in the direction of better understanding of how culture might enhance DRR. Understanding the differing priorities of people 'at risk' requires broadening our frame of reference. It is no longer sufficient to focus narrowly on trying to understand how to improve DRR programmes where these are defined by 'outsiders'. Instead it becomes necessary to invest in understanding the rationalities for different (and seemingly irrational) behaviour. This means moving beyond our conventional frames of reference in the disaster sector and incorporating the perspectives and insights from other disciplines such as social psychology, public health, anthropology, geography, sociology, history, architecture and behavioural economics. Moreover, it requires confronting the (often) difficult questions that different rationalities pose for DRR programming, namely that 'outsider's' priorities are often not the same as those they aim to help. Paying more attention to people's own priorities, perceptions, belief systems and actions, and involving the people affected and their interests in the planning and implementation of DRR and aid measures, may increase the acceptance and sustainability of DRR concepts within societies.

We argue that most DRR interventions either expect people to behave in ways that minimise the same risks as those identified by the outsider, or those intervening are only dimly aware of 'other' cultural framings of threats and disasters and for various reasons do not take them into account. In short, we argue that many DRR organisations or policies are sometimes divorced from the reality, either in their thinking or their doing, or both, of those they are trying to help. DRR staff and institutions are often aware of the gaps between their own goals and those of the people they are claiming to help. But they are often not willing to accept that this is a problem, or find it difficult to challenge the institutions' remit to take it into account. It is very difficult for organisations that are charged with DRR to change their focus and carry out work that supports people in their everyday life, especially when such people are seemingly indifferent to the threat that assails them. Where institutions and their staff are open to the problem, they may still be unable to redefine themselves to reduce the gaps. They are also likely to be dependent on funding from donors who are only willing to support DRR for specific hazards in more or less prescribed ways. By pulling together recent academic work from different disciplines in one volume, we hope to enhance understanding of the complex and complicated linkages between disasters and cultures, and thus contribute to the ongoing debate about how to increase disaster related research and intervention and, perhaps, even challenge some existing approaches and intervention.

Notes

1   International conference 'Cultures and Disasters', 5–6 July, 2011, convened at the Centre of Interdisciplinary Research (Zentrum für Interdisziplinäre Forschung ZEF), University of Bielefeld (Germany), and international conference 'Cultures and Disasters II: Exploring the Links between Disasters and Culture(s): Preparedness, Response, Policies.', 11–12 July, 2013, convened at the Institute of Geography, Friedrich-Alexander-Universität Erlangen-Nürnberg (Germany); both organised by the editors of this volume.

References

Ariely, D., 2009. *Predictably Irrational*. New York: HarperCollins.

Bankoff, G., 2004. In the eye of the storm: The social construction of the forces of nature and the climatic and seismic construction of God in the Philippines. *Journal of Southeast Asian Studies*, 35 (1), 91–111.

Bankoff, G., 2011. Historical Concepts of Disasters and Risk. In: B. Wisner, JC Gaillard and I. Kelman, eds. *Handbook of Natural Hazards and Disaster Risk Reduction*. London, New York: Routledge, 31–41.

Barry, A., 2006. Technological zones. *European Journal of Social Theory*, 9 (2), 239–53.

Beck, U., 1992. *Risk Society: Towards a New Modernity*. London: Sage Publications.

Becker, F. and Geissler, W. P., eds, 2009. *Aids and Religious Practice in Africa*. Leiden: Brill.

Birkland, T., 1998. Focusing events, mobilization, and agenda setting. *Journal of Public Policy*, 18 (1), 53–74.

Cannon, T., 2008. Vulnerability, 'innocent' disasters and the imperative of cultural understanding. *Disaster Prevention and Management*, 17 (3), 350–7.

Cannon, T., Titz, A. and Krüger, F., 2014. The Myth of Community? In: IFRC/T. Cannon, L. Schipper, G. Bankoff and F. Krüger, eds, *World Disasters Report 2014—Culture and Risk*. Geneva.

Casimir, M., ed., 2008. *Culture and the Changing Environment—Uncertainty, Cognition and Risk Management in Cross-Cultural Perspective*. New York, Oxford: Berghahn Books.

Chester, D., 2001. The 1755 Lisbon earthquake. *Progress in Physical Geography*, 25 (3), 363–83.

de Beer, F., 2012. Community-based natural resource management: Living with Alice in Wonderland? *Community Development Journal*, 48 (4), 555–70.

Dewey, J., 1929. *The Quest for Certainty. A Study of the Relation of Knowledge and Action*. London: Georg Allen & Unwin.

Dewey, J., 2005. *Art as Experience*. London: Penguin Publishing Group.

Douglas, M., 1986. *Risk Acceptability According to the Social Sciences*. London: Russel Sage Foundation.

Douglas, M., 1992. Risk and Danger. In: M. Douglas, ed., *Risk and Blame—Essays in Cultural Theory*. London, New York: Routledge, 38–54.

Douglas, M. and Wildavsky, A., 1982. *Risk and Culture*. Berkeley, CA: University of California Press.

Dynes, R., 2000. The dialogue between Voltaire and Rousseau on the Lisbon earthquake: The emergence of a social science view. *International Journal of Mass Emergencies and Disasters*, 18 (1), 97–115.

Etzioni, A., 1996. Positive aspects of community and dangers of fragmentation. *Development and Change*, 27 (2), 301–14.

Geiselhart, K., 2009. Stigma and discrimination: Social encounters, identity and space; a concept derived from HIV and AIDS related research in the high prevalence country Botswana. *Social Science Open Access Repository* [online]. Available from: http://nbn-resolving.de/urn:nbn:de:0168-ssoar-290930 [accessed 23 May 2014].

Hewitt, K., 1997. *Regions at Risk. A Geographical Introduction to Disasters.* Harlow: Longman.

Hulme, M., 2009. *Why We Disagree About Climate Change.* Cambridge: Cambridge University Press.

Johnson, C., Tunstall, S. and Penning-Rowsell, E., 2005. Floods as catalysts for policy change: Historical lessons from England and Wales. *Water Resources Development,* 21 (4), 561–75.

Jones, S., Aryal, K. and Collins, A., 2013. Local-level governance of risk and resilience in Nepal. *Disasters,* 37 (3), 442–67.

Krüger, F. and Macamo, E., 2003. Existenzsicherung unter Risikobedingungen – Sozial-wissenschaftliche Analyseansätze zum Umgang mit Krisen, Konflikten und Katastrophen. *Geographica Helvetica,* 58 (1), 47–55.

Macamo, E., 2013. Translating culture: Everyday life, disaster and policy making. Keynote Lecture held at the Conference '*Exploring the Links between Disasters and Cultures(s): Preparedness, Response, Policies*', University of Erlangen (Germany), 11–12 July 2013 [online]. Available from: www.video.uni-erlangen.de/clip/id/3173.html [accessed 18 July 2014].

Moore, H. E., 1964. *And the Winds Blew.* Austin, TX: Hogg Foundation for Mental Health.

Müller-Mahn, D., ed., 2013. *The Spatial Dimension of Risk.* London, New York: Routledge.

Niehof, A., Rugalema, G. and Gillespie, S., eds, 2010. *AIDS and Rural Livelihoods.* London: Earthscan.

Oliver-Smith, A., 1986. *The Martyred City: Death and Rebirth in the Andes.* Albuquerque, NM: University of New Mexico Press.

Oliver-Smith, A., 2005. Communities after Catastrophe. Reconstructing the Material, Reconstructing the Social. In: S. E. Hyland, ed., *Community Building in the Twenty-First Century.* Santa Fe, NM: SAR Press, 45–70.

Schatzki, T. R., 2003. A new societist social ontology. *Philosophy of the Social Sciences,* 33 (2), 174–202.

Stillwaggon, E., 2006. *AIDS and the Ecology of Poverty.* Oxford: Oxford University Press.

Twigg, J., 2009. *Characteristics of a Disaster-Resilient Community. A Guidance Note.* Version 2. London.

Varley, A., ed., 1994. *Disasters, Development and Environment.* Chichester: Wiley.

Wenger, D. E. and Weller, J. M., 1973. *Disaster Subcultures: The Cultural Residues of Community Disasters, Preliminary Paper No. 9.* Columbus, OH: Disaster Research Centre, Ohio State University.

Wisner, B., Blaikie, P., Cannon, T. and Davis, I., 2004. *At Risk. Natural Hazards, People's Vulnerability and Disasters.* 2nd ed. London, Routledge.

Yates, F., ed., 1992. *Risk-taking Behavior.* Chichester: Wiley.

PART 1

The culture of (de-)constructing disasters

1

FRAMING DISASTER IN THE 'GLOBAL VILLAGE'

Cultures of rationality in risk, security and news

Kenneth Hewitt

Introduction

> *Every culture and every age has its favourite model of perception and knowledge that it is inclined to prescribe for everybody and everything.*
>
> McLuhan 1964, p. 21

In the disasters field, questions of culture fall into two main sets. First, since calamity affects distinctive, often diverse, groups of people, the case for cultural awareness and understanding, or efforts to minimise what are 'lost in translation', can be made. This is usually stressed in encounters between different ethnic groups. However, in modern societies, techno-scientific expertise or agency agendas of, say, disaster risk reduction (DRR) or 'homeland security', hardly speak the languages of lay publics and subcultures, or for multi-cultural cities. For them, cultural awareness and communication can be equally important issues.

Second, the cultural forms and baggage of modernity exert enormous influence in public life and professional disaster ideas or management agencies (cf. Douglas 1992, Steinberg 2000, Klein 2007, Hannigan 2012). Modern ideas shape how danger and responses are addressed (cf. Burchell *et al.* 1991, Ericson and Haggerty 1997, Franklin 1998, Gilbert 1998). Behind accepted forms are histories of social struggle and so-called paradigm shifts (Green 1997, Quarantelli 1998). As discussed below, some entrenched or newly advocated developments in the disasters field have become problematic for thought and practice. They invite cultural critique.

Other chapters, and sources cited below, make the case for involving culture and cultural critiques in DRR. It seems fair to say, however, that modern disaster management largely ignores this in its own concerns. A positivist, 'applied science' or technocratic stance prevails, focused on material realities. Systems are supposed to be rationally devised to work with or towards an exact mirror of environmental phenomena. Arbitrariness is despised. Science is required to respect the sovereign facts, peer review and solutions constrained by clear methods. Even as one can support such values, one must recognise them as just that. The dangers of an implied superiority over alternatives can inhibit engagement with other cultures, generally at risk communities, and awareness of biases in our own socially constructed forms.

Some elements of culture critique

A useful approach is to think of 'culture' as it is in 'agriculture' or 'horticulture'. Humans are likely to cultivate whatever they come into contact with: gardens, clients, friendships and danger. With respect to nature, culture carries a sense of taming or domestication. Lack of cultural debate in the disasters community may reflect, in part, the dubious view of disasters as natural or 'untamed' threats. Discussion of the counter-argument that most, if not all, modern disasters are more appropriately called 'unnatural', even if earthquake or flood trigger them, is as much a cultural as technical and practical debate. It speaks to modern interpretations of disaster and responsibilities of the DRR community (cf. Wijkman and Timberlake 1984, Mileti 1999, Steinberg 2000).

Of cultivation, conversation and meaning

Cultivation has the comfortable ring of practicality but depends on broader cultural matrices and worlds. Behind each situation are many conversations between members of a group and outsiders. To appreciate a culture requires upbringing and training, an ability to share experience with others by speaking the community's language(s). Societies live in webs of symbolic communication that give distinctive meaning to feelings and perspectives, placing a huge burden on brainwork, invention and dialogue. In the modern world, a range of media and institutionalised exchanges add to and complicate more conventional cultural dialogues.

As knowledge workers, culturally given notions need our attention. They include definitions of 'disaster' itself – as Quarantelli (1998) insists. Other core concepts with cultural overtones are hazards, accident, uncertainty, vulnerability and resilience (cf. Green 1997, Bankoff 2001, Alexander and Davis 2012). Troublesome ideas, shared with the larger cultural milieu, include 'nature', land management, sustainability, security and social justice (see Wisner 2012, Hilhorst 2013, Renaud et al. 2013).

Professionals, such as those in DRR, work within modern disciplines and enterprises, and with evidence or messages developed in institutional frameworks that have their own culture milieus. They include safety cultures, perhaps of '(high) reliability' and others that appear 'risk-takers'. DRR activities require us to negotiate with officials and community leaders, pre-existing agencies, laws and protocols. Ideas inherited from our mentors and disciplines are hard to shake or challenge, not least those ideas that some of us helped to set up and want to defend.

However, a culture is unlikely to take shape and persist without some consensus among most members, or the more influential ones. A degree of continuity, conformity and entrenched formality is required. Even so, cultures are rarely, if ever, monolithic. Living cultures are partly transitional, if not provisional, struggling to adapt to new conditions. Some try to return to or re-invent old ones. Cultural concerns are most likely taken seriously when they seem threatened, or basic disagreements and contradictions arise.

Some contradictions

> *Everything is permeated by ambivalence; there is no longer any unambiguous social situation. Just as there are no more uncompromised actors on the stage of world history.*
> *Bauman and Donskis 2013, p. 5*

For its critics, modernity is an odd as well as a compelling and dangerous business. It is said to favour material and secular values, evidence and demonstration over traditional or privileged authority, innovation, growth and progress. As such, it has had a good press. On the negative side, those values are constantly under attack or are sources of conflict. 'Rational', industrialised forms have inspired some of the worst disasters. Stalin and Hitler sit at the head of this table, but are not alone. Discussions of disaster and modernity can hardly ignore the world wars. The greatest calamities in history, they were conflicts between the most modern states and generated many of today's prevailing cultural forms, including how disasters are treated (cf. Hewitt 1997).

For DRR the most troubling contradiction is, perhaps, how an increasing trend in disaster losses coincides with unprecedented growth in disaster-related investment and institutions. Few doubt that disaster numbers and losses have been growing (EM-DAT 2013). Less often mentioned is the great expansion in resources and organisations devoted to disaster concerns. Of late, profitable industries have developed around disaster response in reinsurance, security technologies, relief and reconstruction. According to Calhoun (2004), 'management of emergencies [is] a very big business [. . .] mobiliz[ing] tens of thousands of paid workers and volunteers [. . .]'. Hannigan (2012, p. 22) finds 'natural disasters' to comprise, '[. . .] a global policy field [. . .] becoming considerably more crowded and turbulent [with] the influx of thousands of new NGOs into emergency operations [. . .]'. He identifies further marked expansion – if greater confusion – as disaster management is entwined with climate change adaptation. Meanwhile, some places and activities do seem to enjoy safety standards, reliability and options far beyond anything in the past. They highlight questions of why such improvements and protections are missing where disaster losses increase worldwide for 'at risk' majorities.

The language and practices of disaster management emanate mainly from the wealthiest countries, but they are polarised there too. Some of us emphasise preventive, adaptive and sustainability agendas, a focus on people at risk, and long term prevention (cf. Pelling 2003, Wisner *et al.* 2004, Cannon 2008, Hewitt 2013). Similar concerns affecting DRR are official priorities of the Hyogo Framework for Action (HFA) and agencies such as the United Nations Development Programme (UNDP 2004) and International Federation of Red Cross and Red Crescent Societies (IFRCRCS 2004). However, in our so-called homelands, disaster management is increasingly part of broad security complexes with other priorities and subordinate to other threats. This is the more immediate context of the discussion.

Rationalities of risk, security and news

> *Risk is not assessed exclusively in terms of scientific knowledge [. . .] determinations are made within the prevailing criteria of rational acceptability among knowers in a particular culture.*
>
> *Ericson and Haggerty 1997, p. 90, emphasis added*

For present purposes, aspects and implications of three modern notions highlight some key issues: risk, security and news. These are found in common use, as well as in quite narrow technical and practical domains. Some see risk as the defining notion of modernity (cf. Douglas and Wildavsky 1982, Hacking 1990, Garland 2003). It gives a particular meaning or slant to public and private affairs. In itself, risk is an abstract concept to express the potential for harm. Narrower technical usage looks at the probabilities of adverse outcomes. In this way risk, or risk-benefit, can be used to bracket bio-ecological and economic life. It implies an actuarial framework, more or less well informed, but rooted in value judgements about what constitute gains and losses. In everyday talk, risk refers more to a sense of dangers or expectation of harm. It engages with anxieties, luck and life chances. The usage is broader and more flexible than quantifiable uncertainties, if less precise.

Security identifies concerns and organised actions to provide safety in the social and political realm. Formerly, the modern emphasis was on social security, safety nets for citizens against personal, economic or health misfortune, shared safety in cooperative and unionised groups. These now seem under attack even where they were formerly well entrenched. Security against external and internal 'enemies' receives ever more investment and actions. Where disaster management is housed in 'homeland security', threats dramatise survival of the nation or way of life, variously seen as actual, 'existential' or contrived (cf. Buzan *et al.* 1998). The prospect of harm or disaster is used to justify and legislate extraordinary measures. Below, some parallels are identified with *sécurité* as investigated by Michel Foucault.

For most of us, news means mass media coverage of current events, a source of information of compelling interest for citizens. Disasters are often front page or prime time news. In disaster management, there is limited discussion of media coverage, although quality or accuracy are widely questioned and attributed to a lack of well-trained disaster news hounds (Radford and Wisner 2012). This is unlikely to be an accident. Nevertheless, from a cultural perspective, the role of news media in popular and political notions of disaster is hard to overestimate – hence McLuhan's (1964) 'global village', cited in my title.

'The risk society'

> [. . .] *the production of risks is the consequence of scientific and political efforts to control and minimize them.*
>
> Beck 1998, p. 12

A major concern is to define and assess 'disaster risk', the likelihood of extremely destructive events, or casualties and damage. Discussions see people as being 'at risk' or 'living with risk' (UN/ISDR 2002). As a calculation of exposure to losses, risk is identified with damage statistics, probability functions and actuarial science. Typically, risk expresses a potential, is identified with 'uncertainty' and appears future-oriented. Ideally, it encompasses exposure to natural or technological hazards, vulnerability, resilience, access to and absent protections among other factors of endangerment. Technical projections for the future are mainly based on risk profiles from past events.

An influential idea, proposed by the German sociologist Ulrich Beck (1992), is of 'the risk society'. He sees it as 'a new modernity'; a world beset by 'manufactured risks' and people or policies preoccupied with them. Beck looks especially to dangers from modern industrial and consumer lifestyle, hazards and weapons of mass destruction or, rather, how these are perceived and treated in urban-industrial societies. Seemingly intractable, he believes that is because they overshadow the stereotypical features of modernity, its productive forces and the drive for wealth, growth and advancement. As such, the risk society not only differs from other cultures, but from modern culture before the later twentieth century. From its origins in the West, an emergent 'global risk society' is foreseen (Beck 1999).

While underscoring threats, Beck does not find modern life necessarily less safe than other times and places – although it can be. A counter-intuitive view is how new risk obsessions preoccupy the otherwise main beneficiaries of modernity who seem, from other perspectives, the most affluent, safest populations in history. Yet they are filled with anxieties about threats to health, habitat and social freedoms. He sees techno-scientific initiatives lying behind this as they experiment with public safety and personal health, and threaten ecological survival. He also suggests that, in the risk society 'the *state of emergency* threatens to become the *normal state*'; a '*catastrophic society*' even without disaster as usually understood (Beck 1992, pp. 98–9, emphasis in original). It is as if a pessimistic DRR mindset takes over.

From a cultural perspective, Beck's more radical claim is to trace today's great problems to 'the failure of *techno-scientific rationality*' (Beck 1992, p. 59, emphasis added). Does this also lie behind the increase in disasters? One could readily imagine that New Orleans after 'Katrina', or Fukushima, no less than the Bhopal or 'Deep Water Horizon' disasters, as examples of such failures. However, there are singular difficulties in blaming technocratic culture.

Rationalising failure

> *Disasters often involve regulatory failures. Somebody was responsible for safety and failed to ensure it, through negligence or lack of imagination, or both.*
>
> McLean and Johnes 2000, p. 220

A set of inquiries, looked at for other purposes (Hewitt 2012),[1] engages with Beck's concerns and offers a way to test them. Scientific and organisational efforts were their focus too. Detailed documentation is provided and expert testimony in relevant scientific and engineering fields, as well as from persons directly involved.

On one level they do seem to vindicate Beck. 'Techno-scientific rationality' is present and relevant in all cases. Pertinent risks and failures cannot be separated from 'scientific and political efforts to control and minimize them' (Beck 1992, p. 26). However, the conclusions do not actually state that the disasters were due to this, nor even offer support for such a view. Rather, inquiry conclusions call for more, not less, techno-scientific rationality. They underscore indifference to, or lack of, enforcement of known and mandated technical principles. They highlight failures to accept, adopt or maintain established technical standards and practices. Final recommendations urge that these take priority over other management concerns, notably 'bottom lines' and political influence. Also, unlike much of mainstream disasters work, external initiating hazards – storms, floods, explosions – were not deemed decisive causes (Hewitt 2012).

If anything, the conclusions find administrative 'rationality' more decisive – management priorities and agendas that over-rode or compromised safety concerns. Its failures were present in all cases. Thus:

> [What happened on the afternoon of the power failure] was not a cause in itself. Rather, deficiencies in corporate policies, lack of adherence to industry policies, and inadequate management [were responsible].
>
> (US-Canada Power System Outage Task Force 2004, p. 18)

> [. . .] if the Ministry of the Environment had adequately fulfilled its regulatory and over-sight role, the tragedy in Walkerton *would have been prevented [. . .] or at least significantly reduced in scope.*
>
> (O'Connor 2002, emphasis added)

Rather than faulty techno-scientific measures, the inquiries uncovered prior safety inspections, or concerned operatives, urging measures that could have reduced or prevented damages. Previous inquiries were uncovered into similar events in the same jurisdiction whose recommendations, if implemented, would have prevented or greatly reduced the disaster (cf. Campbell 2003, p. 5, Taylor 1990, p. 4).

Now, the inquiry commissioners and this author may be mistaken; 'mystified', as Beck suggests, by entrenched techno-scientific rationality. It surely has its problems, not least in mixed and cross-cultural contexts. However, scientists and technical staff were found to operate in much more restricted positions and ways than Beck's view suggests; rarely at levels where crucial decisions are made. When they are, findings show them tending to choose bureaucratic, military or business 'rationalities' over technical ones. Arguably, these identify the more critical cultures for modern safety.

Beck's more basic claim also seems challenged: that risk society has replaced the liberal agendas of wealth generation, growth, class and innovation. Since 1986, when his *'Risikogesellschaft'*-book first appeared, productivity in almost all countries, financial transactions and trade, have grown faster than population or disaster losses. So has ownership of modern items such as cars and cell phones as well as military spending (cf. Hewitt 2013). Far from disappearing behind risk preoccupations, a culture of growth and innovation still dominates economic and political priorities. The main source of his 'manufactured risk' surely involves the transformations of international labour, industrial production, resource extraction, finance and commerce associated with globalisation.

Two great risk obsessions of recent years, climate change and the post-2008 financial crisis, may seem to typify the 'risk society'. However, they too are driven by a culture of production and consumption; continued massive investment in and use of fossil fuels, and financial schemes to accumulate ever more wealth. Beck's concerns do seem to be increasingly relevant for a cultural critique of modernity, but related to the priorities of those with decision making powers, rather than scientists or consumers.

Systems of security

> In an age of uncertainty, we are continually facing new and unforeseen threats to our security.
>
> HM Government 2010, emphasis added

The French scholar Michel Foucault preferred to speak of 'systems of thought' or 'history of mentalities', but used 'culture' interchangeably with them (cf. Gordon 1991). His approach was to trace conditions back through historical precedents, in the genealogy of systems such as asylums and prisons (Foucault 1965, 1977). A broad outcome was to find, 'in short, one needed to analyse the series: security, population, government' (Foucault 1977, p. 87). He argued that a society's 'threshold of modernity' arrives when the object of government shifts towards 'life-administration' or 'bio-politics'; primarily the protection and disciplining of the national population (Foucault 1980, Bernauer 1990). Then, institutionalised security assumes a foremost role, one:

> [. . .] inscribed in and definitive of the emergence of the modern West, beginning in the eighteenth century 'Polizeiwissenschaft' or *science of policy* and 'Cameralism' or *science of administration*.
>
> (Gordon 1991, p. 10, emphasis added)

Although sharing Beck's interest in techno-scientific rationalism, Foucault finds the crux of the matter in political and administrative affairs, or 'Governmentality'. His thought inspires a surge of work in cognate areas of welfare, risk, poverty, civil society and insurance (cf. Burchell *et al.* 1991).

For DRR, critical parallels are between his 'apparatuses of security' and developments that place disaster preparedness within Homeland Security (HS). In the United States, 'The vision of homeland security [. . .] is to ensure a homeland that is safe, secure, and resilient against terrorism *and other hazards* where American interests, aspirations, and way of life can thrive' (US/DHS 2012, p. 9, emphasis added). For Britain, 'The National Security Council ensures *a strategic and tightly coordinated approach* across the whole of government to the risks and opportunities the country faces' (HM Government 2010, emphasis added). The approach and priorities make disasters subordinate to other concerns. For the United States, 'ensuring resilience to disasters' is 'Mission 5' (US/DHS 2012). Britain puts 'major accident or natural hazard which requires a national response' third, among four 'priority risks'. The other three concern armed or related aggression with 'International terrorism' the highest (HM Government 2010). Meanwhile, disaster management is integrated with war-preparedness, counterinsurgency, disease control, protection of major infrastructure, weapons inspection or decommissioning, border security, policing of international migrant labour and trafficking. Key elements are ever expanding surveillance, secrecy and carceral systems – to some more a threat to, than guarantees of, public safety. It would be difficult to overstate the implications for DRR in the twenty-first century. Some aspects of HS could exemplify the 'Foucault Effect' (Burchell *et al.* 1991), including the virtual Panopticon of the 'Cyber-Security Industrial Complex' (Talbot 2011), or the disciplining, punishing and profiteering of 'the Prison-industrial Complex' (Whitehead 2012).

The strong cultural underpinnings of HS are indicated by the quote above about 'an age of uncertainty' (HM Government 2010). The sense of a culture of contradictions (and exaggeration?) is reinforced by a Beckian assertion: 'Britain today is both more secure and more vulnerable than in most of her long history' (HM Government 2010, p. 3). However, these developments have a longer 'genealogy' in the modern emergence of 'bio-politics' (Foucault 1980) but, more narrowly and specifically, through a century of civil defence.

Disaster management as civil defence

> *In 1951 NATO established the Civil Defence Committee [and] it soon become apparent that the capabilities to protect our populations against the effects of war could also be used to protect them against the effects of disasters.*
>
> *NATO-OTAN 2001, p. 5*

Civil defence was developed mainly in the two world wars as air raid protection for cities (cf. Ikle 1958, Hewitt 1997). It was extended and modified for nuclear preparedness in the Cold War. Researchers formerly involved in wartime civil defence made important contributions to disaster studies (cf. Wolfenstein 1957, Baker and Chapman 1962, Barton 1969). Until the mid-1960s, Canada's Emergency Measures Organisation (EMO), for example, focused on civil protection in nuclear war, was staffed mainly by ex-military officers and closely linked to NORAD (North American Aerospace Defence Command). The Federal Emergency Management Agency (FEMA) in the USA, and its European and Soviet counterparts, had much the same policies. Peacetime 'mass emergencies' were an add-on, an opportunity to use 'assets' hanging around until nuclear Armageddon.

In the 1980s and 1990s, civil defence acquired a bad odour. It became clear defending civilians or, indeed, the biosphere against nuclear attack was a non-starter. Some saw it as a cruel joke (Piel 1962). However, in the 1980s, terrorist threats were already a major focus of national security (Clymer 2003). Reaction to the 9/11 attacks and military engagements in Iraq and Afghanistan brought an urgent remake of emergency preparedness. Along with 'ensuring resilience to disasters', the whole range of threats, responsible agencies and outsourced responses became situated within the, suitably termed, 'Security-Industrial Complex' (Mills 2004). This has further blurred the lines between disaster aid and military intervention, even giving rise to 'militant humanitarianism' and 'humanitarian wars'.

The idea of the security-industrial complex extends the 'military-industrial complex' of President Eisenhower's 1961 farewell speech. However, it was built on complexes already grown enormously in the Second World War, as indicated, for example, by operational or operations research that laid the foundations of modern management culture (Shrader 2006). Then again, fascist and state socialist dictatorships also co-opted and conjoined industrial, scientific and corporate systems, as in Germany's V-rocket and genocidal programmes, or Stalin's 'Gulag archipelago' (Hilberg 1989, Hewitt 1997). Echoes of Foucault appear in pursuit of political and wealth control through a 'state-within-the-state'.

The justifications for such developments have been built around actual, perceived, or even invented threats. A technical as well as a cultural critique of this could begin with world war civil defence. Recollections rarely go beyond the heroic notions – or enemy 'infamy' – yet in the event, civil defence was never enough and, essentially, failed. When needed most, in attacks from the air, the defences could not protect civilians. No doubt firemen and wardens did heroic work, countless civilians were extraordinarily brave, but no nation invested the resources needed, and nothing like the techno-scientific and industrial assets they had, given their own bomber fleets. The detailed police surveillance and reports on civilian morale suggest protecting them was not, anyway, the main purpose. The 'home front' projects were to control them, ensure they supported the war effort and put up with wartime sacrifices (cf. Stern 1947, Harrisson 1976, Havens 1978). Later, during the cold war, the shortfalls of civil defence against nuclear weapons are legendary.

A cultural effect is observed in socially constructed meanings of disaster. The security-industrial complex reinforces a 'patterns of war' (Gilbert 1998) or 'logic-of-war' paradigm (Buzan et al. 1998). The focus on threats has a similar effect to the hazards or agent-specific determinism that has prevailed in the disasters field (Hewitt 1983). Research, monitoring and reportage tend to foreground spectacular geophysical and other impacts, or 'heroic' external, humanitarian response. The approach has been referred to as a 'securitisation', 'disasterisation' or 'catastrophisation' of disaster policy – ugly terms for an ugly process, perhaps (Warner 2013). These refer to a logic of exception typical of crises or wartime. They favour priorities and actions echoing the security rationales outlined above and commonly lead to a suspension or bypassing of constitutional, stakeholder and market mechanisms (Warner 2013).

This security culture involves top-down hierarchy and centralised control of activities. Funding is pegged to disaster declarations, charitable donations and dedicated relief activities – after the event. 'The emergency' elbows aside long term risk reduction and preventive measures. Environmental risks are subordinated to 'new warfare' threats, victims' concerns overshadowed by the protection of major infrastructure. Inhabitants who are at risk or disaster victims are commonly portrayed as dysfunctional or panic-stricken, even though they typically carry out most of the immediate, life-saving actions, bear the brunt of the losses, hardships and disruptions, and are often side-lined in reconstruction efforts (Auf der Heide 2004). Modern relief operations commonly disenfranchised them. They may even find themselves treated like dangerous persons or 'felons'; likely to be shot at or incarcerated for breaking curfews, in supposed 'looting' or, not least, when demonstrating against inadequate, corrupt and abusive relief (Barsky *et al.* 2006).

The deployment of military forces in declared 'natural' and other disasters is taken for granted in most countries. It has expanded as those forces have grown. They are usually the first and largest to be mobilised, and their role is so important that military establishments have operating manuals for it; military colleges and journals are dedicated to training and preparedness. I cannot say military deployment is never necessary, even exemplary in some cases and, perhaps, inescapable, given today's out-of-control military spending. Excepting a few societies or enclaves with high quality public and private services, only the armed forces have the personnel and equipment readily available for emergency relief. Nonetheless, militant preparedness is troubling.

Contrary to a common impression, military contributions in disasters are not free, nor cheap. Well paid professional forces, ships and planes, helicopters, motor fuel, accidents and the like absorb a large fraction of outlays. Haiti is a rare case where breakdown of costs is available. In the first three years after the earthquake over a fifth, possibly a third, of outlays went to armed forces – less than 10% in total to the Haitian Government, NGOs and businesses, barely 2.5% to victims directly (UN/SC 2010, UN/OSEH 2012).

A blessing in disguise?

The security-industrial complex in the disasters scene resembles the United States prison system where, according to Lilly and Knepper (1993) decisions rest 'within a closed circle or elite of government bureaucrats, agency heads, interest groups and private interests that gain from *the allocation of public resources*' (p. 152, emphasis added). A significant and growing part of the expenditures and organisations involved reflect commercial and geostrategic opportunities. Investments in risk technologies, notably catastrophic risk insurance and commitments of financial institutions, have grown exponentially (Kunreuther and Michel-Kerjan 2010, Swiss Re 2012). The World Bank and Asian Development Bank have funded more than one thousand DRR projects in recent years (World Bank/IEG 2006, pp. 12–14, ADB 2013). Large corporations or their subsidiaries play an ever greater role.

A multi-billion dollar 'emergency' industry claims the lion's share of research, expenditures and institutions that respond to disaster. Reports from Haiti and New Orleans show how, when actions are militarised and secretive, disaster expenditures and benefits are awarded 'in house', increasingly for profit and with little or no accountability. They challenge a presumption that, in disasters, relief of suffering outweighs other agendas or that no one benefits.

Official and charitable humanitarian funds may offer unusual opportunities. They can escape constraints that encumber foreign aid, development bank loans and national debt, taxes and oversight, not least in times of enforced austerity. If this sounds like Klein's (2007) 'shock doctrine', in many ways it is, especially where she documents recent capitalising on disaster. Again, a critical issue is how this all happens with an overall rise in disaster numbers and losses, and whether it is a factor. It does seem contrary to principles promoted by HFA and agencies such as the United Nations Development Programme (UNDP) and IFRCRCS, yet rarely part of popular awareness or media reports.

News lines: disaster as spectacle, opportunity and propaganda

> It has been a long-standing chestnut of terrorist studies that the mass media and terrorism thrive off one another.
>
> Clymer 2003, p. 215

Disasters loom large in media coverage, usually on the front page or in prime time news. Media revolutions have altered the way events are known about and treated. Journalists and, increasingly, survivors with digital electronic equipment help shape the story of disaster-time impacts or crisis response.

New technologies such as cell phones do offer immediate, 'on the-ground' sources. Mainstream media can hardly ignore them, but do they pointedly call the material 'unverified' to contain or diminish them? In fact, most news is 'unverified'. McLuhan's 'the medium is the message' has singular cultural implications for disaster notions. Nevertheless, governments and news corporations go to great lengths to shape messages to favour their own interests, constituencies or advertisers. The news is also about storylines and socially constructing opinion and belief, as well as datelines, spectacle and the makeup of audiences. This does allow editorial and network intervention, with cultural implications for societies, even professions, whose members depend on the media for information and opinion.

Conventional wisdom says the modern progressive state needs independent media to serve an informed public and root out abuses of power. If so, the reality can be another story. From reports in places such as New Orleans, Haiti and Fukushima, it seems the wartime strategy of 'embedded' journalists increasingly prevails in disasters. Independent reporters may be excluded from, detained or killed in crisis zones (Rash 2011).

Media studies themselves have been preoccupied with how the experience of time and space has been transformed. In McLuhan's (1964) 'global village' the latest electronic media are seen to virtually erase time, distance and geography (cf. Virilio 2007). These are compelling changes, but potentially deceptive and, in the present context, misleading. Disasters plunge people into old, even anciently slow speeds and friction of distance. They are thrown back into Stone Age survival, even with working cell phones and helicopters circling overhead! Emergency food or medical intervention may now take hours or days where, a generation or two ago, it required months. Nonetheless, these hours or days are the most life threatening moments, and decisive aspects of trauma, response and relief do not operate instantly or in virtual space. They turn upon how people deal with immediate and terrible realities. We have to ask ourselves, does the news adequately convey this?

The images and storylines are remarkably similar for every disaster – datelines and local, 'cultural' objects excepted. Disaster images are mostly of spectacular destruction and heroic rescue: maps, experts and spokespersons in distant metropolitan and agency headquarters; lorries, planes and ships bringing modern goods and expertise from donor countries and by dedicated agencies. Commentary tends to support a hazards determinism of extreme, 'unscheduled' events view, and an 'expert systems' response. They provide popular culture with the imaginaries of 'Acts of God', 'Mother Nature', 'enemies' and miraculous survivals; awareness and decisions shaped by the emergency. Glimpses of wretched survivors notwithstanding, there is little sense of disasters as sober reminders of public safety failures, of underlying risks and avoidable losses. Most news broadcasts tend to minimise the social relations of loss, despite evidence of major differences by class, wealth, gender, ethnicity or religion. Typically, the newsreels infer that the wretched appearance, thin bodies, poor clothes, shredded homes of most victims are products of the disaster. Research suggests they most often record long term hardship and chronic distress – disasters waiting to happen.

Disasters would not be as common in prime time and headline news if they were not successful 'copy'. Nowadays, movies and TV series about disasters are not far behind crime, war, spy and hospital dramas as steady and lucrative offerings. Back in vogue are 'Forces-of-Nature' documentary-dramas, with the obligatory Social Darwinist/Neo-Malthusian voice-over. The spectacle of Nature's wrath or human error apparently make for better stories than community-level mitigation and victim's concerns. Obviously, modern news media profit from the stories they tell, as they boost sales and advertising revenue. It can increase the value of disaster specialists like us. But cannot the stories and our work do a better job if focused on humanitarian and DRR goals? The question arises as part of cultural critique.

Concluding remarks

These brief observations are intended to illustrate a need for cultural awareness and critique in modern contexts. They underscore the prevailing, better-funded and broadcast aspects of disaster management built around Homeland Security, if I arrive at a rather negative view of that. However, the critique arises specifically from a DRR perspective, mainly with principles advocated in the Hyogo Framework (HFA). These too are artefacts of modern thought and practice. They are strongly supported by research, a range of institutions and leading international agencies. From the discussion so far, they may seem more like a countercultural alternative to the powerful security-industrial complex. Yet, 'poor relation' or not, DRR offers a viable and attractive modern alternative, and will serve as a more positive reprise.

First and foremost, Hyogo principles turn away from threat-driven, aggressive and repressive systems, their crisis-focus and top-down organisation. Militant humanitarianism is countered by calls for greater cultural sensitivity, non-violent, conciliatory and restorative measures. The argument is not against techno-scientific rationality, nor the participation and responsibility of any institutions capable of reducing harm. However, they must be balanced by appropriate, participatory involvement of communities at risk and respect for their everyday concerns.

The 'emergencies first' approach is countered by calls to address, fund and prioritise risk reducing and preventive measures in the longer term. HFA was based on constructive social processes that emphasise the traditions of serving the 'public good', popular needs and stakeholder concerns. Safety is measured according to the wellbeing, resilience and uplift of society at large and, most obviously, for those vulnerable to environmental dangers and least equipped to respond to them. The DRR literature recognises that the latter are pivotal to the ongoing increase in disaster events and losses, as well as persons with the least voice in the institutions that dominate official disaster policy (cf. Davis 2006, Cannon 2008, Hewitt 2013).

Culturally sensitive approaches to preventive and community safety have a different focus and style from the security-industrial complex. In summary they focus on people rather than rules, persons at risk rather than 'systems' or 'indicators'; on enhancing relationships rather than 'processing', participation rather than control, shared decision making rather than top-down, enforced compliance; on non-violence rather than coercion, and reconciliation and compensation rather than policing and punishment. These too reflect cultural values and practices emergent from modernity and widely seen as its more progressive elements.

Finally, it should be stressed that all of this is not so new and was, perhaps, more widely accepted two or three decades ago. A singular inspiration is the work of Gilbert F. White (1911–2006), the distinguished American geographer and peace activist. He developed a conservationist and human ecology approach to environmental hazards. It identified and advocated carefully choosing among the range of possible adjustments (White 1969). His initial concerns with flood hazards countered a prevailing approach based on flood-fighting, dams and other flood control structures; hugely expensive projects, largely the responsibility of the US Army Corps of Engineers, and challenged by the recurrence of destructive floods. White pointed to the benefits of broadening responses to include watershed and flood plain management, multi-purpose projects, insurance and other precautionary or avoidance options – decidedly non-violent methods and working with not against rivers and shorelines. In hazard perception studies, he inspired an effort to listen to and assess the awareness of disaster-prone individuals and their communities. Ultimately his influence has been global (White 1974) and continues in the ongoing work of his many students (cf. Wescoat and White 2003).

To summarise, DRR calls for societal approaches very different from the 'patterns of war' emergency measures and agent-specific paradigms. It looks to approaches that prioritise:

- prudential, preventive and precautionary approaches to dangers;
- risk assessments and measures that recognise and respect the connectedness of people, cultures and habitats;
- living and acting with nature, rather than coercing, abusing or making war against it;
- working towards fairness and justice in social safety, differentiating assistance only according to need;
- enhancing or creating social arrangements that address the needs of all but, especially, those most at risk from and harmed by disasters;
- refusing to invest in structures and cultural norms that separate and privilege some people at the expense of others, or exclude the very people and groups whose risks need to be addressed;
- non-aggressive, participatory responses, cooperative and sharing models.

Such notions have been pioneered in mutualist and peace-making modern thought and through transformative policy initiatives. They have been close to the heart of kindred developments such as the Earth Summit (UNCED) 1992, the Man and Biosphere Programme and the Kyoto Accord among others. If they seem countercultural, or less 'realist' than some would like, they reflect well-developed alternatives within modern cultural and progressive thought.

Note

1 Inquiries reviewed are detailed in the citation.

References

ADB, 2013. *Investing in Resilience: Ensuring Disaster-Resistant Future*. Manila: Asian Development Bank.

Alexander, D. and Davis I., 2012. Disaster risk reduction: An alternative viewpoint. *International Journal of Disaster Risk Reduction*, 2, 1–5.

Auf der Heide, E., 2004. Common Misconceptions about Disasters: Panic, the 'Disaster Syndrome', and Looting. In: M. O'Leary, ed., *The First 72 Hours: A Community Approach to Disaster Preparedness*. Lincoln, NE: iUniverse Publishing, 340–80.

Baker, G. W. and Chapman, D. W., eds, 1962. *Man and Society in Disaster*. New York: Basic Books.

Bankoff, G., 2001. Rendering the world unsafe: 'Vulnerability' as Western discourse. *Disasters*, 25 (1), 19–35.

Barsky, L., Trainor, J. and Torres, M., 2006. Disaster realities in the aftermath of Hurricane Katrina: Revisiting the looting myth. *Quick Response Report* [online], 184. Available from: www.colorado.edu/hazards/qr/qr184/qr184.html [accessed 18 February 2014].

Barton, A. H., 1969. *Communities in Disaster. A Sociological Analysis of Collective Stress Situations*. New York: Doubleday.

Bauman, Z. and Donskis, L., 2013. *Moral Blindness: The Loss of Sensitivity in Liquid Modernity*. London: Polity.

Beck, U., 1992. *Risk Society: Towards a New Modernity*. London: Sage.

Beck U., 1998. Politics of Risk Society. In: J. Franklin, ed., *The Politics of the Risk Society*. London: Polity, 9–22.

Beck, U., 1999. *World Risk Society*. London: Polity.

Bernauer, J. W., 1990. *Michel Foucault's Force of Flight: Towards an Ethics for Thought*. Atlantic Highlands, NJ: Humanities Press International.

Burchell, G. M., Gordon, C. and Miller, P., 1991. T*he Foucault Effect: Studies in Governmentality*. London: Harvester Wheatsheaf.

Buzan, B., Waever, O. and Wilde, J., 1998. *Security. A New Framework*. London: Lynne Rienner.

Calhoun, C., 2004. A world of emergencies: Fear, intervention, and the limits of cosmopolitan order. *The Canadian Review of Sociology and Anthropology*, 41, 373–95.

Campbell, A., 2003. SARS Commission: Interim Report. Toronto: Government of Ontario.

Cannon, T., 2008. Vulnerability, innocent disasters and the imperative of cultural understanding. *Disaster Prevention and Management*, 17 (3), 350–57.

Clymer, J. A., 2003. *America's Culture of Terrorism: Violence, Capitalism and the Written Word*. Chapel Hill, NC: University of North Carolina Press.

Davis, M., 2006. *Planet of Slums*. New York: Verso.

Douglas, M., 1992. *Risk and Blame*: Essays in Cultural Theory. London: Routledge.

Douglas, M. and Wildavsky, A., 1982. *Risk and Culture: An Essay on the Selection of Technological and Environmental Dangers*. Berkeley, CA: University of California Press.

EM-DAT, 2013. *The OFDA/CRED International Disaster Database* [online]. Université Catholique de Louvain, Brussels. Available from: www.emdat.be [accessed 18 February 2014].

Ericson, R. V. and Haggerty, K. D., 1997. *Policing the Risk Society*. Toronto: University of Toronto Press.

Foucault, M., 1965. *Madness and Civilization: A History of Insanity in the Age of Reason*. London: Tavistock.

Foucault, M., 1977. *Discipline and Punish: The Birth of the Prison*. London: Allen Lane.

Foucault, M., 1980. *Power/Knowledge: Selected Interviews and Other Writings, 1972–1977*. New York: Pantheon Books.

Franklin, J., ed., 1998. *The Politics of the Risk Society*. London: Polity.

Garland, D., 2003. The Rise of Risk. In: R. V. Ericson and A. Doyle, eds, *Risk and Morality*. Toronto: University of Toronto Press, 48–86.

Gilbert, C., 1998. Studying Disasters: Changes in the Main Conceptual Tools. In E. L. Quarantelli, ed., *What is a Disaster? Perspectives on the Question*. New York: Routledge, 11–18.

Gordon, C., 1991. Government Rationality: An Introduction. In G. M. Burchell, C. Gordon and P. Miller, eds, *The Foucault Effect: Studies in Governmentality*. London: Harvester Wheatsheaf, 1–52.

Green, J., 1997. *Risk and Misfortune: The Social Construction of Accidents*. London: University College Press.

Hacking, I., 1990. *The Taming of Chance*. Cambridge: Cambridge University Press.

Hannigan, J., 2012. *Disasters without Borders: The International Politics of Natural Disasters*. London: Polity.

Harrisson, T., 1976. *Living Through the Blitz*. London: Collins.

Havens, T. R. H., 1978. *Valley of Darkness: The Japanese People and World War Two*. New York: W. W. Norton.

Hewitt, K., 1983. The Idea of Calamity in a Technocratic Age. In K. Hewitt, ed., *Interpretations of Calamity: From the Viewpoint of Human Ecology*. London: Allen and Unwin, 3–32.

Hewitt, K., 1997. *Regions of Risk: Hazards, Vulnerability and Disasters*. London: Longman/Pearson.

Hewitt, K., 2012. Environmental disasters in social context: Toward a preventive and precautionary approach. *Natural Hazards*, 66 (1), 3–14.

Hewitt, K., 2013. Disasters in 'development' contexts: The contradictions for a preventive approach. *Jàmbá: Journal of Disaster Risk Studies*, 5 (2): 1–8.

Hilberg, R., 1989. The Bureaucracy of Annihilation. In: F. Furet, ed., *Unanswered Questions: Nazi Germany and the Genocide of the Jews*. New York: Schocken Books.

Hilhorst, D., ed., 2013. *Disaster, Conflict and Society in Crisis*. London: Routledge.

HM Government, 2010. *A Strong Britain in an Age of Uncertainty: The National Security Strategy*. London: The Cabinet office, HMSO.

IFRCRCS, 2004. World Disasters Report: Focus on Community Resilience. Geneva: International Federation of Red Cross and Red Crescent Societies.

Ikle, F. C., 1958. *Social Impacts of Bomb Destruction*. London: William Kimber.

Klein, N., 2007. *The Shock Doctrine: The Rise of Disaster Capitalism*. Toronto: Knopf Canada.

Kunreuther, H. C. and Michel-Kerjan, E. O., 2010. Market and Government Failure in Insuring and Mitigating Natural Catastrophes: How Long-Term Contracts Can Help. In W. Kern, ed., *The Economics of Natural and Unnatural Disasters*. Kalamazoo, MI: W. E. Upjohn Institute, 9–38.

Lilly, J. R. and Knepper, P., 1993. The correctional-commercial complex. *Crime and Delinquency*, 39, 150–66.

McLean, I. and Johnes, M., 2000. *Aberfan: Disasters and Government*. Cardiff: Welsh Academic Press.

McLuhan, M., 1964. *Understanding Media: The Extensions of Man*. New York: McGraw-Hill.

Mileti, D., 1999. *Disasters by Design: A Reassessment of Natural Hazards in the United States*. Washington, DC: Joseph Henry Press.

Mills, M. P., 2004. The Security-Industrial Complex. *Manhattan Institute for Policy Research* [online]. Available from: www.manhattan-institute.org [accessed 18 February 2014].

NATO-OTAN, 2001. *NATO's Role in Disaster Assistance.* Brussels: NATO Civil Emergency Planning.

O'Connor, D. R., 2002. *Report of the Walkerton Inquiry.* 1st ed. Toronto: Ministry of the Attorney General, Queen's Printer for Ontario.

Pelling, M., 2003. *The Vulnerability of Cities: Natural Disasters and Social Resilience.* London: Earthscan.

Piel, G., 1962. The illusion of civil defense. *Bulletin of the Atomic Scientists*, 18 (2), 2.

Quarantelli, E. L., ed., 1998. *What is a Disaster? Perspectives on the Question.* New York: Routledge.

Radford, T. and Wisner, B., 2012. *Media, Communication and Disasters.* In: B. Wisner, JC Gaillard and I. Kelman, eds, *The Routledge Handbook of Hazards and Disaster Risk Reduction.* London: Routledge, 761–71.

Rash, J., 2011. Press Under Fire, and Not Just in War Zones. *The Star Tribune* [online]. Available from: www.startribune.com/opinion/118682489.html [accessed 18 February 2014].

Renaud, F. G., Sudmeier-Rieux, K. and Estrella, M., eds, 2013. *The Role of Ecosystems in Disaster Risk Reduction.* Tokyo: United Nations University Press.

Shrader, C. R., 2006. *History of Operations Research in the United States Army: Volume 1: 1942–62.* Washington, DC: United States Army.

Steinberg, T., 2000. *Acts of God: The Unnatural History of Natural Disaster in America.* New York: Oxford University Press.

Stern, J., 1947. *The Hidden Damage.* New York: Harcourt Brace.

Swiss Re., 2012. *Global Insurance Review 2012 and Outlook 2013/14.* Zurich: Swiss Reinsurance Company.

Talbot, D., 2011. The Cyber Security Industrial Complex. *MIT Technology Review* [online]. Available from: www.technologyreview.com/news/426285/the-cyber-security-industrial-complex/ [accessed 19 February 2014].

Taylor, P., 1990. *The Hillsborough Stadium Disaster. 15 April 1989* (Inquiry by Lord Justice Taylor). London: Home Office, HMSO.

UNDP, 2004. *A Global Report, Reducing Disaster Risk: A Challenge for Development.* New York: United Nations Development Programme, Bureau of Crisis Prevention and Management.

UN/ISDR, 2002. *Living With Risk: A Global Review of Disaster Reduction Initiatives.* Geneva: United Nations, International Strategy for Disaster Reduction.

UN/OSEH, 2012. *Can More Aid Stay in Haiti and Other Fragile Settings? How Local Investment can Strengthen Governments and Economies.* New York: United Nations Office of the Special Envoy for Haiti.

UN/SC, 2010. *Report of the Secretary General on the United Nations Stabilizing Mission in Haiti.* Report S/2010/200. New York: UN Security Council.

US-Canada Power System Outage Task Force, 2004. *Final Report on the August 14th Blackout in the United States and Canada.* Ottawa: Government of Canada.

US/DHS, 2012. *Mission Statement.* Washington, DC: Department of Homeland Security.

Virilio, P., 2007. *The University of Disaster.* London: Polity.

Warner, J., 2013. The Politics of 'Catastrophization'. In: D. Hilhorst, ed., *Disaster, Conflict and Society in Crisis.* London: Routledge, 76–94.

Wescoat, J. L. Jr. and White, G. F., 2003. *Water for Life: Water Management and Environmental Policy.* Cambridge: Cambridge University Press.

White, G. F., 1969. *Choice of Adjustments to Floods.* (Vol. 93). Chicago: University of Chicago Press (Department of Geography Research Papers).

White, G. F., ed., 1974, *Natural Hazards: Local, National, Global.* New York: Oxford University Press.

Whitehead, J. W., 2012. Jailing Americans for Profit: The Rise of the Prison Industrial Complex. *The Rutherford Institute* [online]. Available from: www.rutherford.org/publications_resources [accessed 18 February 2014].

Wijkman A. and Timberlake, L., 1984. *Natural Disasters: Acts of God or Acts of Man?* London: Earthscan.

Wisner, B., Blaikie, P., Cannon, T., Davis, I., 2004. *At Risk: Natural Hazards, People's Vulnerability and Disasters.* 2nd ed. London: Routledge.

Wisner, B., 2012. Violent Conflict, Natural Hazards and Disasters. In B. Wisner, J. C. Gaillard and I. Kelman, eds, *The Routledge Handbook of Hazards and Disaster Risk Reduction.* London: Routledge, 71–82.

Wolfenstein, M., 1957. *Disaster: A Psychological Essay.* Glencoe: Free Press.

World Bank/IEG, 2006. *Hazards of Nature, Risks to Development: An IEG Evaluation of World Bank Assistance for Natural Disasters.* Washington, DC: The World Bank, Independent Evaluation Group.

2

CONVERSATIONS IN CATASTROPHE

Neoliberalism and the cultural construction of disaster risk

Anthony Oliver-Smith

Introduction

Until recently, disaster research and management have focused largely on the infrastructural, demographic, political, ecological and socioeconomic aspects of disaster, from pre-event vulnerability to impact and through reconstruction, in the process almost totally ignoring the cultural aspects of disasters. However, it has become clear that neglecting the deep cultural roots of every aspect of the disaster scenario has left troubling insufficiencies in research and tragic deficiencies in disaster praxis (Hoffman 2013). The problem has been exacerbated by our disciplinary compartmentalisation leading us to disengage culture conceptually from economy, society, politics and the environment when in fact all human constructions are based on culturally framed symbolic representations. This shared symbolic meaning is the foundation of culture, which in turn is the basis of moralities and values, and conditions social organisation and social reproduction. It is clear that a fully integrated research methodology on disaster risk reduction must include culture (ICSU 2008). The task of articulating the relationship between disaster and culture can be challenging because neither concept has lent itself to hard and fast definitions to the effect that framing the relationship becomes an exercise in finding links between a set of extremely varied and dynamic phenomena, only remotely 'fixable' in the context of specific cases.

The multidimensionality of disaster

Over the last fifteen years there has been some concern that, partially due to its disparate origins and conceptual and practical diversity, the field of disaster research and management suffers from a lack of consensus on the concept of disaster that potentially undermines both the intellectual integrity of disaster studies as well as its research enterprise. In short, the question 'What is a disaster?' is not considered to have been satisfactorily answered (Quarantelli 1998, Perry and Quarantelli 2005).

What is it about disasters that has made it so problematic to reach a consensual definition? One problem is that a disaster is a collectivity of intersecting processes and events: social, environmental, cultural, political, economic, physical and technological, and it can transpire over varying lengths of time. Disasters are totalising events. As they unfold, all dimensions of a sociocultural formation and the totality of its relations with the environment may become involved, affected, and focused, expressing consistency and inconsistency, coherence and contradiction, cooperation and conflict, hegemony and resistance, and can be expressed through the operation of biophysical and sociocultural systems and their interaction among populations, groups, institutions and practices (Oliver-Smith and Hoffman 2002). Disasters bring about the multiple forces and factors in causal chains of such features as natural forces or agents, the intensification of production, population increase, environmental degradation, diminished adaptability and all their sociocultural constructions. An adequate definition and approach to disaster must be able to encompass this multidimensionality (Oliver-Smith 1998).

Thus, disasters are both culturally constructed and socially experienced differently by different groups and individuals, generating multiple interpretations of an event/process. A single disaster can fragment into different and conflicting sets of circumstances and interpretations according to the experience and identity of those affected. Disasters force researchers and practitioners alike to confront the many and shifting faces of culturally imagined and socially enacted realities (Hoffman and Oliver-Smith 2003).

We are also aware that the culturally imagined realities are themselves extremely dynamic through time. The disaster process, as it evolves through various stages, which while difficult to demarcate exactly for and experienced differently by every affected individual and group, displays characteristic social relations, that are informed by situated knowledge forms and moralities, among the affected population as well as between the affected population and humanitarian and reconstruction aid agents. That is, different forms of knowledge and different moralities become activated as the character of social relations changes in the evolution of the disaster process.

Culture: the human concept

Culture presents its own complexities. Edward Burnett Tylor, often referred to as the father of anthropology in the nineteenth century, said that 'Culture or civilisation, taken in its broad, ethnographic sense, is that complex whole which includes knowledge, belief, art, morals, law, custom, and any other capabilities and habits acquired by man as a member of society' (Tylor 1920). Despite the clear existence of animal 'proto-cultures', culture in the sense of symbolic expression of abstract ideas is the human concept. All animals fear; only humans imagine risk.

However, conceptual or definitional consensus has not been easily come by. In 1952 Alfred Kroeber and Clyde Kluckhohn, two of the most eminent anthropologists of the time, published a book entitled *Culture: A Critical Review of Concepts and Definitions* in which they analysed the distinction properties and components of 164 different definitions of culture (1952). Although the number of definitions to which anthropologists currently ascribe has diminished, total consensus on the concept has not been reached. Since then, debates have struggled over the issues of whether culture was primary (Geertz 1973, Sahlins 1976), or secondary to material relations (Harris 1979), whether culture is subject to objective scientific analysis or humanistic interpretation, and whether cross-cultural generalisations are possible or whether culture is idiosyncratic, accessible only in its own terms (Rosaldo 1989).

Culture in the anthropological sense is generally held to be the grid, lens or frame through which we experience and interpret the material world and the world of our experience. It is through cultural knowledge, belief and attitudes that we generate behaviour or actions. Culture commonly refers to the different ways that people live, including kinship, subsistence, economics, politics, dress, food, exploitation of place, scheduling of time, arrangement of space, classification of humans, social roles and relationships, child raising, rules and laws. Culture is also a people's cosmology: how they see their gods, their ancestors, the universe, earth and stars, their spirituality, beliefs, explanations and their beliefs about their origins and purpose. It guides a people's perception of colours, smells, sounds and touch, and therefore how they filter and format information. Every human absorbs the culture of the group in which they are raised from childhood on up, and by and large live their lives by its grid, but always through the filter of individual experience (Hoffman 2013).

And therein lies one of the sources of change. Culture changes at varying rates, usually slowly, but in some circumstances very rapidly. Because of that change, not all the people of a society live exactly according to their culture's dictates, but the degree of their variance is rather like the stretch of a rubber band: limited. It also does not mean that at times people do not contest their culture. They do, but their contestation emerges from the reflective comprehension of their own life ways (Hoffman 2013). Practices such as speaking one's native tongue or practising one's religion, all constitute elements of culture and cultural heritage. Such elements play a central role in individual and collective identity formation, in the way that time and history are encoded and contextualised, and in interpersonal, community and intra-cultural relations. Space, kin relations, local communities, cosmology and tradition are linked in the construction of what Giddens calls 'an environment of trust' (1990). The relationship between people and their environment – their place – is thus culturally encoded.

Current understandings of culture draw on writers such as Gramsci (1971) and Bourdieu (1977) who frame culture as an array of discourses – essentially a conversation (hence my title) about how life should be lived and how it actually is lived. This perspective is therefore more dynamic. It is culture as a process of production of understandings, as constantly being produced, interpreted, reinterpreted, imposed and resisted. In any community, much less any entire society, there are many different views about how life should be lived. There are no completely homogeneous communities. Since society is not generally the organisation of completely equitable distribution of resources, every society is a dynamic arena of contesting interests organised along some lines of differentiation, whether they be only age and gender, or in more complex situations such as colour, class, caste, religion, ethnicity, kinship or any of the other myriad ways that humans have of grouping themselves and distinguishing others seen as different.

In effect, all the material and economic structures that channel our behaviours are ultimately based on and derived from culturally based values about life and the world, values about what the relationships should be between human beings, and between human beings and the material world. In other words, even physical forces such as earthquakes and hurricanes have to be culturally interpreted before they can be responded to. As Alf Hornborg has recently noted, culture is crucial to socio-ecological processes and relations. It shapes our way of thinking about ecology, economy and technology, and generates our specific kinds of fetishism and consumption, many of which specifically channel the way we inhabit and use our environment (2009, p. 255). And disasters are socio-ecological processes par excellence.

Over the decades anthropologists have explored the depth and nuances involved in human culture, from the overarching to the local level. It is, however, culture in its most discerning sense that has proven the key to understanding why and how various people worldwide deal with risk and disaster. It is central in the calculation of risk, the construction of vulnerability, the experience of catastrophe and recovery or its failure.

Thus, culture is the conversation among all those different people who share minimally the same space and language about life and the way it is or should be lived. Culture is not a single voice, but multiple voices sharing some perspectives, differing in others, all articulating their particular interpretations. All the social characteristics that significantly structure people in a society will play a role in the way those meanings and explanations are constructed. Some voices however can, by virtue of different kinds of power and wealth, come to dominate the conversation. Nowhere is this more apparent than in the conditions of vulnerability and the aftermath of disasters. Everyone in a disaster has a particular perspective and a particular set of interests, perhaps shared with others with whom they share certain characteristics.

Globalisation, culture and disasters

Today, and for quite some time, the process of globalisation has further complicated our understanding of both culture and disasters. Just as the emergence of print media became a major force in the development of national cultures (Anderson 1983), today electronic mediation, transport technologies, multiple forms of mass migration, social movements and rapidly circulating information and imagery breach the localities of village as well as nation (Appadurai 1996). The increasing post-Second World War pace of global capitalist expansion, largely under the rubric of a model of development and economic growth known as neoliberalism, has both underwritten and been intensified by these processes. Although globalisation and neoliberalism are clearly closely related, for present purposes I will refer to globalisation as a set of processes underway in the world and neoliberalism as a general cultural construct, informing both social and environmental relations, that is both producer and effect of globalising processes with profound implications for the construction of risk and vulnerability to natural and anthropogenic hazards.

Globalisation, in my understanding – which draws significantly on Hoogvelt (2001) – is comprised of an array of mutually constitutive economic, social, cultural and ecological flows of energy, information, imagery and material, intersecting densely at various points around the globe, and less densely at others. Metaphors or images of flows or currents are apt because they depict the circulation of capital, the passages of people, and the transfers of energy and material. These flows or currents encircle or encompass the globe certainly, but their courses do not spread them out evenly, particularly in terms of outcomes or impacts. They converge and disperse in accord with very specific logics. The availability of advanced communication and transport technology permits these flows of energy, information and material to intersect or disperse at vastly reduced time and space scales producing a set of impacts that are both the logical outcomes of longstanding systems and relationships, and in some ways fundamentally new.

The contemporary global market is clearly differentiated from that of earlier moments in the first half of the twentieth century by the adoption of geographically dispersed productions systems by multinational corporations with the effect that intra-firm trade now constitutes fully one-third of world trade (Hoogvelt 2001). These integrated production systems have opened the way for a global market principle to exert new forms of discipline on domestic supplies of capital, labour and resources. In effect, national boundaries no longer serve to protect workers and companies – particularly in the core nations – from the competitive discipline of the global market, thus placing the definition and powers of the state in play.

The ability of transnational corporations to organise production through subcontracting or outsourcing of both services and assembly fundamentally alters the capital–labour relation. Companies can now organise themselves as primarily design and marketing operations, outsourcing all their production activities, leading to 'forms of production in which capital no longer needs to pay for the reproduction of labor power' (Hoogvelt 2001, p. 113). Under these conditions of the commodification of production capacity, much employment becomes temporary or casual, the wage paid only on delivery of goods rather than for a period of time during which the entire production process takes place. Under conditions of global competition, not only for markets but for jobs as well, wages are pressured downward and the benefits associated with permanent employment are eliminated.

Today the growth of financial transactions far exceeds the growth of the underlying economic fundamentals of production and trade (Hoogvelt 2001, p. 128). Global financial transactions now constitute at least double the value of global production. More money is now being made by manipulating the circulation of currency rather than commerce in actual goods. Furthermore, flows of capital and commodities are intensifying in the core rather than in peripheral nations that are, in effect, not even worth exploiting in global financial operations (Hoogvelt 2001, p. 86). In the words of John Reed, at the time the chairman of Citicorps (the largest American bank): 'There are five billion people living on earth. Probably 800 million of them live in societies that are bankable and probably 4.2 billion (84%) are living within societies that in some very fundamental way are not bankable' (quoted in Hoogvelt 2001, p. 83).

'Bankable' here refers to the probability of offering a safe return on investment. If societies are not bankable it is unlikely that significant investments of capital for growth and development will be attracted. If more and more secure money is to be made elsewhere in the global economy, the societies of the third world – roughly 85% of the world's population – are essentially structurally irrelevant (Hoogvelt 1997, p. 84). Or in the words of American anthropologist, Sherry Ortner (2011), the global economy is creating populations that are 'dispensable, disposable and replaceable'.

David Harvey has referred to this total process as a system of 'accumulation by dispossession', which he summarises as having four main components: 1) the privatisation and commodification of public goods; 2) 'financialisation', which refers to the conversion of virtually anything, good or bad, into an instrument for economic gain; 3) the management and manipulation of crises; and 4) 'state redistribution' in which the neoliberal state abets and protects the flow of wealth towards privileged sectors (2007, pp. 159–64).

These processes have in turn energised a number of economic, cultural and social flows that are expressed in the ways human actions impact the environment, setting in motion a series of ecological flows that reduce the diversity of ecosystems. This reduction in diversity contributes 'to breaking down nature's bioregional equilibrium of local ecosystems, thereby turning the planet into one big ecosystem, with ecologically (and perhaps long term economically) irrational results' (Murphy 1994, p. 18). The continued expansion of human activities in the globalised world is straining the limits of both human adaptability and the resilience of nature (Holling 1994).

Currently, a spectrum of problems is emerging, caused by human effects on air, land and water that slowly gather momentum until they trigger rapid alterations in local systems impacting the health of populations, the renewability of resources and wellbeing of communities. Furthermore, with the increasing globalisation of trade and migration, this globalisation of biophysical phenomena, coupled with the intensification of linkages, is creating problems across scales in space and in time. In effect, local environmental problems today may have their root causes and triggering agents, and possibly their solutions, on the other side of the globe. This globalisation process means that problems are basically non-linear in causation and discontinuous in both space and time, and therefore inherently unpredictable and fundamentally precluding of the traditional response of observing a signal of change and then adapting to it. These problems place both societies and natural systems in such basically new and unknown terrain that both their social and ecological elements have evolutionary implications. Basically, Holling suggests, people, economies and nature are now in a process of co-evolution on a global scale, each influencing the others in unfamiliar ways and at scales that challenge our traditional understandings of structure and organisation with serious implications for the adaptive capacities of people and societies (Holling 1994, pp. 79–81).

In the field of disaster research in the 1970s and 1980s geographers and anthropologists working in the developing world found there was little in mainstream disaster research that helped them understand why disasters were so much worse in the Global South. When these researchers established the empirical findings that led to the concepts of risk and vulnerability, they created the bases for a powerful model of causation for disaster research and practice (O'Keefe and Wisner 1976, Wisner, Westgate and O'Keefe 1977, Hewitt 1983). In situating causality of disasters in society, or in societal–environment relations as influenced by drivers – both historically embedded and recent, from both near and far – the study and practice of disasters framed local disasters in broader historical and structural frames. The concepts of risk and vulnerability addressed the identifiable social features and their drivers that lead to damages and deaths from specific hazards, but also scrutinise the longstanding trajectories of development and globalisation that produce and sustain such vulnerabilities and risks.

The concept of vulnerability required closer examination of the underlying causes of disaster and was also complexly intertwined with the questions of development. Indeed the concept is foundational for important advances in our understanding of disasters, but unfortunately it has not led to policies or practices that have significantly reduced disaster losses or damages in much of the world. The reasons for our lack of progress towards such goals lie at the core of the cultural and social organisational principles of the dominant states and societies of the world, which, moreover, have now influenced the process of societal development in much of the world.

The question that clearly arises from these findings is why, almost forty years after a major paradigm shift and the resulting huge accumulation of knowledge and data, there has been so little progress in reducing the impacts of socio-natural disasters, particularly, but not exclusively, in the developing world? Why are the problems I saw forty years ago after the Peruvian earthquake of 1970 still with us when we have learned so much about their drivers? Why, after having learned more, are we losing more?

Culture, ideology and policy choices

Men make their own history, but they do not make it just as they please; they do not make it under circumstances chosen by themselves, but under circumstances directly encountered, given and transmitted from the past. The tradition of all the dead generations weighs like a nightmare on the brain of the living.

Marx 1972, p. 437

I begin this section with a quote from Marx because he so succinctly captured the fundamental tension in social change, one that we are especially experiencing at the present historical moment. The trajectory laid down over the last five centuries of global history has expressed itself variously in codified ideological frames over the years as it has led us into a series of both chronic and acute problems and challenges, not the least of which is climate change. These deeply rooted problems all too frequently manifest themselves in the form of the tragic events we understand as disasters. The current iteration of that trajectory is known as neoliberalism.

It is clear that all human cultures have been shaped and transformed in long histories of regional- to global-networks of power, trade and meaning. As such, the expansion of capitalism, science and politics are based on global connections and spread through desires to fulfil universal imaginaries and goals. Nevertheless, despite these ever expanding global connections, it is clear that such universal claims, although they emerge out of different kinds of cultural dialogues, do not make everything uniform everywhere (Tsing 2005). We have not created a global culture, but rather a multiplicity of local discourses about global phenomena. These processes have been uneven, sometimes discontinuous, engendering both accommodation and resistance, coproducing cultures through what Tsing calls the 'awkward, unequal, unstable, and creative qualities of interconnection across difference' (Tsing 2005, p. 4). Despite the overarching characteristics of capitalism and its all-encompassing visions, corporate growth has proven to be nonetheless discontinuous, inefficient and violent, particularly in the developing world where its intensification has resulted in widespread ecological and social transformation. Over the last three decades, capitalism not only expanded its reach, it was essentially transformed by new international rules of trade extending enormous benefits to the globally powerful corporations (Tsing 2005).

The fact that this latest stage of capitalism has become known, in a terminological shift, as neoliberalism suggests that there are two streams in the cultural narrative of capitalism in which various key relationships have been altered (Ortner 2011). The first shift alters the relationship between capital and labour from one essentially grounded in a quasi-collaborative ethos and balance of interests achieved through labour organisations to one in which labour has lost power and capital has achieved an increasingly stringent and strict disciplinary capacity. The second shift involves a change in the relationship between capital and government from one in which government played a role in regulating the economy and providing social programmes for general welfare to one in which the role of government is basically reduced to ensuring strong individual private property rights, the rule of law and the institutions of freely functioning markets and free trade (Ortner 2011).

In cultural terms, neoliberalism is no longer only an economic project embodied in a set of practices, it is an ideology – basically a cultural product that requires careful scrutiny because, while ideologies may reflect material circumstances, they also influence the way people interpret and frame those circumstances as they occur over time. An ideology is the totality of ideas and values or representations common to a society or current in a given social group, a conceptual framework for the way people deal with reality. In effect, it is a social set of representations that could be construed as coterminous with culture. In everyday life it is through the ideology of our society in the first place that we become conscious of everything around us. The basic tenets of an ideology are likely to remain implicit in the sense that fundamental assumptions about the world are so obvious, so omnipresent, that they do not need expression. They are simply part of the definition of the world as it is perceived (Dumont 1977). They are, as Marx put it, a tradition. Within a given tradition an ideology provides, among other things, an implicit guide to the relations between humans and the cosmos, nature and each other. However, few ideologies are monolithic. As previously noted, we understand culture today not as a single voice, but as multiple voices all articulating their particular interpretations. These voices share some perspectives, and differ in others, creating a rich dimension to how culture is defined. All the social characteristics that significantly structure people in a society will play a role in the way those meanings and explanations are constructed. However, as also noted, some voices can, by virtue of different kinds of power and wealth, come to dominate the conversation. Although it is not without contesting views, neoliberalism is now ascendant, particularly as regards to the relationships between humans, and the relations humans have with the natural world, and vice versa.

From an ideological standpoint, what is neoliberalism? In effect, neoliberalism, like its classical predecessors, purports to represent something that is virtually isomorphic with human nature. The discipline of economics has always represented itself and indeed is founded on this proposition. Basically it is asserted that capital and capitalism were always there, just needing the emergence of the market as a dominant institution to claim their rightful universality. In that sense, neoliberalism is an ideology that expresses, largely in economic terms, a more or less complete text about the kind of society it purports to create, one in which efficiency and competitiveness, individualism and freedom from interference in the use of resources are privileged. Neoliberals are faithful to Adam Smith's view that the pursuit of individual self-interest through the market is the best means to benefit society (Harvey 2007), although without Smith's deep concerns about the moral implications of such a position. In fact, in such a society, the values of the private and individualistic domain pervade the public sphere, resulting in the penetration of mercantile and individualistic forms of conduct and values into social life at all levels (Sunkel 2005). For example, financial instruments known as catastrophe bonds – essentially securities that manage the risks of catastrophic natural events – are now traded like other derivatives in stock exchanges (Cooper 2010). In this context, neoliberalism fundamentally becomes a social and cultural policy in which everyone is obliged to become a profit or utility maximiser, downgrading all forms of social cohesion, altruism, solidarity, identity and the natural environment. In such circumstances, individual freedom becomes synonymous with free enterprise (Polanyi 1944). The outcome of such a process is, as we have seen, an enormous concentration of wealth and power achieved through the relatively unrestrained exploitation of human labour and the natural environment.

The role of the state in neoliberal regimes is basically reduced to ensuring strong individual private property rights, the rule of law and the institutions of freely functioning markets and free trade. The freedom of businesses and corporations to engage in free trade in free markets is considered to be essential. The state must seek the privatisation of assets and functions. The transfer of public assets to private interests, as well as the deregulation of social, political and economic processes, is considered fundamental to improve efficiency and productivity, and reduce costs and taxes. Moreover, the state must seek to remove all barriers to the free mobility of capital between sectors, regions and countries (Harvey 2007). The state's contract regarding the security of the citizenry is seen to be achieved most efficiently through the functions of the market.

Neoliberalism, development and disasters

In so far as disasters are the outcome of the intersection between a society and a force from the natural or built environment, the implications of such an ideological system are clear. The fundamental tenets of neoliberalism, like any ideology, set out a number of general frames that guide both our thought and behaviour regarding those interactions of society and environment that we call disasters. It is not accidental that the emphasis of governments (and the funding), despite forty years of research, in disaster management still emphasises emergency management.

Disaster risk and social vulnerability are in large measure the products of historical and existing processes of social and economic development. The energy basis for the development of industrialised societies is the driving force behind global climate change. The ideologies and practices of development, the latest of which is neoliberalism, are deeply implicated in the construction of vulnerability of large segments of the world for the last half century – indeed, for the last half millennium. Are we, as the beginning quote from Marx suggested, locked into a nightmare?

Neoliberal (and other past conventional) models of development with their agendas of economic growth through intensified production stimulated by modern industrial market economies assert that such approaches are the best means to combat poverty and raise standards of living on a global scale. However, many of the processes that drive risk and vulnerability are standard development strategies that only add to climate change drivers (Cannon and Müller-Mahn 2010). Any costs occasioned by productive development have been externalised, to be absorbed either by the environment in terms of resource exploitation and waste processing, or by the general population when social, cultural and economic disadvantages – such as increased risk and disasters – occur. Moreover, the power of the state to limit the processes accentuating vulnerability and disaster risk has been severely curtailed under current neoliberal dictates.

Clearly, among those disadvantages are the conditions that emerge from the inconsistencies, imbalances and inequalities engendered by the dominant development model that compound the social vulnerability of large numbers of people who are increasingly exposed to an expanding number of hazards, now often in a cascade of linked calamities. Despite recent disasters in the industrialised world, it is generally clear that, in terms of mortality, development has reduced vulnerability and enhanced resilience in those nations. It is also clear that social vulnerability to climate change is largely concentrated in the developing world. However, in the United States for example, that reduced social vulnerability is distributed in unequal ways. Hurricane Katrina revealed that much of the flooding, death, destruction and dislocation of minority communities were due largely to strategies of urban development implemented initially as far back as the 1920s that privileged economic gain over environmental security and sustainability. These strategies urbanised highly exposed flood-prone areas to be occupied largely by a vulnerable and discriminated against minority population (Cotton 2006).

Neoliberal disasters: the case of Haiti

Although deeply rooted in Haitian history, the historical construction of Haiti's impoverishment and vulnerability was exponentially compounded by more recent developments during the last quarter of the twentieth century. Following the brutal dictatorship of 'Papa Doc' Duvalier, the ruinous reign of his son 'Baby Doc' left the nation in even greater debt to foreign lenders because of either misappropriation or outright theft by the dictator. The second Duvalier regime, a virtual kleptocracy, coincided with the catastrophic USAID-ordered slaughter of all of Haiti's pigs to limit the spread of African swine flu virus. The loss of the pig population, the source of peasant savings, emergency capital and nutrition left rural people, the majority of the population, even more impoverished and vulnerable (Diederich 1985, Dupuy 1989).

Ever more bereft of resources, rural Haitians were forced to cut down more and more trees to produce charcoal, eventually deforesting almost all of Haiti's territory. United States Agency for Internation Development (USAID) pro-grammes, working with large landowners, encouraged the construction of agro-processing facilities, while IMF-imposed tariff reductions opened Haitian markets to subsidised US rice surpluses, undercutting local production of the nation's staple crop and dismantling the rural economy. The goal of these measures was to develop Haiti's cities into centres of export production for US and multinational companies. The destruction of the rural economy and investment in urban export production stimulated a massive migration to the nation's cities, where impoverished migrants took up residence in festering slums and hillside shantytowns with few services of any sort. The demand for jobs by displaced rural people quickly outstripped the supply, deepening the impoverishment of ever denser populations in vulnerable locations in cities. Political instability during the last twenty years has also led to a reduction of companies available to offer jobs (Chavla 2010, Milne 2010).

Thus, as the year 2010 began, Haiti found itself extraordinarily vulnerable to the natural hazards of its environment. In the previous quarter century, few development efforts, misguided and mismanaged as they were, had privileged the issue of environmental security or hazard mitigation. Moreover, the neoliberal strategies of tariff reductions and Free Trade Zone production sites in the never ceasing search for cheaper labour entrained a number of demographic and economic processes that placed large numbers of people in acute risk. A lack of building codes, together with informal settlements, widespread undernourishment and hunger, disease, poor access to clean water or electricity, inadequate educational and health facilities and services at the national and municipal levels, and crime and corruption led to the construction of extreme vulnerability. In addition, Haitians were largely unaware of the seismic risk on the island, although seismologists had been warning of the possibility of a strong earthquake. Because of this social construction of extreme vulnerability, when the 7.0 magnitude earthquake struck on January 12, 2010, devastating the capital of Port-au-Prince and the surrounding region, more than 300,000 Haitians died, according to Haitian Government estimates. The unregulated and informal housing stock of the city of Port-au-Prince was flattened, its basic service lifelines, inadequate as they were, destroyed.

Neoliberalism, disaster risk reduction and the culture of development

In effect, the challenge of disasters, as well as the impacts of increasing climate change effects – many if not most of which will be experienced as disasters – requires significant alteration of the way we interact with nature and the way we interact with each other. In point of fact, our ability to manage risk and disasters depends on our ability as a society to introduce changes in human thought, behaviour and action when confronted with forces from nature that are beyond our control, or to prevent the construction of new hazards and risks (Lavell 2011, p. 6). To that effect, the disaster research and management community has developed strategies or particular types of instruments for disaster risk reduction. Few significant differences exist between the strategies devised for DRR and those that have been identified for adaptation to climate change effects (Lavell 2011). Cannon *et al.* (2003) locate such strategies within five basic domains:

- improvements in social living conditions;
- livelihood strengthening and increases in resilience;
- self-protection;
- social protection; and
- governance factors.

More specifically this array of strategies can be summarised in the following way:

- environmental management, natural resource and environmental services management;
- territorial organisation and land use planning;

- use of protective infrastructure;
- application of new and traditional technologies and science;
- strengthening and promotion of sustainable livelihoods;
- financial mechanisms–micro credit, insurance etc.;
- integral sector and territorial planning;
- environmental and social monitoring systems and early warning systems;
- education, training, consciousness and participation; and
- mechanisms and processes that increase risk governance in general (Lavell 2011).

It surely does not escape attention that all of the above involve significant participation by the public sector at a variety of levels. However, the continued pressure on and by the state (and the conditionalities of multilateral development institutions) to reduce social services, privatise and deregulate significantly reduces the capacity of the public sector to implement these strategies. Wisner, for example, addressing the post 2001 earthquake reconstruction process in El Salvador notes not only that nothing in the plans dealt with actual drivers of the disaster, but that the increasing privatisation of government functions would likely result in contracts with transnational corporations for implementation (2001).

Despite the significantly publicised attention of international development agencies to risk reduction to natural hazards and to climate change, it may be fairly claimed that contemporary forms of development as enacted generally give it little priority in planning or programmes. Despite continued insistence on the importance of addressing the underlying causes of disasters and disaster vulnerability within or without the context of climate change, political focus and funding are still largely centred on emergency management. In many ways, it may be reasonably claimed that the continued emphasis on emergency management expresses a reluctance to grapple with the issues of disaster risk reduction, which imply confronting the fundamental contradictions at the level of social and environmental relations that characterise our neoliberal systems.

Generally speaking, the development process and specific development projects take place without these inputs because the neoliberal world view informs development policy and planning. Thus, in the name of development, for failing to include disaster risk reduction in our development initiatives, we continue to put more and more people in harm's way, engendering disasters of development and development disasters (Stephen Bender, personal communication). If we want to address the underlying risk factors, called for in the Global Platform for Disaster Risk Reduction in 2013, we must address the inconsistencies and contradictions in current neoliberal policies and models of development as well as the huge imbalances in power. However, the question remains as to whether the political and economic structures of any nation can, even in the face of devastating loss and destruction, truly come to grips with a hegemonic cultural construct that dictates and conditions the terms on which the forms and practices of human and environmental relations are organised.

Conclusion

In locating the cause of disasters in society and societal–environment relations, social vulnerability reveals the fundamental tensions that exist in our environmental and social relations and presents us with a major choice (Oliver-Smith 2013). Do we continue down the path, 'the tradition of all the dead generations, weighing like a nightmare on the brain of the living', in Marx's words? Are we locked in, as some have said, by the iron laws of the market? However, with human thought and action there are no natural laws. The law of gravity has no cultural equivalent. Culture frames and structures even our most basic biological realities. Thus, neoliberal market logics and structural constraints are ultimately cultural products, the outcome of values and norms that guide the decisions and choices made by people. Nonetheless, under such cultural frames, a process of authentic disaster risk reduction becomes extremely difficult. Neoliberal frames of social and environmental relations have come close to being completely naturalised in current discourses, making alternative understandings extremely difficult to empower (Gramsci 1971).

A frequently employed triad of goals for disaster risk reduction strategies states that they must be 1) socially acceptable, 2) economically viable, and 3) ecologically sustainable. However, can these three goals truly be reconciled when their relative cultural power is so unequal in a society whose basic cultural orientations and most powerful institutions are informed by a neoliberal perspective? A truly meaningful adjustment of the relationship among these three goals under such a cultural regime would seem a daunting challenge. That challenge can be voiced in a single question: how can a political ecology of disasters that situates causality in the systemic features of society mobilise significant disaster risk reduction in those entrenched systems? Given current trends of progressively growing densities of increasingly vulnerable populations in exposed regions, the question of effective disaster risk reduction becomes urgent. However, under current global neoliberal economic structures, the capacity of any single society to address such root causes is open to question. Our current failure to diminish disaster risk would seem to confirm that assessment.

References

Anderson, B., 1983. *Imagined Communities*. London: Verso.

Appadurai, A., 1996. *Modernity at Large: Cultural Dimensions of Globalization*. Minneapolis: University of Minnesota Press.

Bender, S., 2009. Personal Communication.

Bourdieu, P., 1977. *Outline of a Theory of Practice*. Cambridge: Cambridge University Press.

Cannon, T. and Müller-Mahn, D., 2010. Vulnerability, resilience and development discourses in context of climate change. *Natural Hazard*, 55 (3), 612–35.

Cannon, T., Twigg, J. and Rowell, J., 2003. Social Vulnerability, Sustainable Livelihoods and Disasters. *Report to DFID Conflict and Humanitarian Assistance Department (CHAD) and Sustainable Livelihoods Support Office* [online]. Available from: www.nirapad.org/admin/soft_archive/1308222298_Social%20Vulnerability-%20Sustainable%20Livelihoods%20and%20Disasters.pdf [accessed February 2014].

Chavla, L., 2010. 'Has the US rice export policy condemned Haiti to poverty?' *Hunger Notes*, 23 April.

Colton, C. E., 2006. *An Unnatural Metropolis: Wrestling New Orleans from Nature*. Baton Rouge, LA: Louisiana State University Press.

Cooper, M., 2010. Turbulent worlds: Financial markets and environmental crisis. *Theory, Culture and Society*, 27 (2–3), 167–90.

Diederich, B., 1985. Swine Fever ironies: The slaughter of the Haitian Black Pig. *Caribbean Review*, 14 (1), 16–17.

Dumont, L., 1977. *From Mandeville to Marx: The Triumph of Economic Ideology*. Chicago, IL: University of Chicago Press.

Dupuy, A., 1989. *Haiti in the World Economy: Race, Class and Underdevelopment since 1700*. Boulder: Westview Press.

Geertz, C., 1973. *The Interpretation of Cultures*. New York: Basic Books.

Giddens, A., 1990. *The Consequences of Modernity*. Cambridge: Polity Press.

Gramsci, A., 1971. *Selections from the Prison Notebooks of Antonio Gramsci*. Edited and translated by Q. Hoare and G. Nowell-Smith. New York: International Publishers.

Harris, M., 1979. *Cultural Materialism: The Struggle for a Science of Culture*. New York: Random House.

Harvey, D., 2007. *A Brief History of Neoliberalism*. Oxford: Oxford University Press.

Hewitt, K., 1983. *Interpretations of Calamity*. Boston: Allen & Unwin, Inc.

Hoffman, S. M., 2013. The anthropological perspective on disaster and the key concept of culture. *International Conference on Anthropological Studies on Hazards and Disasters*, Yunnan Academy of Social Sciences, Kunming, China, 21–22 August 2013.

Hoffman, S. M. and Oliver-Smith, A., 2002. *Catastrophe and Culture: The Anthropology of Disasters*. Santa Fe, NM: The School of American Research Press.

Holling, C. S., 1994. 'An Ecologist View of the Malthusian Conflict'. In: K. Lindahl-Kiessling and H. Landberg, eds, *Population, Economic Development, and the Environment*. New York: Oxford University Press, 79–103.

Hoogvelt, A., 1997. Globalization and the Post-Colonial World: The New Political Economy of Development. Baltimore, MD: The Johns Hopkins University Press.

Hoogvelt, A., 2001. *Globalization and the Post-Colonial World*. Baltimore, MD: The Johns Hopkins University Press.

Hornborg, A., 2009. Zero-sum world: Challenges in conceptualizing environmental load displacement and ecologically unequal exchange in the world system. *International Journal of Comparative Sociology*, 50 (3–4), 237–62.

ICSU, 2008. *A Science Plan for Integrated Research on Disaster Risk: Addressing the Challenge of Natural and Human-Induced Environmental Hazards*. International Council for Science, Paris.

Intergovernmental Panel on Climate Change (IPCC), 2007. *Climate Change 2007: Climate Change Impacts, Adaptation and Vulnerability, Summary for Policy Makers*. Contribution of Working Group II to the Fourth Assessment Report of the Intergovernmental Panel on Climate Change, Brussels.

Kearney, M., 1996. *Reconceptualizing the Peasantry: Anthropology in Global Perspective*. Boulder, CO: Westview Press.

Kroeber, A. L. and Kluckhohn, C., 1952. *Culture: A Critical Review of Concepts and Definitions*. New York: Vintage Books.

Lavell, A., 2011. *Unpacking Climate Change Adaptation and Disaster Risk Management: Searching for the Links and the Differences: A Conceptual and Epistemological Critique and Proposal*. IUCN-FLACSO Project on Climate Change Adaptation and Disaster Risk Reduction.

Marx, K., 1972. The Eighteenth Brumaire of Louis Bonaparte. In: R. C. Tucker, ed., *The Marx-Engels Reader*. New York: W. W. Norton, 436–525.

Milne, S., 2010. Haiti's suffering is a result of calculated impoverishment. *The Guardian*, 20 January.

Murphy, R., 1994. *Rationality and Nature*. Boulder, CO: Westview Press.

O'Keefe, P. and Wisner, B., 1976. Taking the naturalness out of natural disasters. *Nature* 260 (5552), 566–7.

Oliver-Smith, A., 1998. Global Challenges and the Definition of Disaster. In: E. L. Quarantelli, ed., *What is a Disaster? Perspectives on the Question*. New York: Routledge, 177–95.

Oliver-Smith, A., 2013. A matter of choice. *International Journal of Disaster Risk Reduction*, 3, 1–3.

Oliver-Smith, A. and Hoffman, S. M., 2002. Why Anthropologists Should Study Disasters. In: S. M. Hoffman and A. Oliver-Smith, eds, *Culture and Catastophe: The Anthropology of Disaster*. Sante Fe, NM: The School of American Research Press, 3–22.

Ortner, S., 2011. On Neoliberalism. *Anthropology of this Century* [online], May 2011. Available from: http://aotcpress.com/articles/neoliberalism/ [accessed February 2014].

Perry, R. W. and Quarantelli, E. L., 2005. *What is a Disaster? New Answers to Old Questions*. *International Research Committee on Disasters*. Available from: http://higheredbcs.wiley.com/legacy/college/perry/0471920770/chap_resource/ch01/what_is_disaster.pdf [accessed February 2014].

Polanyi, K., 1944. *The Great Transformation*. New York: Farrar and Rinehart.

Quarantelli, E. L., 1998. *What is a Disaster? Perspectives on the Question*. New York: Routledge.

Rosaldo, R., 1989. *Culture and Truth: The Remaking of Social Analysis*. Boston, MA: Beacon Press.

Sahlins, M., 1976. *Culture and Practical Reason*. Chicago, IL: The University of Chicago Press.

Sunkel, O., 2005. The Unbearable Lightness of Neoliberalism. In: B. A. Roberts and C. H. Wood, eds, *Rethinking Development in Latin America*. University Park, PA: Pennsylvania State University Press.

Tsing, A. L., 2005. *Friction: An Ethnography of Global Connection*. Princeton, NJ: Princeton University Press.

Tylor, E., 1920. *Primitive Culture*. New York: J. P. Putnam's Sons.

Wisner, B., 2001. Risk and the neoliberal state: Why post-Mitch lessons didn't reduce El Salvador's earthquake losses. *Disasters*, 25 (3), 251–68.

Wisner, B., Westgate, K. and O'Keefe, P., 1977. Global systems and local disasters: The untapped power of people's science. *Disasters*, 1 (1), 47–57.

3

DESIGN BY DISASTERS

Seismic architecture and cultural adaptation to earthquakes

Greg Bankoff

It is nearly 15 years since Dennis Mileti published *Disasters by Design: A Reassessment of Natural Hazards in the United States*, a title by which he somewhat ironically inferred that many disasters were the predictable result of interactions stemming from the physical environment including hazards, the social and demographic characteristics of the communities that experienced them, and the buildings, roads, bridges and other components of the constructed environment. Climate change, the number of people living in hazard-prone regions, especially coastal locations, and the sheer density of the urban fabric, he argued, were increasing the potential losses from natural forces (Mileti 1999). Yet through history, if the hazard was frequent enough, many peoples have built structures that were quite literally 'designed' to accommodate such an event and minimise its consequences. Rather than being 'disasters by design', these buildings were designed so to speak by disasters, a 'seismic architecture' that represented a cultural adaptation by peoples to earthquakes. It is difficult, at times, to clearly identify specific instances when a better understanding of culture materially contributes to disaster risk reduction (DRR). Yet this 'seismic architecture' provides a very visible manifestation of how culture is ineluctably intertwined with disaster risk management (DRM) not only in terms of practical adjustments to hazards but also as a form of long-term educational transmission through heritage.

Hazards and cultures

Sociologists and psychologists have long been intrigued by the idea that people who are frequently exposed to hazards adapt their behaviour to accommodate risk, and that these patterns can become embedded to some extent in cultures over time. Interest, however, has not been with the one-in-one hundred year 'big one' as with the smaller scale, more frequent hazards that can be regularly anticipated. The constant recurrence of such events, it is argued, can give rise to alternative normative structures that replace the routine social norms with ones more in accord to a time of crisis: the very frequency of hazard can engender cultural adaptation. The degree of cultural adaptation can vary enormously from specific forms of risk mitigation strategies such as seismic architecture designed to counter a repeated threat, to a geographically bounded subcultural accommodation to a single risk, to a more wholesale societal adjustment to one or more hazards. None of these categories are mutually exclusive in that risk-mitigating strategies may form an aspect of both subcultural and societal adaptations, and one or more subcultures may be embedded within the larger society.

Harry Moore is credited with first coining the term 'disaster subculture' to describe 'those adjustments, actual and potential, social, psychological and physical, which are used by residents of such areas in their efforts to cope with disasters which have struck or which tradition indicates may strike in the future' (Moore 1964, p. 195). He was writing about how people who lived in coastal communities acted as Hurricane Carla hit Texas in September 1961. Moore concluded by explaining their behaviour as pertaining to a 'hurricane subculture'. A few years later, William Anderson similarly identified a 'flood disaster subculture' in Cincinnati to designate a social-physical environment of repeated inundations along the Ohio River Valley (Anderson 1965). Moore never went on to fully formulate his ideas; this was left to a research paper by Dennis Wenger and Jack Weller written nearly a decade later. Based on research carried out in a number of communities across the United States regularly affected by flooding and hurricanes, they sought to uncover the existence of disaster subcultures through indices that showed 'the perpetuation of successful patterns of adaptation to the disaster context through socialization' (Wenger and Weller 1973, p. 1).

Wenger and Weller distinguished between 'instrumental' elements and 'expressive' ones. The former include those normative, technological, evaluative, knowledge and resource components that are related to predicting, preparedness, management and responding to the physical impact of a hazard. In particular, these elements are primarily concerned with individual and organisational behaviour and intra- and inter-organisational patterns. Wenger and Weller, however, also recognised the importance of what they termed expressive elements. These traits consist of the norms, values, beliefs, legends and myths that people hold about hazards and their impact on communities. To a large extent, such elements define whether an event is regarded as more of a nuisance within the routine of daily life or as disaster. Rather than viewing disaster subcultures as special arrangements apart from the general culture of the community that are resorted to in times of need – more in the way Moore argued – Wenger and Weller view the subcultural traits that evolved out of disaster experiences as being fully integrated into the dominant non-disaster culture. Moreover, these traits constitute a residue of learning that is both passed down from one generation to the next and is transmissible to new community members (Wenger and Weller 1973, pp. 2–5). Drawing on their study of Midwestern North American communities and the variations in the extent to which they possess distinctive cultural characteristics, they were able to identify three factors that appear to be crucial to the emergence of a disaster subculture within a community: that a hazard has to be repetitive, it is of such a nature as to allow a period of forewarning, and that its consequences inflict significant damage to human and material resources (Wenger and Weller 1973, p. 9).

In some instances, however, the disaster subculture may not be confined to only a single geographical area but more generally permeates the entire society, raising the question of the former's relationship to the latter. In the Netherlands, for example, the boundaries between the two are not clear and one might speak, as Petra van Dam does, of an 'amphibious culture' that encompasses the entire society or state (van Dam 2012, pp. 1–10). The Philippines, on the other hand, is beset by so many and so varied natural hazards that the archipelago's many different ethnicities share a distinctive and common cultural range of adaptation to disasters (Bankoff 2003). The threat that might otherwise give rise to regional or local subcultures is so frequent and widespread in these instances that they have come to constitute part of the wider culture. Floods and flood-related events in the Netherlands and the sheer variety and magnitude of seismic and meteorological threats in the Philippines influence social and material culture on a truly national scale to create 'disaster cultures'.

Significantly, too, these studies add a historical dimension to the notion of disaster culture, recognising that people are made both vulnerable and resilient over time. Culture can be regarded as the outcome of physical hazards acting upon a society whose past actions and present practices both expose people to risk and motivate them to develop capacities to learn from their past experiences to make themselves more resilient in the future (Wisner *et al.* 2004, Lorenz 2010). Central to this perspective is the notion that history prefigures disasters; that populations are rendered either powerless or capable by particular social orders that, in turn, are often modified by that experience to make some people even more vulnerable or resilient in the future. Not only does history reveal that disasters may take centuries in the making but also that culture reflects an adaptation to risk over time. Societies, destructive agents and cultures are mutually constituted and embedded in natural and social systems as unfolding processes. Anthony Oliver-Smith refers to this as a historically produced pattern of vulnerability and resilience where the 'life-history of a disaster begins prior to the appearance of a specific event-focused agent' (Oliver-Smith 1999, pp. 29–30).

Earthquakes, in particular, satisfy the three factors identified by Wenger and Weller as crucial to the emergence of a disaster subculture within a community: they occur at frequent if irregular intervals; people living in seismically active areas know they are at risk; and the consequences inflict significant losses in terms of mortality and material damage. The historical frequency of earthquakes in Japan, for instance, have had such an enduring impact on people's daily lives that they have shaped the material and symbolic culture of that society to form what Gregory Clancey calls an 'earthquake nation' (Clancey 2006). Writing on 17 August 1999 Kocaeli (Izmit) earthquake along the North Anatolian fault in western Turkey, Jacqueline Homan and Warren Eastward identify very specific adaptations to the hazards of the natural world around them that are built up over time – a 'seismic culture' that has become part of the discourse of the place, that is constructed by and known to those who live there, and that makes use of local materials, skills and resources. They define seismic cultures as those having: 'The knowledge (both pragmatic and theoretical) that has built up in a community exposed to seismic risks through time' (Homan and Eastward 2011, p. 624). In particular, they were interested in how a tradition of architecture that had developed over the centuries from the Byzantine and Ottoman periods to mitigate the worst effects of earthquakes had subsequently been lost.

Seismic architecture

The key to the emergence of a seismic culture is the development of construction techniques that have proven effective in minimising loss of life and mitigating damage to buildings during earthquakes. As 1.2 billion people or 20 percent of the world's population in 2000 lived on land estimated to have at least a 10 percent probability of a medium sized earthquake (pga of <2 m/s^2) in a fifty year period, seismic cultures have not only emerged in Asia Minor but also in many geographically dispersed areas throughout history (Dilley et al. 2005, p. 43). However, not all earthquakes produce seismic cultures. On the one hand, it is difficult to distinguish the most effective anti-seismic techniques in devastating earthquakes when nearly all buildings are destroyed. Similarly, if the death toll is high or the settlement is abandoned, then local practices are likely to be lost. On the other hand, minor earthquakes do not cause enough damage to buildings for local inhabitants to acquire the necessary construction knowledge and know-how to enable techniques to be improved. Instead, very much in accord with Wenger and Weller's criteria, the creation of a seismic culture depends upon earthquakes that do not cause complete devastation to the built environment, and so allow for the identification of structures that are better able to withstand ground movements, and yet occur often enough so that the intervals between events are not long enough for people to forget their effects. Where earthquakes are frequent and not catastrophic, construction techniques are tested and improved upon after each event and their effectiveness becomes well-known and widely adopted within the community. The European Centre for Cultural Heritage distinguishes between a 'seismic culture of prevention', or the development of long term earthquake resistant building techniques, and a 'seismic culture of repairs' where people are responsive to disaster mitigating measures in the immediate aftermath of an event but then tend to revert back to pre-disaster construction practices and lifestyle (CUEBC 1993, p. 5). The style of architecture that evolved in seismically active zones in the past often fused taste and aesthetics with practical earthquake resistance measures: they were designs, so to speak, 'inspired' by disasters. Often the threat of frequent earthquakes underlies the foundations of particular engineering practices, upon which are subsequently wrought the fancies of devotion, power and ostentation.

The origins of seismic engineering date from the 'classical' period, where the designation is used simply to denote pre-modern architecture rather than a purely Western European connotation of antiquity. Evidence of adaptations in structures that rendered buildings better able to accommodate sudden, violent earth movements have been suggested from as early as the Bronze Age Minoan civilisation of Knossos between the 27th and 5th centuries BCE (Langenbach 2007, p. 35). In particular, the classical temples of Ancient Greece and Rome pose some intriguing questions, given the construction techniques employed and the frequency of earthquakes in those areas. The archetypal temple façade of columns constitutes a segmental (multi-block) rocking system that works to re-centre the building's axial load during violent ground movements, although it is unclear whether the design was purposeful or accidental (Pampanin 2008, p. 118). The construction of columns, however, seems more certain. Each column is joined together with an iron pin in a lead or wood surround to absorb shock and allow the structure to flex in earthquakes (Konstantinidis and Makris 2005). Archaeological remains from a house in Herculaneum suggest that the *insulae*, or seven to eight-storey tenements that housed much of Rome's urban population, utilised timber frames and infill masonry techniques designed to make the buildings more resistance against earthquakes (Langenbach 2007, p. 35). The Roman *angora* (market) at Perge near Antalya in present-day Turkey dating from the third century CE also exhibits clear signs of the use of 'proto-reinforced beams'. All these examples point to a diffusion of knowledge about earthquake resistant construction techniques in seismically active areas around the Mediterranean. In fact, seismic engineering techniques may have evolved independently in many civilisations around the world though knowledge about these adaptations has now been lost. For example, in South America, remaining Inca buildings exhibit many of the characteristics of an earthquake resistant architecture. Structures employ carefully bonded corners and alternate rows of stone headers (laid with the short side exposed) and stretchers (laid with the long, narrow side exposed), interlocking monumental stonework and trapezoidal archways – though some doubts have been cast upon the extent of their seismic properties (Cuadra, Karkee and Tokeshi 2008).

Over the centuries, at least three zones of identifiable earthquake resistant architecture have emerged that correspond to a single seismically active region, the Alpine-Himalayan belt: the Eastern Mediterranean, Southern Europe and the Himalayan arc. Though the peoples of these regions are culturally distinct, they share a similar risk from earthquakes and have evolved in many ways a common response to that threat through seismic architecture. In fact, earthquakes may provide an underlying environmental unity that provides a metanarrative or means to understanding a wider region whose history and present problems are usually divided by continental designation, ethnicity or religion (Schwartz 2005, pp. 382–4). The Alpine–Himalayan belt, which stretches from the Atlantic Ocean to the Himalayan Mountains, was formed when the Arabian, African and Indian continental plates collided with the Eurasian Plate (Erdik 2000, p. 724). Earthquakes and the seismic cultures that they gave rise to provide another tool for understanding the history of this broader region and the important cultural impact of hazards on society.

The Eastern Mediterranean

The Eastern Mediterranean appears to be one of the cradles of seismic engineering dating from at least Roman times. In this area there are many examples of the reinforced beam structures that persisted through the Byzantine Empire to the late Ottoman period. In particular, the modern nation-states of Greece and Turkey are located in one of world's most seismically active zones caused by the complex interaction of numerous tectonic microplates juxtaposed between colliding major plates. Greece is located at the boundary of the Africa–Eurasia convergence with subduction along the Aegean or Hellenistic arc. Here, minor earthquakes are common but major ones rare (a return period of about 1,000 years). In Turkey, the North Anatolian Fault Zone extends 1,400 km from the east to the Sea of Marmara in the west, along which major earthquakes are regular events (Ambraseys and Jackson 2000, Kouskouna and Makropoulos 2004). Given the nature of seismicity in this region, people have developed common engineering techniques and building practices over time in response to the frequency of earthquakes.

The seismic architecture that developed progressively from the third century CE involved the use of reinforced beams placed intermittently around the walls of buildings known as *hatillar* (singular *hatil*), a common feature of Byzantine and Ottoman structures. They were made either from thin boards of wood laid into the walls so that they overlapped at the corners, or as a narrow course of masonry different from the material used in the rest of the structure. *Hatillar* have the following seismic engineering functions: they work as vertical and horizontal shock absorbers in the case of wood because they are more compressible than the surrounding masonry; they operate as slip planes to prevent any frictional drag in the superstructure from foundations and walls below a *hatil*; they provide a horizontal tie member all around a building at a common level to resist tensile forces; and they act as a rigid horizontal girder in such a manner to maintain the structure's overall configuration even if the walls undergo a degree of lateral shift (Duggan 1999). *Hatillar* give the characteristic stripes to be seen in the design of many Byzantine and Ottoman buildings and have been proven scientifically to prevent damage to buildings during earthquakes. Unfortunately, their seismic purpose ceases to function properly from wear, natural erosion, lack of maintenance or the recycling of materials. *Hatil* techniques continued to be used in construction into the twentieth century, though its function became less widely understood and increasingly only applied as a fashionable decoration to the outside of walls. A case in point is the reconstruction of sections of the Theodosian city walls of Istanbul where strips of red brick have been attached merely as a veneer and serve no seismic purpose (Langenbach 2007, p. 37). In rural areas of Anatolia, *hatil* techniques were a common feature in local housing till recently (Hughes 2000a, p. 2).

The use of timber in buildings has long been noted for its resistance to earthquakes. While some 'log cabin' types of wooden structures were common in small areas of the West of Turkey, the use of infilled timber frameworks was much more common (Hughes 2000a, p. 1). Known as *himiş* or *bağdadi*, the construction method involved a timber framework tied by studs to a general timber armature and then infilled with a single thickness of masonry in the former case or short, rough pieces of timber in the latter technique (Gülkan and Langenbach 2004, pp. 4–5). Structures are held together by using weak rather than strong mortar bonds to encourage sliding along the bed joints instead of cracking the masonry units when panels are deformed by violent ground movements. Infill masonry walls respond to the stress of earthquakes by 'working' along joints to defray the strain and so dissipate much of the seismic energy. The only visible manifestation of these internal movements is usually cracks in the interior plaster along the walls and at the corner of rooms. The timber studs that subdivide the infill frameworks mean that the loss of one or even several masonry panels does not lead to the progressive destruction of the rest of the wall. Timber frameworks are often found in the same building with *hatillar*, the latter used to reinforce the weight-bearing ground floor and *himiş* used for upper storeys. These buildings survive earthquakes by not fully engaging with them: structures may not have much lateral strength but they do have lateral capacity. Buildings respond to seismic forces by swaying with them rather than attempting to resist them with rigid materials and connections (Gülkan and Langenbach 2004, pp. 10–11). Aesthetically, too, they can be made to please the eye with the infilled use of masonry wall bricks arranged at angles to form decorative patterns.

Given the shared history under both Byzantine and Ottoman domination, Greek architecture reflects a similar seismic cultural origin but exhibits a wide diffusion of practices. Studies of the few extant traditional post-1500 houses reveal a similar use of embedded-in-masonry horizontal timber reinforcements designed to improve the tensile and bending strength of brittle masonry to accommodate seismic forces (Tsakanika-Theohari and Mouzakis 2010). A common feature of domestic architecture in both countries is that the height of buildings is usually restricted to one or two storeys. A particular feature of post-Byzantine Greek architecture is for structures to have a dual load-carrying system consisting of a three-dimensional timber frame where the ground floor is encased in thick masonry and upper storeys are infilled (Makarios and Demosthenous 2006, p. 266). The timber frame acts to carry the load should the masonry fail. The downward force exerted by the roof is carried by diagonally braced wall-posts that are independent of the masonry walls (i.e., not secured by nailing or dowels) to the foundations (Porphyrios 1971). Greek buildings also often have a sub-foundation composed of three layers of specially treated tree trunks to accommodate ground movement where the soil is poor and/or where there is a high level of underground water. The development of an earthquake resistant architecture is a particularly prominent feature of buildings dating from the nineteenth century on the Adriatic islands of Lefkás and Théra known to have the highest seismicity in the country (Makarios and Demosthenous 2006, pp. 264–6).

Southern Europe

Forming the western part of the Alpine-Himalayan belt, seismicity in southern Europe is mainly confined to southern Portugal, coastal regions in the south and south-east of Spain and Italy, especially Calabria and Sicily. Most shocks correspond to shallow events with magnitudes, in general, less than 5.5, but this region is periodically shaken by much larger earthquakes that have precipitated major changes in architectural design and practice on several historical occasions (Buforn *et al.* 2004, pp. 623–5). Most notably the Lisbon earthquake of 1755 provides a documented instance of cause and effect as the scale of destruction convinced the chief minister, the Marquis of Pombal, to mandate by law the use of seismic techniques and practices in the rebuilding of the city centre, the Baixa (Mullen 1992). The subsequent style that became known as pombaline, though mainly implemented by the military architect Manuel de Maia, is a structure of four or five storeys with wood framed masonry walls (frontals) that, together with the timber floor beams, form a type of three-dimensional wooden 'bird-cage' or *gaiola*. The *gaiola* helps stabilise a building during an earthquake as the frames of the cage flex at each corner and so absorb some of the lateral force of the earthquake to reduce the stress on walls. Within the frames, a diagonal 'x' brace known as the 'Cross of St Andrew' stiffens the walls against lateral movement. The timber framed design was said to have been inspired by Portuguese expertise in ship construction (Cardoso, Lopes and Bento 2004).[1] The triangular spaces between the crosses were then 'nogged' or filled with a mixture of stone rubble and broken brick and covered with plaster to hide the infill and timber frame. For protection from fire, probably the most devastating factor in the destruction of the city in 1755, Maia advised the adoption of high masonry firebreaks between roofs of the new building blocks to prevent flames from spreading (Tobriner 1980, pp. 13–14). Exterior façades of Baixa buildings are clad with load bearing masonry walls, some of which also have timber frames on the inside face. The pombaline style remained the standard building practice in Lisbon until the 1920s and gives central Lisbon its characteristic architecture (Langenbach 2007, p. 48). After 1880, however, as the memory of 1755 began to fade, a decline in the standard of execution particularly in the use of the Cross of St Andrew meant that subsequent buildings were less seismically resistant.[2] Interestingly, the adoption of the pombaline style is said to have come from the observation by contemporaries that half timbered structures had withstood the earthquake better than more solid masonry buildings (Langenbach 2007, p. 48). It also shares many construction techniques and practices with earthquake resistant construction methods in the Eastern Mediterranean.

The reconstruction of Lisbon after 1755 to improve the city's ability to withstand earthquakes was not simply a matter of seismic engineering but also of urban planning. The new urban layout was designed to provide open spaces and wide boulevards to expedite evacuation, as well as facilitate traffic flow and maximise air and sunlight for public health purposes. Urban planning to minimise damage was certainly a feature of southern Italy, too, where a similar seismic culture was 'mandated' as a consequence of a major earthquake. After the 1693 earthquake in south-eastern Sicily, many of the cities and towns damaged during the event were rebuilt to incorporate seismically appropriate designs or even refounded on new, safer sites. Catania, for example, which one commentator observed had been flattened 'like the palm of your hand', was replanned with straight, wide streets so that householders could leave their homes in safety and with inordinately large piazzas so that people could flee to open spaces (Tobriner 1980, pp. 12–13). After the 1783 earthquake in Calabria, the Bourbon Government decided to reconstruct the entire region utilising traditional seismic engineering techniques and practices known as the *baraccata* system. It had been noted that the elasticity of wood framed, infilled structures better resisted lateral forces (Tobriner 1983, pp. 132–3). With its origins in the fourteenth century, but undoubtedly inspired by pombaline building standards, the *casa baraccata* (the word means wooden barracks) consisted of a masonry structure with one large interior inner timber frame that linked the entire building together from roof to foundations (Dutu *et al.* 2012, p. 621). Structural timbers were intersected by transverse reinforcements, diagonal members and x-shaped braces in the interior walls that were sometimes also visible on the exterior of the building. The walls were infilled with small stones, rubble or brick held together using a quality blend of Neapolitan lime based mortar. Balconies and exterior decoration were limited to the lightest possible type and a maximum height of around 10 m prescribed. These guidelines became law in 1785 and remained in effect till 1854. The basic principles of the *casa baraccata* continued to influence manuals, construction practices and patent applications on seismic architecture up until the first few decades of the twentieth century when the techniques sadly fell into disuse (Tobriner 1983, p. 135).

The Iberian conquest and settlement of the New World in the Americas and the Pacific led to encounters with entirely new environments, many of which proved very seismically active – as Europeans discovered to their cost. On the one hand, the European invasion precipitated a total eclipse of indigenous monumental seismic architecture and, on the other, it gave rise to the development of entirely new seismic cultures over time. These new earthquake resistant architectures were often hybrids born out of southern European and indigenous traditions. Many of the buildings still to be found in the central districts of Peru's major cities such as Lima, Trujillo and Arequipa founded in the aftermath of the Spanish conquest in 1535 exhibit a traditional method of construction. The *adobe-quincha* building technique is a blend of indigenous materials and the 'timber framed' seismic architecture of Southern Europe. Buildings made in this style comprise a ground floor fashioned of *adobe*, or sun-dried bricks, and upper storeys made of a timber frame and a mesh of cane and mud plaster, or quincha. Canes are fixed to the wooden frame by means of cords or ropes that are anchored at the bottom by means of heavy materials, such as *adobe* (Cuadra, Saito and Zavala 2012). As such, these structures function according to much the same principles as the pombaline or *baraccata* house type but with the substitution of local materials. The widespread adoption of this type of building in Lima was prompted, as in the case of Sicily in 1693 and Lisbon in 1755, by the devastating earthquake of 1746. In the wake of this disaster, streets were widened, a limitation was imposed on the heights of buildings, arch towers were prohibited, adequate plazas were planned to act as open public spaces in the event of a similar misfortune, and stone structures were replaced with ones utilising quincha (Walker 2003, pp. 68–9).

Much the same developments took place elsewhere in the New World. The *bahareque* vernacular architecture of much of Latin America substitutes bamboo (particularly *Guadua Angustifolia Kunth*), the poor man's timber, for the wooden frame. Bamboo is used for the structural poles and as beams for the walls, floors and roofs. It is split and flattened into longitudinal sections (*esterillas*) by removing the softer interior of the culms, and used to fashion the exterior of the walls, which are then rendered with a mixture of mud and horse dung. Depending on the climate, the walls may be filled with a mixture of mud and broken tiles. The roof structure is also usually made out of bamboo and covered with clay tiles. The bamboo frame provides the necessary structural integrity to resist seismic forces (Gutierrez 2004, pp. 19–21). Across the Pacific, in the Spanish Philippines, another hybrid variant of seismic architecture emerged in Manila following the earthquake of 1645. The so-called 'earthquake baroque' style of colonial architecture consisted of the extensive use of massive buttresses, low body structures and squat bell towers in public and religious buildings. Domestic buildings also underwent a similar radical transformation: heavy roof beams were supported by trusses and rested on struts planted deep into the ground to provide more flexibility; and the number of storeys was reduced to two with the upper one increasingly constructed from lighter materials such as wood and latticework (Bankoff 2012, p. 173).

The Himalayan arc

Seismicity along the Himalayan arc at the easternmost extent of the Alpine–Himalayan belt is mainly attributable to the processes of collision between the Indian and Eurasian plates as the former thrusts underneath southern Tibet at a speed of 16–18 mm a year, a rate that is assumed to have remained constant for many thousands of years. The region experiences a large number of great earthquakes, four recorded events over magnitude 8 in the last 500 years though the historical record is possibly incomplete (Bilham and Ambraseys 2005). While no earthquake is known to have struck the same site twice in this period, about 40 percent of all earthquakes in the Himalayas occur in clusters along the 2,500 km long arc that extends through the present-day nations of Afghanistan, Bhutan, China, India, Nepal and Pakistan (Roy and Mondial 2012, p. 353). Much of the vernacular architecture in this zone reflects cultural adaptation to living in this highly volatile region. In particular, the Vale of Kashmir, a heavily populated area at the western end of the mountain chain, is distinguished by its great monuments as well as by its characteristic residential architecture. Buildings not only have to contend with frequent earthquakes but also with soft soils as Srinagar, the capital, and many of the other towns and villages in Kashmir are located on the site of a former prehistoric lake. The dual nature of the danger has given rise to a characteristic type of house able to undergo a certain amount of inelastic deformation without losing vertical load-carrying capacity. As the well-known architectural historian and building conservator, Randolph Langenbach, observes: in this type of environment 'rigidity in a construction carries the potential for destruction' (Langenbach 2009, p. 1).

Two types of seismic engineering have developed in Kashmir that share many of the characteristics of the use of reinforced beams or *hatillar* of the Eastern Mediterranean or the three-dimensional wooden-framed cage or *gaiola* system of Southern Europe (Khurana 2011). Indeed, it is thought that many of these construction traditions followed the paths of migration and conquest as Islamic culture spread from the Middle East across Central Asia and parts of India (Langenbach 2009,

p. 6). *Taq* refers to buildings made from unreinforced masonry embedded with horizontal timber lacing and configured into a modular layout (or *taq*) of masonry piers and window bays. These masonry piers are thick enough to carry the vertical loads and are tied together by a ladder-like set of horizontal timber beams located at the base of the structure above the foundation and at each floor level and window lintel level. The timbers are deliberately unbonded and held in place by the full weight of masonry. The building is then faced with small, rough-surfaced fired bricks that serve as a weather-resistant facade over the sun-dried brick or rubble. Often three or more storeys in height, the buildings sway during earthquakes while the timber ladder bands keep the individual masonry piers from separating and causing the house to collapse. *Taq* gives buildings a flexibility to withstand considerable deforming movements either from seismic forces or subsidence and has the added advantage of being constructed from easily sourced local materials (Langenbach 2009, pp. 8–12).

Dhajji dewari is a variation of the timber frame and masonry construction method previously described. The term is Persian meaning 'patchwork quilt wall' and may indicate its provenance. Essentially it consists of the same timber framed infilled masonry as the Turkish *himiş*. Walls are commonly only one half brick thickness so that masonry/rubble and timber are flush on both sides. Each storey generally constitutes a distinct platform framed separately on the one below. While the strength of these disparate materials are relatively low, the ability of the house to resist collapse in an earthquake depends to a large extent on the interaction of wood and masonry, though houses with smaller wall panels generally perform better than those with larger ones. Houses are often built with an unreinforced masonry *taq* construction on the ground floor and *dhajji dewari* construction for the upper stories (Langenbach 2009, pp. 13–16). Such building practices continue to be used into the present and underwent something of a revival in the wake of the 7.6 magnitude Kashmir earthquake of 2005, when traditional houses proved to be surprisingly seismically resistant (Schacher and Ali 2009, p. iii).

Other variants of seismic architecture have developed along the Himalayan arc. The traditional *Koti Banal* architecture of the Indian state of Uttarakhand, lying on the south-eastern slopes of the mountain range and where two of India's largest rivers – the Ganga and Yamuna – originate, is perhaps the most visually spectacular. Some of its four- or five-storey buildings are reportedly over 800 hundred years old (Chakraborty 2008). A simple, rectangular tower-like construction, these buildings are raised on a dry masonry platform upon which alternate layers of wooden logs and dressed flat stones are laid. At the corners, pairs of logs are secured by thick wooden nails or dowels. The structure is further reinforced by wooden beams that run from the middle of the walls of one side to the other intersecting at the centre and so dividing the interior space into four parts, which provide the joists for supporting the floor beams in each storey. The fourth or fifth floor of the building has a balcony running around all four sides with a wooden railing. Openings are kept to a minimum and are relatively small with strong wooden empanelment to compensate for loss of strength (Rautela and Joshi 2009, pp. 306–7). These buildings are constructed along principles somewhat similar to that of the timber framed structure already described. While their characteristic architecture evolved out of their proven safety in earthquakes, the design also served other purposes such as security, defence and protection against heavy snowfall. Similar construction techniques are found to the west of Kashmir in the Hunza Valley of northern Pakistan. In particular, hill forts – such as that at Baltit – employ a form of timber lacing infilled with stone or lose rubble that may have been introduced by Alexander the Great from the Eastern Mediterranean. Known as cator and cribbage, lengths of timber cators are laid out in a rectangular pattern and infilled with loose rubble or stone with the overlapping corners laced together by 'cribbage columns' (vertical timber box frames found typically at wall corners) comprised of shorter wooden pieces (Hughes 2007, p. 102). Seismicity was by no means the only factor governing design and construction methods. The British Army campaigning in the Northern Areas of Pakistan in 1891 discovered that artillery shells had little effect on the walls of such hill forts (Hughes 2000b).

Design by disasters

As this brief historical overview indicates, seismic architecture has evolved in many different places and at many different times. Though the examples cited here may indicate a common origin in the construction techniques dating back to the classical civilisations of the Eastern Mediterranean, other forms of earthquake resistant buildings have evolved quite independently in other societies, as on the island of Nias off the Sumatran coast or around Yogyakarta in Java (Weichart and Herbig 2013).[3] Several questions, however, still remain to be answered: To what extent were earthquakes solely responsible for the design of buildings? Does the construction of earthquake resistant houses by themselves constitute a seismic culture in a wider sense? And what relevance, if any, has these past construction techniques for DRR in the present?

Amos Rapoport, in his foundational text on the relationship between house form and culture, identifies the principal determining factors as physical – climate and the need for shelter, materials and technology – and site as well as social – economics, defence and religion. As such, he does not include earthquakes as a determinant. However, his main argument is that form cannot be attributed to a single cause 'but is the consequence of a whole range of socio–cultural factors seen in their broadest terms' (Rapoport 1969, p. 18). As the historical case examples have shown, seismicity is usually one of many considerations that determine structural forms and construction techniques over time. Defence was clearly an imperative in the building of the hill forts of northern Pakistan as cator and cribbage constructions could not be easily breached by ballistics nor toppled and subsided by mining the foundations. Moreover, materials were easily dismantled and reused in a 'lego-like' modular fashion or buildings expanded and contracted as needs demanded (Hughes 2000a, p. 47).

Nor was earthquake alone the only hazard determining the shape and form of structures, as evidenced by the importance of attic firewalls in the pombaline houses of post-1755 Lisbon or the vernacular domestic architecture of the Spanish

Philippines where the evolution of a characteristic architecture was the product again of a combination of pyro-seismic factors (Tobriner 1980, p. 14, Bankoff 2012). The *fachwerk* or *blockbau* stone and wood buildings of South Tyrol, a region not known for earthquakes, may have had perfect seismic properties but the construction techniques also worked well to offset subsidence and to withstand heavy snowfalls and strong seasonal winds (Lanner and Barbisan 2000). The main point here perhaps is not so much the original factor or factors that led to the development of an architectural style, which were likely to be multiple and varied, but why, in seismically active areas, a particular method of construction was retained – often largely unchanged – for centuries. *Dhajji dewari* evolved in the Kashmir Valley for a variety of economic and cultural reasons that led to the development of similar forms of construction elsewhere around the world, but its retention up until the present was probably due to its performance during earthquakes especially for structures built on soft soils (Langenbach 2009, p. 15). Such reasons may explain the development of infilled timber framed construction techniques that are characteristic of mediæval and early modern housing in most of Europe even in regions with low seismicity, such as the 'half timbered' style of Tudor and Elizabethan houses in England and *colombage* in France. It may also help to explain why these fashions were subsequently replaced by less earthquake resistant forms (Dutu *et al.* 2012, pp. 622–5).

If seismic architecture does not necessarily produce a seismic culture, then what does? At least two other factors need consideration. The first is more a definitional matter and regards whether what constitutes seismic architecture is confined only to the way in which a building is constructed or whether it includes the wider built environment as well, the urban or community setting in which a structure is situated. Again, the evidence suggests that in many cases the entire fabric of the urban environment was refashioned to better accommodate the risk of earthquakes: How Sicilian planners rebuilt Catania after the 1693 earthquake with wide straight streets and inordinately large piazzas, or the symmetrical blocks of pombaline houses that still dominate the layout of post-1755 central Lisbon (Tobriner 1980, pp. 12–13, Langenbach 2007, p. 48). More to the point, how subsequent to the 1783 Calabrian earthquake Italian architectural treatises such as those written by Nicola Cavalieri di San Bertolo (1831) and Antonio Favaro (1883) widely publicised and promoted the application of seismic engineering methods mainly based on traditional building techniques tried and tested over the centuries (Lanner and Barbisan 2000, p. 3). How the *gaiola* and *casa baraccata* construction were mandated by law as standard building practice in Portugal and Naples are both examples of this. The development of a local seismic culture depends on the frequency with which an earthquake recurs though its initiation may have been attributed to an infrequent event of devastating proportions: for example, the 1505 earthquake in the Himalayas, the 1509 and 1766 earthquakes in Istanbul, the 1755 earthquake in Portugal and the 1783 earthquake in Calabria. Scholars at the European University Centre for Cultural Heritage have even tried to calibrate the frequency with which high intensity but not catastrophic events must occur so that they give rise to a seismic culture. Any generation, they argue, must be affected at least twice in a lifetime; that is an interval of between forty and sixty years (CUEBC 1993, p. 5).

The second factor to consider when looking at the the existence of a seismic culture is a perceptual matter and concerns how people regard a disaster. The critical point here is not whether a disaster took place but whether it caused cultural change. Modern Disaster Risk Management gauges an event in terms of the number of deaths caused or property damaged. Only if the mortality figures or economic losses exceed a certain threshold is the event considered a disaster. The hallmark of a seismic culture, however, is whether such an event had a significant effect on a social group and resulted in cultural change, and whether this change was directly attributable to the disaster or acted more as a catalyst for change in the longer term. Archaeological evidence, again, suggests that the key variable in this respect is not magnitude but frequency. Disasters that occur frequently over relatively long period of time can instigate adaptation and engender cultural change or, alternatively, they may hinder any such developments. In this respect, disasters are perceived not so much as natural events but as social ones that are explained or refuted according to how a threat is interpreted through a specific cultural and temporal lens (Torrence and Grattan 2002, pp. 11–13). Viewed historically, a noticeable feature of seismic cultures is how they are eclipsed over time. Vernacular seismic architecture had disappeared by the early twentieth century first in Southern Europe and then in the Eastern Mediterranean. It persisted longer in the Himalayan arc, but is now under threat there too. The loss of traditional construction expertise is rarely compensated by the introduction of modern seismic building codes that are often carelessly modified in their application or simply ignored (CUEBC 1993, p. 2). A 'catch-22' situation ensues in which local communities are no longer familiar with past practices and more modern anti-seismic regulations have not yet been adopted or become part of the culture. As one local Turkish woman summarised the state of affairs in the aftermath of the 1999 Izmit earthquake: 'It is the old people that know how to live with earthquakes; they live in wooden frame buildings of one-two storeys—we must learn from the older generations if we want to survive earthquakes' (Homan 2004).

So what relevance does this tradition of seismic architecture and seismic culture have for contemporary DRR? Why is culture important for DRR? The past challenges are notions that contemporary ways are always better and that techniques and practices developed by peoples and communities centuries ago to cope with the hazards that beset them have no bearing on how to deal with such events in the present. Indeed, science and technology are often wrongly credited as the product of the modern, largely Western age. There are innumerable reports detailing how properly maintained vernacular seismic architecture performed well during recent earthquakes.[4] Given the apparent superiority of these structures' performance in earthquakes, it must be asked what led to the decline of vernacular construction techniques in so many seismically active regions of the world? A number of related factors led to the gradual disappearance of seismic architecture during the twentieth century that were as true of the Kashmir earthquake as they have been of many communities and regions worldwide. Population pressure is the underlying causal agent compounded by the age profile of many developing countries whose young populations share no knowledge of traditional construction practices. Fashion and status, too, give a marked preference towards more modern looking houses that do not incorporate earthquake resistant features. Finally, the deforestation of many areas in recent decades and the significant amount of timber required in vernacular architecture has driven lower socioeconomic groups to use alternatives such as cheap concrete (Halvorson 2010, pp. 198–9). A scarcity of wood was partly responsible for the decline in the use of *hatil* construction in rural parts of Turkey in the early twentieth century (Hughes 2000b). Other factors unrelated to actual construction techniques may also have had bearing on people's decisions such as the lack of personal comfort in traditional *Koti Banal* architecture, whose design is purely utilitarian (Rautela and Joshi, p. 314).

Conclusion

Most people in the developing world live in non-engineered structures, whether these are rural domestic dwellings or, increasingly, urban low-rise. Formal building codes in developing countries rarely consider housing in the formal and low income sectors, despite the fact that between 60 to 70 percent of the population may live in such settlements (Lewis 2003). As in housing and many other features of the built environment, a fuller understanding and appreciation of past technologies, combined with modern science, are essential components of DRR. Most damage to structures and loss of life results from either disrespect of the tried and tested rules of the past or non-adherence to the seismic codes of the present.

Over the centuries, societies and the environment they have built have adapted to earthquakes. Where earthquakes are frequent and of a magnitude to inflict regular damage to property and cause loss of life, people in the past acquired the pragmatic and theoretical knowledge of learning to live with threat on a day-to-day basis – an accommodation that is reflected in their architecture styles and building methods. Many of these seismic cultures persisted well into the twentieth century, even in Europe. They continue to influence housing in rural areas of some parts of the developing world up until the present, though they are fast losing ground to the 'economics of the cinderblock', especially where timber has grown scarce and expensive. Modern building codes regularly ignore traditional seismic technologies and dismiss their scientific value. Unfortunately, local people are often active agents in this process tempted by the appeal of the modern and the symbols such structures convey of outlook, status and wealth. Effective DRR reduces vulnerability by encouraging a community to rediscover, in a critically appropriate manner, their 'own' seismic architecture. Apart from the direct advantages of reducing injuries and minimising damage, such an endeavour has intangible heritage value in terms of identity, memory, history and tradition and more tangible economic returns in terms of tourist potential. Far from just being disasters by design, the built environment is just as much designed by disasters and may even be enriched by them as well.

Notes

1 Initially, a maximum height of three storeys was prescribed but proved unenforceable as house owners wanted to derive more economic benefits from their properties.

2 The subsequent variation in construction technique is known as gaioleiros (Dutu *et al.* 2012).

3 See also the Earthquake Engineering Research Institute and International Association for Earthquake Engineering, Encyclopaedia of Housing Construction Types in Seismically Prone Areas of the World, available from: www.world-housing.net.

4 The high death tolls from the earthquakes in Bam (Iran) in 2004 and Izmit in 1999 were due to the failure of contemporary buildings not the historic ones (Gülkan and Langenbach 2004, p. 2). In a detailed statistical study of the damage district in the latter case, a wide difference was found in the percentage of modern reinforced concrete buildings that failed and those constructed in a traditional manner. In the hills above Gölcük, 60 of the 814 reinforced concrete structures were heavily damaged or collapsed accounting for 287 deaths and only four of the 789 two-three storey traditional buildings collapsed or were heavily damaged resulting in three fatalities (Doğangün *et al.* 2006, p. 982). A similar outcome was observed by N. N. Ambraseys in the aftermath of the 1963 Skopje (former Yugoslavia) earthquake where the old *adobe* houses with timber bracing resisted the shock much better than the modern brick or brick and reinforced concrete structures (Langenbach 1989). The much older buildings in the historic town centre of San Guiliano di Puglia suffered little damage in the Molise (Italy) earthquake of 2002 when the larger buildings of reinforced concrete in the newer part of town were devastated beyond repair (Langenbach and Dusi 2004, p. 343). The crucial point, however, on how well a building performs is the state of repairs, especially the state of the timber bracing. Houses built in a variation of the infilled timber frame system known as taquezal in Nicaragua performed well during the earthquake that shook Managua in 1931 but had been allowed to badly deteriorate in a tropical climate and mainly collapsed in the 1972 earthquake (Langenbach 1989). Older, well-maintained masonry houses in Bhuj, constructed with timbers in the walls, all survived the Gujarat earthquake of 2001 (Langenbach 2003). Again, well maintained *taq* and *dhajji dewari* structures were largely left intact by the 2005 Kashmir earthquake that destroyed nearly half a million houses made from modern building materials (Schacher and Ali 2007, p. 25).

References

Ambraseys, N. N. and Jackson, J. A., 2000. Seismicity of the Sea of Marmara (Turkey) since 1500. *Geophysical Journal International,* 141 (3), F1–F6.

Anderson, W. A., 1965. *Some Observations on a Disaster Subculture: The Organizational Response of Cincinnati, Ohio, to the 1964 Flood, Research Note 6.* Columbus, OH: The Disaster Research Centre, Ohio State University.

Bankoff, G., 2003. *Cultures of Disaster: Society and Natural Hazard in the Philippines.* London: Routledge.

Bankoff, G., 2012. A Tale of Two Cities: The Pyro-seismic Morphology of Nineteenth Century Manila. In: G. Bankoff, U. Luebken and J. Sand, eds, *Flammable Cities: Urban Fire and the Making of the Modern World.* Madison, WI: University of Wisconsin Press.

Bilham, R. and Ambraseys, N., 2005. Apparent Himalayan slip deficit from the summation of seismic moments for Himalayan earthquakes, 1500–2000. *Current Science,* 88 (10), 1658–63.

Buforn, E., Bezzeghoud, M., Udías, A. and Pro, C., 2004. Seismic sources on the Iberian–African plate boundary and their tectonic implications. *Pure and Applied Geophysics,* 161 (3), 623–46.

Cardoso, R., Lopes, M. and Bento, R., 2004. Earthquake resistant structures of Portuguese old "pombalino" buildings. Paper no 918. *13th World Conference on Earthquake Engineering*, Vancouver, BC: Canada, 1–6 August 2004 [online]. Available from: www.iitk.ac.in/ nicee/wcee/article/13_918.pdf [accessed 9 August 2013].

Chakraborty, T., 2008. 1000-year-old quake-proof architecture: Science seal on ancient houses. *The (Calcutta) Telegraph*, Monday 22 September 2008 [online]. Available from: www.telegraphindia.com/1080922/jsp/frontpage/story_9869079.jsp [accessed 11 August 2013].

Clancey, G., 2006. *Earthquake Nation: The Cultural Politics of Japanese Seismicity, 1868–1930*. Berkeley, CA: University of California Press.

Cuadra, C., Karkee, M. B. and Tokeshi, K., 2008. Earthquake risk to Inca's historical constructions in Machupicchu. *Advances in Engineering Software*, 39, 336–45.

Cuadra, C., Saito, T. and Zavala, C., 2012. Dynamic characteristics of traditional adobe-quincha buildings in Peru. Paper no 2653. *15th World Conference on Earthquake Engineering*, Lisbon: Portugal, 24–8 September 2012 [online]. Available from: http://ares.tu. chiba-u.jp/peru/pdf/output/2012/201215WCEE_Cuadra_2.pdf [accessed 8 August 2013].

CUEBC (European University Centre for Cultural Heritage), 1993. ATLAS of local seismic cultures: How to reduce the vulnerability of the built environment by re-discovering and re-evaluating local seismic cultures. *STOP Disasters: The United Nations International Decade for Natural Disaster Reduction Newsletter*, 12, March/April: Supplement.

Dilley, M., Chen, R. S., Deichmann, U., Lerner-Lam, A. L., Arnold, M., Agwe, J., Buys, P., Kjekstad, O., Lyon, B. and Yetman, G., 2005. *Natural Disaster Hotspots: A Global Analysis*. Washington, DC: United Nations.

Doğangün, A., Tuluk, Ö. İ., Livaoğlu, R. and Acar, R., 2006. Traditional wooden buildings and their damages during earthquakes in Turkey. *Engineering Failure Analysis*, 13 (6), 981–96.

Duggan, T. M. P., 1999. The hatil and the lessons of history. *Turkish Daily News*, 25 August [online]. Available from: www.hurriyetdailynews.com/default.aspx?pageid=438&n= the-hatil-and-the-lessons-of-history-1999–08–25 [accessed 2 August 2013].

Dutu, A., Gomes-Ferreira, J., Guerreiro, L., Branco, F. and Gonçalves, A. M., 2012. Timbered masonry for earthquake resistance in Europe. *Materiales de Construcción*, 62 (308), 152–65.

Erdik, M., 2000. Earthquake risk in Turkey. *Science*, 341 (6147), 724–5.

Gülkan, P. and Langenbach, R., 2004. The earthquake resistance of traditional timber and masonry dwellings in Turkey, Paper no 2297. *13th World Conference on Earthquake Engineering*, Vancouver, BC: Canada, 1–6 August 2004 [online]. Available from: www. conservationtech.com/rl's%20resume&%20pub's/RL-publications/Eq-pubs/2004– 13WCEE/GULKAN-LANGENBACH.pdf [accessed 3 August 2013].

Gutierrez, J., 2004. Notes on the seismic adequacy of vernacular buildings, Paper no 5011. *13th World Conference on Earthquake Engineering*, Vancouver, BC: Canada, 1–6 August 2004 [online]. Available from: www.curee.org/architecture/docs/13WCEE-GUTIERREZ-5011.pdf [accessed 10 August 2013].

Halvorson, S. J., 2010. In the aftermath of the Qa'yamat: The Kashmir earthquake disaster in northern Pakistan. *Disasters*, 34 (1), 184–204.

Homan, J. and Eastward, W. J., 2001. 17 August 1999 Kocaeli (Izmit) earthquake: Historical records and seismic culture. *Earthquake Spectra*, 17 (4), 617–34.

Homan, J., 2004. Seismic Cultures: Myth or Reality? [online]. Available from: www.grif. umontreal.ca/pages/papers2004/Paper%20-%20Homan%20J.pdf [accessed 14 August 2013].

Hughes, R., 2000a. Cator and cribbage construction of northern Pakistan. *Proceedings of the International Conference on the Seismic Performance of Traditional Buildings*, Istanbul: Turkey, 16–18 November 2000 [online]. Available from: www.icomos.org/iiwc/seismic/Hughes-C.pdf [accessed 12 August 2013].

Hughes, R., 2000b. Hatil Construction in Turkey. *ICOMOS International Wood Conservation Conference*, Istanbul: Turkey, November 16–18 [online]. Available from: www.icomos.org/iiwc/seismic/Hughes-H.pdf [accessed 3 August 2013].

Hughes, R., 2007. Vernacular Architecture and Construction Techniques in the Karakoram. In: S. Bianca, ed., *Karakoram: Hidden Treasures in the Northern Areas of Pakistan*. Turin, Italy: Umberto Allemandi & Co.

Khurana, M., 2011. Response to climate: Vernacular architecture and sustainability. *Boloji*. [online]. Available from: www.boloji.com/index.cfm?md=Content&sd=Articles&ArticleID= 10561 [accessed 9 August 2013].

Konstantinidis, D. and Makris, N., 2005. Seismic response analysis of multidrum classical columns. *Earthquake Engineering and Structural Dynamics* 34, 1243–70.

Kouskouna, V. and Makropoulos, K., 2004. Historical earthquake investigations in Greece. *Annals of Geophysics*, 47 (2/3), 723–31.

Langenbach, R., 1989. Bricks, mortar, and earthquakes: Historic preservation vs. earthquake safety. *Journal of the Association for Preservation Technology*, 21 (3–4), 30–43.

Langenbach, R., 2003. Survivors among the rubble: Traditional timber-laced masonry buildings that survived the great 1999 earthquake in Turkey and the 2001 earthquake in India, while modern buildings fell. *Proceedings of the International Congress on Construction History*, Madrid: Spain, 20–24 January 2003 [online]. Available from: http://gilbert.aq.upm.es/sedhc/biblioteca_digital/Congresos/CIHC1/CIHC1_120.pdf [accessed 15 August 2013].

Langenbach, R. and Dusi, A., 2004. On the Cross of Sant' Andrea: The response to the tragedy of San Guiliano Di Puglia following the 2002 Molise, Italy, Earthquake. *Earthquake Spectra*, 20 (S1), 341–58.

Langenbach, R., 2007. From 'Opus Craticium' to the 'Chicago Frame': Earthquake-resistant traditional construction. *International Journal of Architectural Heritage* 1, 29–59.

Langenbach, R., 2009. *Don't Tear It Down! Preserving the Earthquake Resistant Vernacular Architecture of Kashmir*. New Delhi: UNESCO.

Lanner, F. and Barbisan, U., 2000. Historical Anti-Seismic Building Techniques: Wooden Contribution. *Estatto dalla memoria presentata al Convegno Internazionale Seismic Behaviour of Timber Buildings*, Venice: Italy, 29 September 2000 [online]. Available from: www.tecnologos.it/index.php?option=com_content&view=article&id=197:historical-antiseismic-building-techniques-wooden-contribution&catid=84:01&Itemid=95 [accessed 13 August 2013].

Lewis, J., 2003. Housing construction in earthquake-prone places: Perspectives, priorities and projections for development. *Australian Journal of Emergency Management*, 18 (2), 37.

Lorenz, D., 2010. The diversity of resilience: Contributions from a social science perspective. *Natural Hazards* [online]. Available from: http://dx.doi.org/10.1007/s11069–010–9654-y [accessed 4 December 2012].

Makarios, T. and Demosthenous, M., 2006. Seismic response of traditional buildings of Lefkas Island, Greece. *Engineering Structures*, 28 (2), 264–78.

Mileti, D. E., 1999. *Disasters by Design: Reassessment of Natural Hazards in the United States*. Washington, DC: Joseph Henry Press.

Moore, H. E., 1964. *And the Winds Blew*. Austin, TX: Hogg Foundation for Mental Health.

Mullen, J. R., 1992. The reconstruction of Lisbon following the earthquake of 1755: A study in despotic planning. *Planning Perspectives*, 7 (2), 157–79.

Oliver-Smith, A., 1999. "What is a Disaster?" Anthropological Perspectives on a Persistent Question. In: A. Oliver-Smith and S. M. Hoffman, eds, *The Angry Earth: Disaster in Anthropological Perspectives*. New York: Routledge, 29–30.

Pampanin, S., 2008. Development in Seismic Design and Retrofit of Structures: Modern Technology Built on 'Ancient Wisdom'. In: L. Bosher, ed. *Hazards and the Built Environment: Attaining Built-in Resilience*. London and New York: Routledge, 96–123.

Porphyrios, D. T. G., 1971. Traditional earthquake-resistant construction on a Greek island. *Journal of the Society of Architectural Historians*, 30 (1), 31–9.

Rapoport, A., 1969. *House Form and Culture*. Englewood Cliffs, NJ: Prentice-Hall.

Rautela, P. and Joshi, G., 2009. Earthquake safety elements in traditional Koti Banal architecture of Uttarakhand, India. *Disaster Prevention and Management*, 18 (3), 299–316.

Roy, P. N. S. and Mondial, S. K., 2012. Identification of active seismicity by fractal analysis for understanding the recent geodynamics of Central Himalaya. *Journal of the Geological Society of India*, 79 (4), 353–60.

Schacher, T. and Ali, Q., 2007. *Dhajji, Construction for One and Two Story Earthquake Resistant Houses. India-UNESCO Country Programming Document 2008–2009, IN/2007/PI/11*. New Delhi: United Nations Educational, Scientific and Cultural Organization.

Schacher, T. and Ali, Q., 2009. *Dhajji, Construction for One and Two Story Earthquake Resistant Houses: A Guidebook for Technicians and Artisans*. Scuola Universitaria Professionale Della Svizzera Italiana.

Schwartz, S. B., 2005. Hurricanes and the shaping of circum-Caribbean societies. *Florida Historical Quarterly*, 83 (4), 381–409.

Tobriner, S., 1980. Earthquakes and planning in the 17th and 18th centuries. *Journal of Architectural Education*, 33 (4).

Tobriner, S., 1983. La casa baraccata: Earthquake-resistant construction in 18th-century Calabria. *Journal of the Society of Architectural Historians*, 42 (2), 131–8.

Torrence, R. and Grattan, J., 2002. The Archaeology of Disasters: Past and Future Trends. In: R. Torrence and J. Grattan, eds, *Natural Disasters and Cultural Change*. London and New York: Routledge.

Tsakanika-Theohari, E. and Mouzakis, H., 2010. A post-Byzantine mansion in Athens: The restoration project of the timber structural elements. *Proceedings of 11th World Conference on Timber Engineering*, Trentino: Italy, 20–4 June 2010 [online]. Available from: www.ewpa.com/Archive/2010/june/Paper_173.pdf [accessed 3 August 2013].

van Dam, P., 2012. Denken over Natuurrampen, Overstomingen en de Amfibische Cultuur. *Tijdschrift voor Waterstaats Geschiedenis*, 1/2, 1–10.

Walker, C. F., 2003. The upper classes and their upper stories: Architecture and the aftermath of the Lima earthquake of 1746. *Hispanic American Historical Review*, 83 (1), 53–82.

Weichart, G. and Herbig, U., 2013. Time for rebuilding, time for heritage? Case studies from Indonesia. *7th European Association for Southeast Asian Studies*, Lisbon: Portugal, 2–5 July 2013.

Wenger, D. E. and Weller, J. M., 1973. *Disaster Subcultures: The Cultural Residues of Community Disasters, Preliminary Paper No. 9*. Columbus, OH: Disaster Research Centre, Ohio State University.

Wisner, B., Blaikie, P., Cannon, T. and Davis, I., 2004. *At Risk: Natural Hazards, People's Vulnerability, and Disasters*. 2nd ed. London: Routledge.

4

'LEARNING FROM HISTORY'?

Chances, problems and limits of learning from historical natural disasters

Gerrit Jasper Schenk

Problem outline, modelling and framing the question

Can one learn from history? Many people these days would spontaneously reply, 'Yes, of course!' In many countries, lessons from history are regarded for the good of the state – for instance in Germany, where resisting anti-democratic tendencies and racism in society is a result of the twentieth century experience of two totalitarian dictatorships and the Holocaust (cf. Wehler 1988, Herzog 1997, pp. 13–16, Brumlik 2004). The question of learning from the history of natural disasters is frequently answered in the affirmative in research writings, appealing to the old belief in history as a schoolmaster. In the context of historical floods and climate history, 'knowledge of the past is a key for understanding present and future; this is especially true for climate history' (Glaser and Stangl 2004, p. 485). Learning from historical disasters is said to take place within research on disastrous flooding, landslides, avalanches, storm tides and other such events (cf. Kempe 2007, Poliwoda 2007, Bankoff 2012, p. 39, Dix 2012). Processes of learning and adaptation are so strongly understood as a force for shaping cultures and nation building that Greg Bankoff speaks of the Philippines as a 'culture of disaster' and Gregory Clancey refers to Japan as an 'earthquake nation' (Bankoff 2003, Clancey 2006). Since there is consensus among researchers that disasters are the outcome of complex, historically induced and causally connected processes at the interface of 'nature' and 'culture', culture becomes a key factor for understanding disasters – and is thereby something that has evolved throughout history.

However, there are very few systematic investigations on whether or even how it is possible to learn from historical natural disasters, and how this learning works in practice. Christian Pfister has developed a functionalist model of adaptive interaction with disasters that defines three phases of response (Figure 2). They are

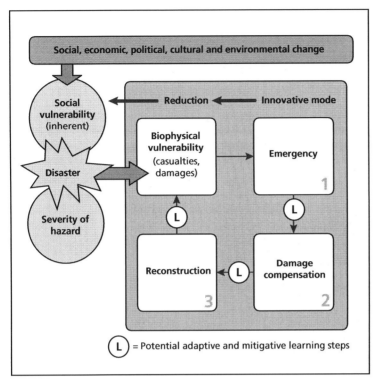

FIGURE 2 A functional model of disaster response

Source: adapted from Pfister 2009b, p. 22.

associated with possible learning steps that result from the disaster experience and are capable of contributing to the reduction of a society's vulnerability (Pfister 2009b, pp. 22–33).

The first of Pfister's response phases is the emergency phase immediately before and after damage occurs, where the response in disaster management is to improve early warning. The second is the damage compensation phase. This begins in the aftermath of the initial disaster and covers, for example, the organisation of first aid and the mobilisation of the affected community and the larger society. The local frequency of disasters and sociocultural setting largely determine whether, and then how, innovative learning processes are triggered during this phase. The third is the reconstruction phase, in which the disastrous experience can be used for reconstruction to reduce the vulnerability of society through either more of the same interventions (higher dams against flooding, for instance) or innovations (such as legislation or reforestation of the upper reaches of rivers).

Despite its abstractness, Pfister's model offers a rather good summary of the possible learning processes of historical and recent societies (for alternative linear and cyclical models, cf. Wisner *et al.* 2004, pp. 87–124, Felgentreff 2008, pp. 284–6). However, it presupposes a human behaviour that largely ignores a number of sociocultural factors, such as religion or emotion, and gives more weight to the functionalist rationality of practical know-how than to cognitive and mental processes (Pfister 2009b, pp. 19–20). Another question is worthy of discussion: to what extent do knowledge and experience of historical disasters going back several generations, or even centuries, still have a practical learning effect? Pfister sees in *longue durée* a quite practical dimension of learning from disasters. One example is the case of choosing a low-risk location for buildings based on high-water marks. In areas at risk of flooding, these marks are still seen on bridges and house walls, enabling an everyday risk assessment through visible and concrete indications of the height and frequency of flooding (Pfister 2009a, p. 242). In view of the specific regional cultures of memory and the importance of disaster awareness and disaster risk reduction over the centuries, Pfister stresses: 'The importance of historical research results lies [. . .] in the fact that they can bring together the myriad of experience gained with respect to certain questions, present them to society as communication about risks or stimulus for reflection and, in so doing, make the unthinkable thinkable' (Pfister 2009a, p. 240).

Pfister's fairly optimistic view of learning from disasters has been met with scepticism from some experts. In view of recent disasters and subsequent reconstruction processes, critics observe that people learn little or nothing from disasters for material needs and sociocultural reasons, such as the need for rapid reconstruction and the effect of biased interests (Felgentreff 2008, pp. 289–92; Glade and Felgentreff 2008, p. 447). Heike Egner has offered an explanation for this observation from the perspective of system theory (Egner 2008, pp. 429–32): societies as functionally differentiated systems can only be regulated to a limited degree due to their closed subsystems. After a disaster, the tendency of subsystems to maintain themselves can thus hamper or prevent a necessary structural change in the whole of society, and even in individual subsystems. The development of social systems (also of organisations and institutions) is therefore not planned and intentional, but evolutionary.

According to Niklas Luhmann, a natural disaster can only be perceived as a signal from the 'environment' of the system if it can be built into the internal communication of the system and cause 'resonance' there (Luhmann 2004, pp. 97–8, 210, 218–26). The systemic environment (as 'nature') must be regarded not as static but as dynamically evolving. This theory could therefore explain why risks are hard to control in complex modern societies and disasters occur surprisingly and unpredictably (Renn 2008, pp. 61–3). It would, however, militate decisively against the possibility of being able to learn at all from recent or past disasters.

There are even more fundamental objections to the possibility of learning from history. Based on considerations from methodology and historical theory, these critiques question in general whether a forecast for the future can be derived from the historical explanation of a past event. Such arguments cite fundamental considerations concerning the nature of scholarly historical knowledge. This knowledge is subject to a double limitation on epistemological grounds (Gabriel 2013): the historian first gains knowledge of the past with the assistance of a methodological constructivism by constituting statements from incomplete sources derived by chance and selection by means of what Kant (1793: XXVI) terms 'reflective judgement' in an heuristic process. Second, this historical knowledge always relates to a specific case, whose relationship with the general is hard to define (Pohlig 2013). This means that historical knowledge is epistemically indeterminate. Its statements are always based on an incomplete and one-off set of data that permits few conclusions about rules or principles and can only make general statements that seem plausible or probable. History does not repeat itself and has no laws. Hence it was recently stated: 'Accordingly, one can learn nothing for the future even from a successful explanation of past history – neither as confirmation or warning' (Gabriel 2013, p. 18). However, due to the challenges of having to make statements about ever more complex problems and the future, for example in global climate models, the epistemic character of statements from the social and natural sciences are growing closer to those from the humanities. This particularly applies when the natural sciences issue information about weather conditions, climatic processes or similar topics with the aid of statistical methods and probability calculations.

The discussion about the use of historical knowledge for the future therefore fluctuates between cautious affirmation, sceptical qualification or even negation. At the same time, it has become clear that the opportunities and problems of useful historical knowledge are dealt with at quite different levels. On the one hand, it is a question of the general constitution of methodologically guaranteed historical knowledge about the past (Gabriel 2013, Pohlig 2013). On the other, the usefulness of this historical knowledge for the present or future is up for debate in another area such as (climate) policy, disaster management or the specific prevention of disasters (Egner 2008, Pfister 2009a and b). The respective types of knowledge, their validity and their methodological status are different, however, and alternate between theoretical and practical knowledge, or between knowledge for planning, action and orientation, or between reflection and application (Mittelstraß 1986, pp. 64–8 '*Verfügungswissen*' and '*Orientierungswissen*', Koselleck 1992, pp. 38–66, 154–7, Detel 2003, pp. 129–30). Furthermore, timescales of the learning processes vary greatly and can range from days to centuries. It thus makes sense for us to focus our notion of 'learning from history': it is not about whether people generally learn from history, but whether they learn, and have learned, from the history of natural disasters. We must ask specific questions about the knowledge that has played a part in the learning process: what epistemic status does it have (uncertain, certain, probable) and what is its sociocultural value as such (expert, general, recognised or esoteric knowledge)? What area of validity (social, local, global) and what use is attributed to it? The investigation becomes difficult as it involves a variety of forms of knowledge that are hard to categorise and a combination of many (historical, sociological, numeric–scientific) methods and insights. Two temporal levels of investigation are dealt with in the next two sections: simultaneous learning from disasters and present learning from past disaster in *longue durée*.

Simultaneous learning from disasters – then and now

Regardless of epistemological and methodological doubts about the possibility of learning from the past, it is beyond dispute in research circles that contemporaries (envisaged as one generation, approx. 30 years) in many epochs and cultures have learned from disasters (in general, cf. Bankoff 2012, p. 41; for Europe, cf. Schott 2012, Poliwoda 2007; for Asia, cf. Bankoff 2007, Bhargava 2007, pp. 199–205). However, the manner, scope and thus benefit of this implicit or explicit 'learning' varies greatly and depends on epoch and culture. This variation reveals the role played by history and culture in the learning process. Determinations of what knowledge is useful are subject to the local value system and worldview prevalent at a given time; hence, they are hardly reflective of present understanding. Anyone who applies present-day criteria to learning processes of the past renders these processes useless for reducing social vulnerability towards natural risks. Learning processes to improve disaster management and prevention should therefore be assessed twofold: on the one hand in the context of their respective age and culture, and on the other regarding our own present-day value criteria and conceptions.

An essential role for dealing with disasters is played by the knowledge available to culture in all its forms. Cultural memory (Assmann 1999) as a knowledge repository or archive of historical experience provides terms, concepts, patterns of interpretation and response schemas, without which natural hazards cannot be perceived as possible disasters and thus processed. This has a far-reaching and comprehensive application to the terms that stand for natural hazards, to the concepts underlying such terms, and to ways of explaining and reacting to disasters.

Let us study some examples from the European Middle Ages. The term *disastro* arose for the first time in fourteenth-century Italy and goes back to the idea of an unfavourable star constellation (disaster) having a negative impact on an earthly situation, causing earthquakes and other such events (Schenk 2013, pp. 192–4). This idea is based on an understanding of nature, which is itself based on ancient ideas of a physical relation between the events in heavens and on earth. The term *disastro* thus highlighted an astro-meteorological explanation pattern of disaster and served the forecasting of disasters by observing the stars and thus reading nature as God's book. Furthermore, in most cases, God was understood to be the first mover of the stars. A disaster could therefore be understood as a natural sign from God, mostly as moral warning or as punishment for sins (Groh *et al.* 2003, pp. 20–1). This idea also characterises the reaction patterns triggered by disasters. When a severe earthquake destroyed thousands of homes in the Mugello valley north of Florence in 1542, the senate strongly condemned sodomy and blasphemy in view of this sign of God's apparent anger as a preventive measure against imminent future disasters (Schenk 2010b, pp. 221–2). Similar and frequently used measures against potential disasters during the European Middle Ages and the Early Modern period were processions and bans on luxury (Martín-Vide and Barriendos Vallvé 1995, Barriendos Vallvé 2005, Schenk 2010c, pp. 516–7, Labbé 2011, pp. 175–6). Following the works of cultural anthropologist Mary Douglas, one can presume that such bans and the search for a scapegoat for the disaster, such as homosexuals in the case of Florence in 1542, are based on the same cultural model: one that is a culturally specific idea of purity that becomes endangered by sin expressed as deviant behaviour or by non-conformist groups of persons (Behringer 1999, Douglas 2008, pp. 117–40). Naturally, these measures had no preventive effect against natural hazards such as earthquakes, floods or droughts. However, they were able to strengthen both the mental resilience of (parts of) the population and the power of the ruling government.

Complex cultures are able to offer different and even contradictory explanation patterns, and this fact has consequences for disaster response. When the Florentine people wanted to hold a religious procession because of the ominously continuous rain in October 1496, the city council hotly debated it because, as one speaker put it, 'it is not sufficient to address oneself to God if we do not make an effort to help ourselves' (Fachard 2002, p. 327). Around 1510, the Strasbourg council asked the cathedral chapter to organise a supplicatory procession against the Black Death, but clerics warned against this idea, stating it could be counterproductive by increasing the risk of infection (Schenk 2010a, p. 73). Contradictory patterns of explanation and conflicting patterns of reaction are dynamising factors in the cultural handling of disasters because they can lead to better preventive measures. Hence, the sociocultural structure of a society can favour or prevent the evolution of learning processes.

Specific local knowledge has comprehensively shaped the handling of natural hazards and potential disasters (Dekens 2007). This encompasses both practical and material considerations, such as the state of embankments at coasts that are threatened by storm tides, as well as cognitive, religious, social and political considerations. In order to allow for embankments to be successfully constructed and maintained, people have to join together and supply resources by investing ideas, and offering labour, time and money. The prevention of disasters can create a sense of community. Taking the example of embankments at the German North Sea coast, this effect can be observed in the centuries-long relations between cooperative and governmental structures (Mauelshagen 2009, pp. 49–58).

The exploration of disasters from the past indicates that societies have mainly learned from disasters when disasters have been recurrent phenomena. Above all, minor disasters have produced a pressure to adapt and have generated behavioural changes in individuals, civic planners and even governments. This has led to long term constructional and institutional adaptations at many levels (Schott 2012, Poliwoda 2007, Schenk 2012a), which have operated in the context of what was usual and conceivable for their age and culture. The need to learn from a disaster and the resultant measures to do so are frequently and explicitly described by contemporary writers in connection with, for instance, floods (Rohr 2007a, pp. 353–84, Schenk 2007, pp. 372–4). However, learning effects can also be understood implicitly by researchers as slow institutional processes. From a present-day viewpoint, the benefit of these learning effects, measured by casualty and damage reduction, is often overlooked or underestimated. Whether whole societies collapsed because learning seemingly did not take place or had little effect, as in Jared M. Diamond's popular argument (Diamond 2005), is doubtful and improbable (McAnany and Yoffee 2010). We should therefore look into the effectiveness of 'learning from disaster'.

What do the above findings reveal about the generally limited effectiveness of learning processes by societies in the high modern and postmodern eras (cf. Egner 2008, pp. 429–32, Felgentreff 2008, pp. 289–92, Glade and Felgentreff 2008, p. 447)? It is hard to make definite statements of global validity: the differences between, for instance, Switzerland and Bangladesh, or between agricultural and industrialised societies are too great. The disaster gap from 1882 to 1976 that Pfister diagnosed in Switzerland supplies an argument for the loss of traditional risk awareness due to the low frequency and intensity of natural disasters during such a period (Pfister 2009a). The scientific and technological worldview, with its specific patterns of perception and interpretation of modern industrialised societies, has probably played a role in this process of suppressing risk awareness (Borst 1981, p. 532). Recent research confirms this assumption regarding landslides in the Swabian Alps, where the advent of the Enlightenment and the expert culture of the modern age gradually forced traditional local knowledge to recede, to the detriment of the population (Dix and Röhrs 2007, pp. 229–31).

So do modern societies have a 'blind spot' about recognising their own ways of dealing with disasters, one which leads them to resort primarily to recent expert knowledge, technologies and, as a system theoretician would say, the inherent logic of societal subsystems when seeking prevention tools? More recent approaches to disaster management have realised there has indeed been a blind spot regarding this issue and have been taking a bottom-up approach. These approaches are beginning to reconnect the population with traditional local knowledge on disaster prevention measures (Wisner *et al.* 2004, pp. 367–74). Can this be characterised as a learning effect, owed to the orientation knowledge that results from analysing past and present disasters, and thus constitute 'learning from history'?

Learning today from the disasters of the past: opportunities, problems and limits

The question of 'learning from history?' becomes more acute when we inquire into present learning from disasters that occurred at least three generations in the past. This limit, which is also cited to distinguish communicative from cultural memory (Assmann 1999, pp. 48–56), at the same time marks a boundary between the personal experience still accessible through stories and memories and a history methodically analysed by a professional historian. Beyond this limit we can no longer truly speak of a common area of experience and must consequently question the legitimacy of drawing analogies between the past and the present.

Disasters, too, belong to the world of 'social facts' that Émile Durkheim wanted to see explained by *faits sociaux antécédents* (Durkheim 1895, p. 135). Anyone who examines disasters using scholarly historical methods will first use the established techniques of the historian or the cultural historian from the social sciences and cultural studies (Lorenz 1997, Jordan 2009). Doing so allows for solid statements, those which are substantiated by strong arguments with a specific scope of validity and insightfulness. Here it is no less true than with other subjects of investigation that historians can and should argue about structures, (causal) connections, interpretations and evaluations. This knowledge has a fundamental value as an intersubjectively verifiable statement on the past. The decisive question is only whether we can still distil statements for understanding the present or the future from this knowledge.

Since questions addressed to the past necessarily spring from an interest in the present, their replies have a relationship with the present through the heuristic 'representation' of the historian (Gabriel 2013, pp. 16, 23–4). The comparison of past and present has always been inherent to this heuristic exercise in order to shape concepts or develop models. This generally takes place through simple logical operations, such as analogies with similar situations or contrastive contouring with divergent outcomes, framework conditions and development parameters. Modelling becomes easier if the present is related to the past in a structure of interlocking effects, however complex and broken this structure might be (Crumley 1994, p. 413). Here, we might think of the landscape of dykes on the North Sea coast that by its very nature has for centuries been the consequence and precondition of 'natural' disasters (storm tide flooding) and at the same time presupposes and requires certain mental attitudes, social activities and societal structures to be maintained.

Even if history does not repeat itself because the starting conditions and contexts of events are always unique, we can still gain structural knowledge by utilising the methods of an historian. Such knowledge has a certain predictive potential because it allows for the identification of long term 'conditions of a possible future' (Koselleck 1992, pp. 156–7). Historical analysis through case studies (Pohlig 2013, pp. 304–6) makes it easier to identify far-reaching and long term social mechanisms (e.g., risk awareness versus security-mindedness), recurrent structures (model) or process types of disaster that can serve as self-enlightening orientation knowledge. Knowledge of the past contributes to understanding present conditions and situations. The example mentioned above of the model 'hydrographic society' of the North Sea coast shows how a sociocultural structure guaranteed the maintenance of a danger-resistant cultural landscape that has crossed national borders and centuries (Mauelshagen 2009, pp. 49–55).

The applicability of structural knowledge that describes an open field of potential, instead of supplying a set of timeless rules, has its limits owing to its basic 'epistemic indeterminism' (Gabriel 2013, p. 17). We cannot make a case for a compelling causal link between, for instance, a specific interaction with natural hazards and a certain type of society (or vice versa). Anyone who would like to deduce concrete policy advice for the present or future from structural knowledge about the past can simply extend the horizon of the present and give guidance by pointing to plausible analogies and probably ongoing constellations, connections and developments. In my view, however, the benefit of analysing historical disaster can in some cases go beyond a conceptual broadening of the scope of possibilities and the generating of 'orientation knowledge' (Mittelstraß 1986, pp. 64–8; 'Orientierungswissen'). Such analysis can also, at least partially, provide application-related knowledge. The study of natural disasters represents a fascinating special case of acquiring knowledge through historical study, theory and method.

This is due to the 'natural factor' in disaster events that, being a physical shock, follows rules different from social rules. In contrast to Durkheim's view mentioned above, here a social fact can thus follow from a preceding physical fact, namely, an extreme natural phenomenon. The specific objects of study and intentions of historical disaster research lead to a situation where, from the start, it is not possible to rely merely on the methods and insights of the humanities. To reply to certain questions, it not only makes sense but is imperative to draw on the findings and methods of both the social and natural sciences.

The historical-critical review of statements from the sources on extreme weather conditions and their duration can best be clarified in cooperation with disciplines such as geophysics, geography, hydrology, biology, meteorology and archaeology. These fields tell us about the height of floodwaters, the location of landslides and the run-off of precipitation as a function of variations in elevation, vegetation, tillage, etc. In other words, like environmental history or soil archaeology (Crumley 1994), historical disaster research must be open to a mix of methods and interdisciplinary exchange. Yet what does the combination of hermeneutic, qualitative and statistical methods mean for the epistemic status of the findings obtained? And what does the argumentative linking of knowledge from different origins mean in terms of the validity of the findings obtained, for its predictive power and for the much-mentioned notion of 'learning from history'?

Through combining methods and findings from the humanities with the social and natural sciences, we increase the possibility of perceiving analogies between historical conditions and those of the present and future. If, for example, knowledge about past natural phenomena can be obtained with the methods of the historian and these findings have a certain probability of recurrence at a certain place, this knowledge can be used for application-related diagnoses and forecasts in the present (Bankoff 2012, pp. 39–40). A number of examples quickly reveal what is at stake.

In the last few decades, historical seismology has developed a sophisticated collection of methods and tools, starting with the historical earthquake catalogues from the fifteenth to the mid-twentieth century (Guidoboni and Ebel 2009; for criticism cf. Waldherr 1997, pp. 18–29). These tools enable a relatively accurate quantitative reconstruction of past earthquakes thanks to the wealth of historical sources (texts, pictures, buildings, maps and photographs), geological finds and geophysical calculations. Historical seismology relies on findings owed to both historical-critical source analysis and to research in scientific fields (including geology, geography and geophysics). For example, damage to buildings allows researchers to draw conclusions about the approximate force of an earthquake.

A comparison of historical reports, drawings or *in situ* remnants of building damage enables the force and epicentre of historical earthquakes to be determined (on methods and scales, cf. Gasperini and Ferrari 1997, Grünthal 1998, Guidoboni 2000, Guidoboni and Ferrari 2000). Geological conditions hardly change at all across historical time, thus forming a permanent material foundation for the world on the earth's crust, which is subject to comparatively rapid change. The geophysical phenomena of the earth's crust and interior follow natural laws and were not noticeably influenced by human activities until the industrial age with its major technological inventions and interventions. The findings of historical reconstructive analysis can thus be checked against scientific findings about the bowels of the earth: can the earthquake have happened as reconstructed from the historical sources? Interdisciplinary findings are available in the form of lists, databases and maps, on which the earthquakes of the last 500 to 2,500 years can be localised and marked with their frequency and force (cf. Ambraseys, Melville and Adams 1994, Guidoboni and Comastri 2005, Catalogue of Strong Earthquakes in Italy 2014, Sismicitè de la France 2014). The historical seismological data from these are admittedly not as exact as those gained instrumentally in the last few decades, as they have been reconstructed from estimates and reports by people living at the time and by post hoc scientific analysis, but they cover a much greater period of time.

This temporal range is the key point, however, since in order to assess present seismic risks it makes sense to start with a large scale time frame to be able to examine as many earthquakes as possible, seeing that they often only occur at long intervals (Quenet 2012, p. 96). Knowledge gained in this way is application-related and can be used directly – to determine the building site for nuclear power plants or for dams, for example. If research work has followed the *lege artis* (rules of the art), a relatively concrete application-related knowledge can be derived from such historical experience. On the other hand, this only produces simple data about past earthquakes in certain zones. Other data are consequences and assessments that result from their present context.

This example from historical seismology has been explained in some detail. It can also be observed in a similar way in other areas of historical disaster research. The findings of historical hydrology, meteorology and climatology are particularly relevant. Here is a brief outline of their contribution to learning from the history of ('natural') disasters.

Historical hydrology attempts to reconstruct historical flooding patterns and events (Brázdil and Kundzewicz 2006, Rohr 2007b). Here, too, different types of sources such as contemporary reports, weather diaries, ledgers and maps are analysed on the basis of source critique and assessed in light of what information they supply on the place, time and severity of a flood. In addition, it provides an evaluation of sediments, high-water marks and the state of buildings, among other physical evidence (Baker 2008). This information is quantified via a multi-level classification system according to the flood's degree of gravity. Details about a flood can, in principle, be reconstructed on the basis of information about landscape relief, the absorption capacity of the soil, the shape of the riverbed, and the river's hydrological details. Compilations of long series of flood (and drought) events reveal the basic breadth of fluctuation over the centuries, and even millennia (Glaser 2012, pp. 105–11, Chalyan-Daffner 2013, pp. 379–588).

Historical meteorology (Luterbacher *et al.* 2002) attempts the reconstruction of historical climate conditions and weather phenomena. Historical climatology reconstructs details on historical macro-weather systems, atmospheric conditions and climatic parameters (cf. Glaser 2001, pp. 13–56, Brázdil *et al.* 2005, Brázdil *et al.* 2010, Mauelshagen 2010, pp. 36–59). The two related sub-disciplines draw on an enormous number of textual source types, even including dendro-chronological data, and proceed both qualitatively and quantitatively, thus drawing on a combination of methods from the humanities and the social and natural sciences. A particularly important role is played by proxy data, which give indirect indications of weather or climate conditions of the past and originate from the 'archives of nature' (such as pollen layers or ice cores) and from the 'archives of society' (e.g., a record of the date of the first apple blossom of the year). It is clear that such complex reconstruction processes using statistical methods and huge, sometimes georeferenced, databases are vulnerable to error at many points (cf. NOAA Databases 2014, TAMBORA 2014, Euro-Climhist 2014). Consequently, there has been no lack of criticism regarding such research. Using source-critical methods, Pierre Alexandre was able to show errors in up to 50 percent of cases of many older collections of source texts relevant to climate and weather history published from 1858 to 1976 – a horrifyingly high rate of error (Alexandre 1987, pp. 13–23, Liebscher *et al.* 1995, pp. 10–13). If the rules of the art are followed, however, the findings gained from this research are very valuable for reconstructing information on all historical extreme events connected to precipitation, temperature and air pressure: flooding and droughts, cold and hot spells, storms and sometimes even hail and thunderstorms.

Yet what is the present value of these findings about historical conditions? Unlike in historical seismology, we cannot start from a largely unaltered material substrate. Almost everything on the earth's surface is subject to change through natural dynamics (such as erosion) and above all through human intervention (agriculture, buildings, dams, canals, emissions, etc.). If the material substrate has considerably changed, drawing analogies from historical to present circumstances becomes considerably more complicated (Seidel *et al.* 2012, pp. 290–2). Nevertheless, the historical findings are valuable since they supply reliable values for extreme events, such as the maximum height of a flood or maximum duration of a drought, in the *longue durée* (cf. Wetter *et al.* 2011, With 2013). As a benchmark used for tasks such as estimating risk for construction zones or calculating minimum quantities of cooling water for power stations, these findings are of very practical utility. However, because of the fundamental change in sociocultural parameters in potentially affected societies, very few far-reaching and application-related conclusions can be drawn from the past regarding the sociocultural interaction with disasters today.

Conclusion: an application-related 'material turn' in historical disaster research

What can we finally conclude from these findings? First, it has become clear that simultaneous learning processes can only take place in time- and culture-specific societal frameworks and are thereby limited. Second, present-day learning from natural disasters from the more distant past is an interesting and special case. Through analysing historical disasters, we can identify long term societal mechanisms, specific types of processes and recurrent structures for dealing with disasters. This structural knowledge hints at the potential scope for future development and can give helpful guidance in decision making.

The 'material turn' in the present-day interdisciplinary exploration of historical disasters has unveiled fascinating prospects. Multidisciplinary and multi-method subdisciplines of historical seismology, hydrology, meteorology and climatology show that it is possible to distil application-related knowledge from experiences of historical disasters and make practical suggestions to reduce societal vulnerability. The knowledge resulting from natural disasters about the intensity, probability and intervals of recurrence of extreme events is helpful when it comes to estimating actual risk potential. These findings are of significance for town planning and the insurance industry, to name just two examples, and enable societies to focus on disaster risk reduction. However, sociocultural factors remain crucial for effectively turning this knowledge into action in modern-day society.

References

Alexandre, P., 1987. Le Climat en Europe au Moyen Âge. Contribution à l'Histoire des Variations Climatiques de 1000 à 1425, d'Après les Sources Narratives de l'Europe Occidentale. *Recherches d'histoire et de sciences sociales; Studies in History and the Social Sciences* (24). Paris: Éd. de l'École des Hautes Études en Sciences Sociales.

Ambraseys, N. N., Melville, C. P. and Adams, R. D., 1994. *The Seismicity of Egypt, Arabia and the Red Sea. A Historical Review.* Cambridge: Cambridge University Press.

Assmann, J., 1999. *Das kulturelle Gedächtnis. Schrift, Erinnerung und politische Identität in frühen Hochkulturen.* 2nd ed. München: Beck.

Baker, V., 2008. Paleoflood hydrology: Origin, progress, prospects. *Geomorphology*, 101 (1/2), 1–13.

Bankoff, G., 2003. *Cultures of Disaster. Society and Natural Hazard in the Philippines.* London, New York: Routledge.

Bankoff, G., 2007. Fire and quake in the construction of old Manila. *The Medieval History Journal*, 10 (1/2), 411–27.

Bankoff, G., 2012. Historical Concepts of Disaster and Risk. In: B. Wisner, JC Gaillard and I. Kelman, eds, *The Routledge Handbook of Hazards and Disaster Risk Reduction*. London, New York: Routledge, 37–47.

Barriendos Vallvé, M., 2005. Climate and Culture in Spain. Religious Responses to Extreme Climate Events in the Hispanic Kingdoms (16th–19th Centuries). In: W. Behringer, ed., *Kulturelle Konsequenzen der Kleinen Eiszeit* (Veröffentlichungen des Max-Planck-Instituts für Geschichte 212). Göttingen: Vandenhoeck & Ruprecht, 379–414.

Behringer, W., 1999. Climatic change and witch-hunting. The impact of the little ice age on mentalities. *Climatic Change*, 43 (1), 335–51.

Bhargava, M., 2007. Changing river courses in North India: Calamities, bounties, strategies. Sixteenth to early nineteenth centuries. *The Medieval History Journal*, 10, 193–208.

Borst, A., 1981. Das Erdbeben von 1348. Ein historischer Beitrag zur Katastrophenforschung. *Historische Zeitschrift*, 233, 529–69.

Brázdil, R., Pfister, C., Wanner, H., von Storch, H. and Luterbacher, J., 2005. Historical climatology in Europe – the state of the art. *Climatic Change*, 70 (3), 363–430.

Brázdil, R. and Kundzewicz, Z. W., eds, 2006. Special Issue: Historical hydrology. *Hydrological Sciences Journal / Journal des Sciences Hydrologiques*, 51 (5), 733–985.

Brázdil, R., Dobrovolný, P., Luterbacher, J., Moberg, A., Pfister, C., Wheeler, D. and Zorita, E., 2010. European climate of the past 500 years: New challenges for historical climatology. *Climatic Change*, 101 (1/2), 7–40.

Brumlik, M., 2004. *Aus Katastrophen lernen? Grundlagen zeitgeschichtlicher Bildung in menschenrechtlicher Absicht.* Berlin, Vienna: Philo.

Catalogue of Strong Earthquakes in Italy, 2014. *461 BC–1997* [online]. Available from: http://storing.ingv.it/cfti4med/ [accessed 30 November 2013].

Chalyan-Daffner, K., 2013. *'Natural Disasters' in Mamluk Egypt (1250–1517): Perceptions, Interpretations and Human Responses.* Heidelberg: PhD Thesis, unpublished.

Clancey, G., 2006. *Earthquake Nation. The Cultural Politics of Japanese Seismicity, 1868–1930.* Berkeley, CA: University of California Press.

Crumley, C. L., 1994. Historical ecology. A multidimensional ecological orientation. In: C. L. Crumley, ed., *Historical Ecology. Cultural Knowledge and Changing Landscapes.* Santa Fe, NM: School of American Research Press, 1–16.

Dekens, J., 2007. *Local Knowledge for Disaster Preparedness. A Literature Review.* Kathmandu: International Centre for Integrated Mountain Development.

Detel, W., 2003. Wissenskulturen und epistemische Praktiken. In: J. Fried and T. Kailer, eds, Wissenskulturen. *Beiträge zu einem forschungsstrategischen Konzept (Wissenskultur und gesellschaftlicher Wandel 1).* Berlin: Akademie, 119–32.

Diamond, J., 2005. *Collapse. How Societies Choose to Fail or Succeed.* New York: Viking.

Dix, A. and Röhrs, M., 2007. Vergangenheit versus Gegenwart? Anmerkungen zu Potentialen, Risiken und Nebenwirkungen einer Kombination historischer und aktueller Ansätze der Naturgefahrenforschung. *Historical Social Research*, 32, 215–34.

Dix, A., 2012. Forgotten Risks: Mass Movements in the Mountains. In: A. Janku, G. J. Schenk and F. Mauelshagen, eds, *Historical Disasters in Context. Science, Religion, and Politics* (Routledge Studies in Cultural History 15). London, New York: Routledge, 140–52.

Douglas, M., 2008. *Purity and Danger. An Analysis of Concept of Pollution and Taboo. With a New Preface by the Author.* Reprint of the ed. 1966. London, New York: Routledge.

Durkheim, É., 1895. *Les Règles de la Méthode Sociologique.* Paris: Felix Alcan.

Egner, H., 2008. Warum konnte das nicht verhindert werden? Über den (Nicht) Zusammenhang von wissenschaftlicher Erkenntnis und politischen Entscheidungen. In: C. Felgentreff and T. Glade, eds, *Naturrisiken und Sozialkatastrophen.* Berlin, Heidelberg: Springer, 421–33.

Euro-Climhist, 2014 [online]. Available from: www.euroclimhist.unibe.ch/de/ [accessed 30 November 2013].

Fachard, D., ed., 2002. *Consulte e Pratiche della Repubbli Fiorentina 1495–1497* (Publications de la Faculté des Lettres – Université de Lausanne 35). Genève.

Felgentreff, C., 2008. Wiederaufbau nach Katastrophen. In: C. Felgentreff and T. Glade, eds, *Naturrisiken und Sozialkatastrophen.* Berlin, Heidelberg: Springer, 281–94.

Gabriel, G., 2013. Fakten oder Fiktionen? Zum Erkenntniswert der Geschichte. *Historische Zeitschrift,* 297 (1), 1–26.

Gasperini, P. and Ferrari, G., 1997. Stima dei Parametri Sintetici: Nuove Elaborazioni. In: E. Boschi *et al.,* eds, *Catalogo dei Forti Terremoti in Italia del 461 a.c. al 1990 Vol. 2.* Rome, Bologna: Istituto Nazionale di Geofisica, 56–64.

Glade, T. and C. Felgentreff, 2008. Naturereignisse sind unausweichlich, Katastrophen nicht!? In: C. Felgentreff and T. Glade, eds, *Naturrisiken und Sozialkatastrophen* Berlin, Heidelberg: Springer, 443–8.

Glaser, R., 2001. *Klimageschichte Mitteleuropas. 1000 Jahre Wetter, Klima, Katastrophen.* Darmstadt: WBG.

Glaser, R., 2012. Klima- und Erdbebenkatastrophen – historische Dimension und Erkenntnisgewinn. In: U. Wagner, ed., *Stadt und Stadtverderben. 47. Arbeitstagung in Würzburg, 21.–23. November 2008* (Stadt in der Geschichte, Veröffentlichungen des Südwestdeutschen Arbeitskreises für Stadtgeschichtsforschung 37). Ostfildern: Thorbecke, 97–121.

Glaser, R. and Stangl, H., 2004. Climate and floods in central Europe since AD 1000: Data, methods, results and consequences. *Surveys in Geophysics,* 25 (5/6), 485–510.

Groh, D., Kempe, M. and Mauelshagen, F., 2003. Einleitung. Naturkatastrophenwahrgenommen, gedeutet, dargestellt. In: D. Groh, M. Kempe and F. Mauelshagen, eds, *Naturkatastrophen. Beiträge zu ihrer Deutung, Wahrnehmung und Darstellung in Text und Bild von der Antike bis ins 20. Jahrhundert* (Literatur und Anthropologie 13). Tübingen: Gunter Narr, 11–33.

Grünthal, G., ed., 1998. *European Macroseismic Scale 1998* (Cahiers du Centre Européen de Géodynamique et de Séismologie 15). Luxembourg: Centre Européen de Géodynamique et de Séismologie.

Guidoboni, E., 2000. Method of investigation, typology and taxonomy of the basic data: Navigating between seismic effects and historical contexts. *Annali di Geofisica,* 43 (4), 609–868.

Guidoboni, E. and Comastri, A., eds, 2005. *Catalogue of Earthquakes and Tsunamis in the Mediterranean Area from the 11th to the 15th Century.* Rome: Istituto Nazionale di Geofisica.

Guidoboni, E. and Ferrari, G., 2000. Historical variables of seismic effects. Economic levels, demographic scales and building techniques. *Annali di Geofisica,* 43 (4), 687–705.

Guidoboni, E. and Ebel, J. E., 2009. *Earthquakes and Tsunamis in the Past. A Guide to Techniques in Historical Seismology.* Cambridge: Cambridge University Press.

Herzog, R., 1997. *Kann man aus der Geschichte lernen? Rede des Bundespräsidenten Roman Herzog zur Eröffnung des 41. Deutschen Historikertages am 17. September 1996 in München mit einer Vorbemerkung von Eberhard Schmitt* (Übersee. Kleine Beiträge zur europäischen Überseegeschichte 30). Hamburg: Abera-Verlag Meyer.

Jordan, S., 2009. *Theorien und Methoden der Geschichtswissenschaft.* Paderborn: Schöningh.

Kant, I., 1793. *Critik der Urtheilskraft.* 2nd ed. Berlin: F. T. Lagarde.

Kempe, M., 2007. Mind the next flood! Memories of natural disasters in northern Germany from the sixteenth century to the present. *The Medieval History Journal,* 10 (1/2), 327–54.

Koselleck, R., 1992. *Vergangene Zukunft. Zur Semantik geschichtlicher Zeiten.* 2nd ed. Frankfurt a.M.: Suhrkamp.

Labbé, T., 2011. Essai de réflexion sur la réaction aux inondations en Milieu urbain au XVe siècle: Du seuil de tolérance catastrophique des sociétés riveraines à la fin du moyen âge. *Revue du Nord. Hors série. Collection Art et Archéologique,* 16, 173–81.

Liebscher, H.-J., Krahe, P. and Witte, W., 1995. *Rekonstruktion der Witterungsverhältnisse im Mittelrheingebiet von 1000 n.Chr. bis heute anhand historischer hydrologischer Ereignisse* (Internationale Kommission für die Hydrologie des Rheingebiets, Bericht II Nr. 9). Lelystad: Eigenverlag.

Lorenz, C., 1997. *Konstruktion der Vergangenheit. Eine Einführung in die Geschichtstheorie* (Beiträge zur Geschichtskultur 13). Böhlau, Köln, Weimar and Wien.

Luhmann, N., 2004. *Ökologische Kommunikation. Kann die moderne Gesellschaft sich auf ökologische Gefährdungen einstellen?* 4th ed. Wiesbaden: Verlag für Sozialwissenschaften.

Luterbacher, J. *et al.*, 2002. Reconstruction of sea level pressure fields over the Eastern North Atlantic and Europe back to 1500. *Climate Dynamics*, 18 (7), 545–61.

Martín-Vide, J. and Barriendos Vallvé, M., 1995. The use of rogation ceremony records in climatic reconstruction: A case study from Catalonia. *Climatic Change*, 30, 201–21.

Mauelshagen, F., 2009. Disaster and Political Culture in Germany 1500–2000. In: C. Mauch and C. Pfister, eds, *Natural Disasters, Cultural Responses. Case Studies toward a Global Environmental History*. Lanham, MD: Rowman & Littlefield, 41–75.

Mauelshagen, F., 2010. *Klimageschichte der Neuzeit 1500–1900*. Darmstadt: WBG.

McAnany, P. A. and Yoffee, N., eds, 2010. *Questioning Collapse: Human Resilience, Ecological Vulnerability, and the Aftermath of Empire*. Cambridge: Cambridge University Press.

Mittelstraß, J., 1986. Wissenschaft als Kultur. *Heidelberger Jahrbücher*, 30, 51–71.

NOAA Databases, 2014 [online]. Available from: www.ncdc.noaa.gov/data-access/paleoclimatology-data/datasets [accessed 30 November 2013].

Pfister, C., 2009a. Die "Katastrophenlücke" des 20. Jahrhunderts und der Verlust traditionalen Risikobewusstseins. *GAIA*, 18 (3), 239–46.

Pfister, C., 2009b. Learning from Nature-Induced Disasters. Theoretical Considerations and Case Studies from Western Europe. In: C. Mauch and C. Pfister, eds, *Natural Disasters, Cultural Responses. Case Studies toward a Global Environmental History*. Lanham, MD: Rowman & Littlefield, 17–40.

Pohlig, M., 2013. Vom Besonderen zum Allgemeinen? Die Fallstudie als geschichtstheoretisches Problem. *Historische Zeitschrift*, 297 (2), 297–319.

Poliwoda, G., 2007. Learning from disasters: Saxony fights the floods of the River Elbe 1784–1845. *Historical Social Research*, 32 (3), 169–99.

Quenet, G., 2012. Earthquakes in Early Modern France. From the Old Regime to the Birth of a New Risk. In: A. Janku, G. J. Schenk and F. Mauelshagen, eds, *Historical Disasters in Context. Science, Religion, and Politics* (Routledge Studies in Cultural History 15). London, New York: Routledge, 94–114.

Renn, O., 2008. *Risk Governance. Coping with Uncertainty in a Complex World*. London: Earthscan.

Rohr, C., 2007a. *Extreme Naturereignisse im Ostalpenraum. Naturerfahrung im Spätmittelalter und am Beginn der Neuzeit* (Umwelthistorische Forschungen 4). Böhlau, Köln, Weimar and Vienna.

Rohr, C., 2007b. Historische Hochwasserforschung. Die Überschwemmungen an der Traun im 15. und 16. Jahrhundert. In: D. Gutknecht, ed., *Extreme Abflussereignisse. Dokumentation, Bedeutung, Bestimmungsmethoden* (Wiener Mitteilungen. Wasser, Abwasser, Gewässer 206). Vienna: Technische Universität Wien, 29–42.

Schenk, G. J., 2007. . . . prima ci fu la cagione de la mala provedenza de' Fiorentini . . . Disaster and 'life world' – Reactions in the commune of Florence to the flood of November 1333. *The Medieval History Journal*, 10, 355–86.

Schenk, G. J., 2010a. Dis-Astri. Modelli interpretativi delle calamità naturali dal medioevo al Rinascimento. In: M. Matheus *et al.*, eds, *Le calamità ambientali nel tardo medioevo europeo: Realtà, percezioni, reazioni. Atti del XII convegno del Centro di Studi sulla civiltà del tardo Medioevo S. Miniato, 31 maggio–2 giugno 2008* (Centro di Studi sulla Civiltà del Tardo Medioevo San Miniato, Collana di Studi e Ricerche 12). Florence: Firenze University Press, 23–75.

Schenk, G. J., 2010b. Human security in the Renaissance? Securitas, infrastructure, collective goods and natural hazards in Tuscany and the Upper Rhine Valley. *Historical Social Research*, 35 (4), 209–33.

Schenk, G. J., 2010c. Lektüren im "Buch der Natur". Wahrnehmung, Beschreibung und Deutung von Naturkatastrophen. In: S. Rau and B. Studt, eds, *Geschichte schreiben. Ein Quellen- und Studienhandbuch zur Historiographie (ca. 1350–1750)*. Berlin: Akademie, 507–21.

Schenk, G. J., 2012a. Managing Natural Hazards: Environment, Society, and Politics in Tuscany and the Upper Rhine Valley in the Renaissance (1270–1570). In: A. Janku, G. J. Schenk and F. Mauelshagen, eds, *Historical Disasters in Context. Science, Religion, and Politics* (Routledge Studies in Cultural History 15). London, New York: Routledge, 31–53.

Schenk, G. J., 2012b. Politik der Katastrophe? Wechselwirkungen zwischen gesellschaftlichen Strukturen und dem Umgang mit Naturrisiken am Beispiel von Florenz und Straßburg in der Renaissance. In: U. Wagner, ed., *Stadt und Stadtverderben. 47. Arbeitstagung in Würzburg, 21–23 November 2008* (Stadt in der Geschichte, Veröffentlichungen des Südwestdeutschen Arbeitskreises für Stadtgeschichtsforschung 37). Ostfildern: Thorbecke, 33–76.

Schenk, G. J., 2013. Vormoderne Sattelzeit? Disastro, Katastrophe, Strafgericht – Worte, Begriffe und Konzepte für rapiden Wandel im langen Mittelalter. In: C. Meyer, K. Patzel-Mattern and G. J. Schenk, eds, *Krisengeschichte(n): 'Krise' als Leitbegriff und Erzählmuster in kulturwissenschaftlicher Perspektive* (Beihefte der Vierteljahrschrift für Sozial- und Wirtschaftsgeschichte 210). Stuttgart: Steiner, 177–212.

Schott, D., 2012. Resilienz oder Niedergang? Zur Bedeutung von Naturkatastrophen für Städte in der Neuzeit. In: U. Wagner, ed., *Stadt und Stadtverderben. 47. Arbeitstagung in Würzburg, 21–23 November 2008* (Stadt in der Geschichte. Veröffentlichungen des Südwestdeutschen Arbeitskreises für Stadtgeschichtsforschung 37). Ostfildern: Thorbecke, 11–32.

Seidel, J., Dostal, P. and Imbery, F., 2012. Analysis of Historical River Floods – A Contribution Towards Modern Flood Risk Management. In: J. Emblesvåg, ed., *Risk Management for the Future – Theory and Cases*. Rijeka, Shanghai: InTechEurope, 275–94.

Sismicitè de la France, 2014 [online]. Available from: www.sisfrance.net/ [accessed 30 November 2013].

TAMBORA, 2014 [online]. Available from: www.tambora.org/ [accessed 30 November 2013].

Waldherr, G. H., 1997. Erdbeben. Das aussergewöhnliche Normale. Zur Rezeption seismischer Aktivitäten in literarischen Quellen vom 4. Jahrhundert v. Chr. bis zum 4. Jahrhundert n. Chr. *Geographica Historica* (9). Stuttgart: Steiner.

Wehler, H.-U., 1988. Aus der Geschichte lernen? In: H.-U. Wehler, ed., *Aus der Geschichte lernen? Essays*. München: Beck, 11–18.

Wetter, O. *et al.*, 2011. The largest floods in the High Rhine Basin since 1268 assessed from documentary and instrumental evidence. *Hydrological Sciences Journal*, 56 (5), 733–58.

Wisner, B., Blaikie, P., Cannon, T. and Davis, I., 2004. *At Risk: Natural Hazards, People's Vulnerability, and Disasters*. 2nd ed. London: Routledge.

With, L., 2013. *Approche Interdisciplinaire des Inondations Historiques dans le Rhin Superieur. Programme Junior 2007–2009*. Alsace, Mulhouse: Maison Interuniversitaire des Sciences de l'Homme.

5

DISASTERS, CLIMATE CHANGE AND THE SIGNIFICANCE OF 'CULTURE'

Terry Cannon

If we are to understand disasters as being a result of people's vulnerability, then it is essential to understand all factors and processes that cause that vulnerability. This chapter argues that while economic and political processes that generate vulnerability have been fairly well understood and integrated with vulnerability analysis, there is a significant gap relating to culture. The idea that disasters (those related to natural hazards) must be regarded as being socially constructed and not 'natural' has been well established for many decades. It is widely accepted by most academics and organisations that operate in disaster research and risk reduction (DRR) (cf. Hewitt 1983, Blaikie *et al.* 1984, Wijkman and Timberlake 1984, Wisner *et al.* 2004). Using this perspective, a disaster happens only when the people who experience the hazard are made vulnerable to it by a range of social factors. These factors can also create varying levels of vulnerability for different groups of people that leads to greater or lesser impacts of the hazard.

In this social construction approach, the disastrous effects of a hazard are explained by the various processes and factors that generate vulnerability. But in mainstream DRR organisations, culture has been almost completely ignored as a part of this process of social construction, despite its great significance in making people vulnerable. Equally lacking is any appraisal of what can be called the organisational culture of DRR. It is not only the people who 'have culture' (here taken to involve beliefs, attitudes, values and behaviours) but also the organisations that support disaster risk reduction, relief and reconstruction. Their own cultures are rarely examined, especially by the organisations themselves. In a book that makes very cogent analysis of 'organizational culture', Ramalingam argues that in aid organisations (his analysis can easily be mapped onto DRR) '*there is not nearly enough reflection on the way we think and act.* The focus has been on technical fixes instead of behavioural changes, on bolt-ons instead of changed business models, on spin instead of substance' (2013, p. 15, his italics).

In short, culture, as a concept to be reflected upon later in this chapter, is missing from much DRR work in the practice of most non-government organisations (NGOs), governments, donors and international organisations. In academic work, inclusion of culture (in relation especially, but not only, to developing countries) as a factor affecting disaster vulnerability is restricted almost entirely to anthropology. Here there has been very significant work that has not been assimilated by DRR organisations (cf. Torry 1979, Oliver-Smith 1986, Oliver-Smith and Hoffman 1999, Oliver-Smith and Hoffman 2002). Another problem relates to the significant differences between the culture of the people affected by hazards and organisational cultures, which means that there is scope for a 'clash of cultures' that reduces the effectiveness of DRR.

This chapter explores the neglect of culture as a factor in vulnerability and as part of the process of social construction of disasters. This is done by using a people-centred and organisation-centred vulnerability framework into which are added different ways that culture can take effect. Its goal is to assess how culture affects people's vulnerability, using a framing of vulnerability that disaggregates it into five components that refer to individuals, households and organisations. Although the argument is made mainly in relation to disasters and DRR, much of it is also relevant to climate change adaptation and the influence of culture on people and organisations in relation to that.

Factors in the social construction of vulnerability and disasters

The factors normally taken into account in most versions of social construction are economic, social and sometimes political. Economic and political factors are rarely framed in relation to culture. Social factors related to vulnerability normally include some that are inherently cultural (especially those relating to gender and ethnicity), and yet they are not often discussed in terms of cultural analysis. Some NGOs also emphasise social factors relating to age groups (for instance, specific vulnerabilities of children or older people) and disability. In other words, key political and economic determinants of vulnerability are rarely analysed in relation to culture. Where social factors are included that should encompass culture, it is rarely looked at in depth. Specific social groups are identified as being distinct in relation to vulnerability, but the cultural processes that generate the divisions between groups are not the main focus of the analysis. Instead, the framing used regards their vulnerability as being group-specific and not culturally constructed.

The main exception is where gender appears to have a significant difference in disaster mortality: here some cultural factors (such as cultural attitudes that restrict women leaving the house for safety, e.g., after cyclone warnings) are discussed by some organisations and academic authors. But the inclusion of gender issues such as inequality in access to education, nutrition and health care is not often included as a *cultural* vulnerability factor, but rather as an issue of rights or inequality – in an economic or political context. For example, gender issues related to disasters are rarely understood in relation to male behaviour (e.g., the significance of *machismo* as a factor in risk taking behaviour) and culturally constructed differences in male and female attitudes to risk. And yet it is the cultural construction of 'maleness' in societies where men are dominant that is important. It is not enough to frame this in terms of rights, and empowerment (of women) only works if it also changes men. To understand gender in relation to risk requires understanding how maleness (and its counterpoint of male domination) embody beliefs, attitudes and behaviours in relation to hazards and disasters, and towards other people in their responses to risk.

Linking vulnerability to cultural factors

In this chapter, vulnerability is framed in a people-centred and organisation-centred approach, and examines the factors that generate different levels of vulnerability as they affect individuals and households. Vulnerability is considered in relation to five components, three of which are largely characteristics of individuals and households, and the remaining two are mainly organisational (i.e., they refer to the actions of entities at scales above the individual and household). A key part of the analysis is the linkages between the five components (Figure 3), shown as arrows.

The first three vulnerability components are those that relate to livelihoods (especially the ability of a livelihood and its relevant assets to withstand hazards), and the income, nutrition and health status (sometimes known as developmental or *baseline status*) that is available from that livelihood, and the level of self-protection and exposure to risks (living and working in a safe building and location). These three are almost always included in local level vulnerability assessments by DRR organisations, for instance when using the many 'community vulnerability assessment' methods such as the Vulnerability and Capacity Assessment and other participatory methods used by the Red Cross Red Crescent and many NGOs (IFRC n.d.).

The remaining two 'organisational' vulnerability factors operate outside the household and individual scale. The first involves the level of 'social protection' from risk (for example in the form of warning systems, flood preparedness, building codes for earthquakes, etc.). The last key factor involves the power relationships ('governance') that largely determine other aspects of vulnerability. Governance is here defined as the systems of power, control and ownership that determine the allocation of resources, income and welfare. Power relations determine how assets (e.g., land), income and welfare are distributed in a society (who owns what and why), how taxes are (or not) collected (and from whom) and how they are used (or not) for the wellbeing of the people. Power is also involved in relation to disaster preparedness, for example whether or not it supports good warning systems, building codes, freedom for civil society and media to act as vulnerability 'watchdogs'. Key issues here include whether power relations curb corruption so that

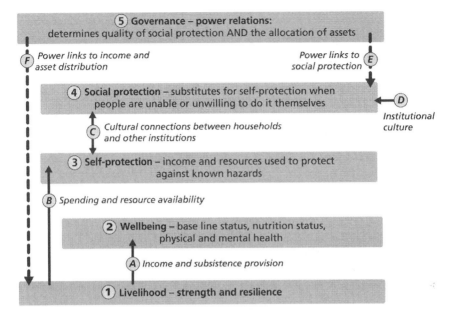

FIGURE 3 The five components of vulnerability and their linkages

buildings are constructed according to seismic codes, and whether a government ensures that where relevant there are early warning systems in operation.

These five key components of vulnerability are defined here in a people-centred way that incorporates all aspects relevant to a person or household. This is different from some approaches to vulnerability where social, physical and economic vulnerability are treated as being distinct types. The five components can be summarised (as in Figure 3).

Vulnerability components at personal and household level

Figure 3: (1). The quality and adequacy of a person or household's livelihood and its strength in relation to hazards likely to affect them. Will the assets be lost or damaged because of the hazard that is likely to affect them, and will it take a long time or not to recover? Will the infrastructure ('physical capital') on which they depend be damaged by a given hazard impact (e.g., access to a road, electricity or water supply; it can be seen that there is an overlap with the five 'capitals' of the Sustainable Livelihoods approach: a person or household's assets in terms of physical, human, natural, financial and social capital) (Carney n.d., Scoones 1998).

Figure 3: (2). The livelihood determines the level of the *person or household's wellbeing*, which is the first line of defence against a shock. These initial conditions (prior to a hazard impact) can be considered the baseline status: how well-nourished and healthy, how educated and resourceful is the person or household. Health includes mental wellbeing and factors such as indebtedness and other sources of stress are likely to predispose people to be more vulnerable to hazards. Wellbeing is primarily determined by the success of the livelihood in providing subsistence and income.

Figure 3: (3). A household's level of income is also very significant in determining their exposure to hazards and whether they have adequate ability to self-protect. Do the people have to live and work in a dangerous place because that is all they can afford? How safe is the construction of their homes and workplaces in relation to anticipated hazards? This can be regarded as self-protection, in which people are faced with a number of options related to income, for example whether they have enough to build a safe house and pay to live in safer places. If they do not, they are likely to be highly vulnerable when a hazard strikes. Do they have enough income but choose not to use it for that purpose? Is it necessary for them to live in a dangerous location in order to have access to livelihood assets (e.g., fertile land on a flood plain or volcanic slope, fishing by a river or the ocean)?

Vulnerability components operating at organisational level

Figure 3: (4). When people do not have adequate income to live in a safe house or location, or when they choose not to do that, they will be reliant on outsiders to prepare for hazards and provide social protection specific to those needs. These can include extended family, a community based organisation (CBO), NGOs, Red Cross Red Crescent, local or national government and international agencies. Such support can include improvements to houses (e.g., roof reinforcements for storms) and workplaces (ensuring structural safety), setting up warning and evacuation systems, setting and implementing earthquake building codes, or flood protection measures.

Figure 3: (5). All of these depend on the wider *power systems* operating in society. These determine the willingness and capacity of institutions beyond the household to provide social protection, and the freedom for civil society to advocate for social protection measures when governments are reluctant. The amount and quality of the hazard social protection is determined then by the type of governance – the power relations and how they affect the allocation of resources and determine priorities within the country. And there is a key link (Arrow F) back to component 1 as well: the power relations also determine how assets and income are distributed and how resources are allocated (or not) for welfare and income redistribution measures to reduce vulnerability and poverty.

It is evident that these five components of vulnerability are predominantly related to economics and politics: the strength and stability of livelihoods, the power systems that allocate resources and social protection, and the operation of organisations that are affected by funding and political will. Conventional DRR approaches fit with this type of framing to a large extent: they usually accept that vulnerability is determined by such political and economic factors and understand that in many cases vulnerability is correlated with (and often largely caused by) poverty. However, what is almost always missing in the conventional approaches is the fundamental effects of culture: both the culture of the people and the culture of the organisations that deal with disasters. Each of these five components of vulnerability is now examined to assess the significance of culture that affects them and the influence of culture on the linkages between them.

By analysing the cultural aspects of the five components, specific attributes of culture can be allocated to the causation of vulnerability. Culture (a complex concept with many definitions) is here considered to involve four interrelated aspects: beliefs, attitudes, values and behaviours and how these relate to risks that arise from natural hazards. Beliefs are often rooted in non-verifiable but deeply held ideas about what controls the world, including some that relate to ideas about the role of 'nature' in relation to risk. Beliefs embody peoples' explanations of themselves and their place in the world and how the world operates in relation to them. For the vast majority of people this includes a religious belief, which is important as so many cultures are also inherently linked to religions. Religions also enable people to have explanations of disasters and ways of behaving that they believe can reduce risks. Attitudes towards (or perceptions of) risk are the outcome of beliefs in the sense that people may decide that a particular aspect of nature is beyond their control and therefore has to be accepted. Values influence how people consider that risk should be shared or distributed among different types of people, for instance that an ethnic group deserves more or less hurt in relation to risk. Behaviours are the composite outcomes of belief, attitudes and values: they are the 'operationalised' expression of a particular culture in relation to a risk. This includes how people act on their own behalf, in relation to other people, and in regard to outside institutions that may play a role in disaster.

Behaviours can involve a culturally determined attitude to risk such as a person being affected by the expected attitudes of their peer group: a man may prefer to take a risk if he is fearful of being considered unmanly by taking preventive measures. A behaviour does not necessarily conform completely with a belief: other factors may intervene to affect the actual outcome so that it is diverted away from complete compliance with the supposed belief. For example, people may claim to have a particular belief in relation to a hazard (e.g., that they should pray to reduce the likely effect), but this does not prevent them from finding scientific explanations useful and being willing to accept practical reductions of the risk (e.g., using a warning system).

Because of the complexity of the concept of culture and wide range of different definitions, the approach to culture here is functional – it examines the way that culture is related specifically to risk. Elsewhere in this volume, Oliver-Smith describes culture in this way:

> Culture in the anthropological sense is generally held to be the grid, lens, or frame though which we experience and interpret the material world and the world of our experience. It is through cultural knowledge, belief, attitudes, that we generate behaviour or actions [. . .] Culture is also a people's

cosmology, how they see their gods, their ancestors, the universe, and stars, their spirituality, beliefs, explanations and their beliefs about their origins and purpose.

(p. 39)

Culture therefore involves a wide range of interactions between people and their (natural) environment, interactions between people themselves, self-awareness and sense of place, and identity in relation to self (affecting what an individual considers is or is not acceptable behaviour). It is not static and a range of processes operate within and from outside a people's culture that leads it to change (IFRC *et al.* 2014, Chapter 7). Of course, not all behaviours are attributable to culture: other factors are relevant as well. Some may be inherently personal or individualistic and related to psychology, heredity and genetic disposition, including epigenetics (the interaction of environment with genetic makeup). These factors are not considered in this chapter, which only examines aspects of culture that are likely to have relevance to risk. It is not a comprehensive assessment of all aspects of culture, nor of all factors that determine human behaviour in relation to risk. It is focused on culture because that is one area that is most obviously missing and ignored, although much can also be learned from psychology, neuroscience and behavioural economics (Ariely 2009, World Bank 2014). The chapter also makes some general-isations, although where possible specific illustrations are given. In addition, DRR organisations themselves have beliefs, attitudes, values and behaviours that constitute an 'organizational culture' that must be analysed. In particular, the 'clash of cultures' (when the beliefs, attitudes and values of the people do not match those of the organisations) must be understood as the site of complications or failure in effective DRR where this clash leads to behavioural outcomes that are contradictory.

Understanding culture as a determinant of vulnerability

As each of the first three components is examined, the amount of detail relating to culture declines. This is to avoid repetition, because there is overlap. Cultural factors affect each of the vulnerability components, but since the same culture is likely to be relevant across the first three, the assessment given at the start has relevance in the subsequent components.

(1): Livelihoods, their strength and resilience in the face of extreme events: what cultural factors affect people's livelihoods and how resilient they are to hazards?

Many hundreds of millions of people live in places that have or will experience severe natural hazards, including floods, cyclones, earthquakes, tsunamis, drought, volcanic eruptions, wildfires and landslides. Although poverty is a primary driver of having to live in danger, it is also evident that significant choice may be made to live in danger zones. This is because people weigh the risks of a hazard occurring against the benefits that are gained from the livelihood on a daily basis. In effect, people are 'discounting' the damage that may happen if and when a hazard strikes in order to justify the benefits of living in that location. Typical rural livelihood benefits include soil fertility (valleys, some coastal zones, volcano slopes), fishing (coasts, lakes, rivers) and water supply in arid areas (sometimes present at the surface in relation to the fault systems that cause earthquakes). Urban places are beneficial mainly for employment or informal sector income.

Many people in both rural and urban locations give priority to their livelihood opportunities and are less concerned about disaster risks (Cannon 2008a, IFRC *et al.* 2014). Culture plays a significant part in enabling people to live with the risks, especially where people have religious beliefs that have arisen in the context of these activities and locations. Attitudes to (perceptions of) the risks people face are embedded in those cultures. For example, many people ascribe to God the dual function of determining both benefits and harms arising in nature: 'God gives and God takes away'. Many millions of people around the world exhibit related 'fatalistic' versions of what happens in a disaster. Others consider that God exercises control over who lives and dies, and they display a readiness to accept God's will in this way: it is a part of the pattern of human existence. In some countries, Westerners are regarded as having unrealistic (and ungodly?) attitudes to the control of risk and nature: death is regarded as natural and related to god's purpose (IFRC *et al.* 2014, chapter 2, Schipper in this volume). In other cultures, people consider that those who suffer are being punished for their sins. After the Indian Ocean tsunami in 2004, some of the Muslim people of Aceh (Indonesia) said that the disaster was a result of god's anger at the disturbance caused by oil drilling, while others blamed the decadence of tourism (Paul and Nadiruzzaman 2013). Similar views were expressed after the Hurricane Katrina in the United States (and were believed by some of the victims in New Orleans) (IFRC *et al.* 2014).

In other cultures, people believe they can influence the behaviour of god(s) by prayer and offerings that make the deity or deities more benevolent towards them. This is especially prevalent in locations that have a specific hazard that is metaphorically identified with the god(s). Crowley (Donovan *et al.* 2012) has found that many people on the slopes of Merapi volcano (Java, Indonesia) accept that there is a mountain deity that can be interpreted by a local 'spirit guide', who also issues advice on when to evacuate from an eruption. Similar beliefs are known on Mount Cameroon in Africa and other volcanoes in Latin America. Crabtree (2010, see also this volume) describes how many of the victims of a devastating flood in Bihar (North India) believed that they had displeased the Goddess Kosi, after which the river Kosi is named. Their idea for future disaster prevention was to pray more and make offerings.

Religions, in particular, often have something significant to say about risks. Beliefs and attitudes to risks have evolved in order to enable them to live with those risks. If people believe that a disaster is a punishment of god(s) for some aspect of their behaviour, then this at least makes some sense of an otherwise unexplainable event. Many people retain beliefs about disasters that are linked with religion even when there are very adequate scientific explanations. This may represent an emotional inertia and the need for a deeper meaning rooted in religious explanations. Where people believe in god(s) that by definition has supernatural powers, it does not make any sense to exclude god(s) from the causal process of disasters, since otherwise that is hardly a god(s) worth believing in. People are also constrained in changing their beliefs by peer pressure: when people live together in villages, small towns or attend a particular religious establishment, it is difficult to give up a belief without running the risk of disapproval or even exclusion.

But we also know that disasters (ranging from personal tragedy up to very large scale events such as the Lisbon earthquake and tsunami of 1755) are often a significant factor in people starting to have doubt in their belief in god(s) (Dynes 2005). This is also the focus of major theological discussion. A belief that sins are the reason that god(s) may cause or permit a disaster allows people to avoid thinking that god(s) simply visits random violence on humanity. The significance of maintaining good relations with god(s) was observed in Ethiopia, where some Coptic Christians will leave ripe crops to be damaged in the fields on days of religious festivals when they instead attend church (Schipper 2010), to invest in a deeper, longer-term risk reduction (and possibly avoid condemnation by their peers).

People's livelihoods are not always chosen, but are determined by culture in a broad sense. As an example, this applies to a very large proportion of people in India and Nepal, and to some extent in Bangladesh, Sri Lanka and Pakistan. The Indian and Nepali caste system is integrated with Hindu religious beliefs and, although it has been altered by modernity (and especially by urbanisation), it still determines the lives of hundreds of millions of people. Originally thought to be a system of division of labour in rural India, castes are arranged in a hierarchy that ranges from higher position Brahmins (who traditionally cannot do manual labour but often own land) down to the lowest level and those who are outcastes (literally beyond the caste system). Outcaste people (previously also known as untouchables because of barriers of contact between them and higher caste people) often now identify themselves as Dalits. This term became popular when it was adopted by a political movement for untouchable people's rights that emerged in the 1960s, when it became clear that the illegality of caste discrimination declared in the Constitution of independent India (1947) was not producing much change.

One of the key aspects of caste is the belief within Hinduism that a person's current caste status is a reflection of the status of the person's soul, which is determined by previous incarnations and behaviours. This has the outcome of culturally justified discrimination against low caste and outcaste people, whose poverty is regarded as legitimate. Outcastes are by tradition forbidden from owning land, have to draw water from different sources than caste Hindus and are prevented from gaining access to other services. Higher caste groups may also object to disaster preparedness for lower status groups and deprive them of disaster relief and reconstruction. Historically, some outcastes have converted to Christianity and Islam in order to sidestep the status accorded to them by the Hindu majority. All of these cultural features mean that a very large proportion of the population is vulnerable to hazards through processes that are actually justified and perpetuated by culture.

This issue of status and hierarchy as a factor in livelihoods, wellbeing and discrimination is similar in other versions of caste-type systems that exist in other parts of the world, especially parts of Africa and the Middle East – including some that legitimise slavery from a cultural perspective (i.e., that involve racial hierarchy or religious superiority). In all these cases, it is reasonable to infer that people's vulnerability in relation to relevant hazards is higher because of the inability to choose livelihoods and safe locations. This is linked to (2) (wellbeing and baseline status) and (3) (ability to self-protect).

Gender relations are another key cultural aspect relevant in most of the world, which determines livelihoods through culture. Sometimes it is embedded in religious beliefs: almost all mainstream religions ascribe different status to females and males, and while some of them claim that this is based on difference and not male superiority, most outsiders (and some of the women affected) find this difficult to substantiate. Gender differences often relate to a division of labour (with women working much longer hours), rights of ownership of assets (especially land) and access to health care, food, education and job opportunities. All of these can have an effect in raising the vulnerability of females in relation to hazards, although most of the effects will be considered under (2) and (3).

(2): Wellbeing and baseline status: what aspects of people's culture affect their capacity or ability to support their baseline status?

Low levels of wellbeing (especially hunger, malnutrition, poor health, low educational attainment and low esteem) are strong correlates with vulnerability to hazards (although these are often subsumed within generic 'poverty'). There is a clear causal relationship between people's livelihood and level of wellbeing, and therefore vulnerability, in this component. Nutritional level, access to health services, ability to pay for education and other services (sometimes including water) are all largely determined by income from employment, sales of produce, self-provisioning or subsistence from farming, fishing and other natural resources. This causal relationship is well understood, and its significance for vulnerability to natural hazards is highly relevant.

People with good levels of wellbeing that are provided by adequate livelihoods are likely to be less vulnerable (and able to recover more quickly) from hazard impacts. This includes the effects of weak, uncertain or declining livelihoods on mental health, including stress, related ill-health and sometimes problems of addiction (to alcohol or drugs). Suicide rates are often high in places where inadequate livelihoods lead to indebtedness, and mental health problems are highly likely to raise levels of vulnerability to hazards.

Some people may benefit from outside support, for example through NGOs, religious institutions or from the government through pensions and other welfare payments, subsidised food and farm inputs, cash payments through social protection programmes or employment guarantee schemes. These welfare transfers may add to the aggregate level of wellbeing for some households as part of what in Western countries used to be called the 'social wage', but do not cover all people in need.

But how is wellbeing and a person's baseline status affected by culture? Some intra-household behaviours derive from beliefs and values that affect gender, age group and other relations between household members. These are cultural attributes that determine the allocation of income, subsistence and welfare (including external aid). People have differential access to the outputs of a livelihood and welfare and social protection provided by the state or outsiders according to their culturally determined status. For access to the assets that enable livelihoods to function, there are usually very significant gender differences within households that reflect the wider culture. Unequal gender relations are embedded in culture in most parts of the world, and normally result in females having less right to own land and other production resources, poorer nutrition, less access to medical care and often reduced access (or even no right) to education.

In many countries (as was also common in the past in the West), girls are discriminated against in many ways and preference for male children leads to low esteem and harsh working lives for many females (and significant imbalance in the male–female ratio, especially in India and China). As a result, females are likely to be more vulnerable in relation to hazards. There may also be culturally determined feeding hierarchies. The head of household and other adult males may be fed first (and with more and better food), leaving less for females and elderly people. Recent research in India suggests that the widely observed stunting of children is an attribute of reduced food for mothers once they have given birth to the first son: stunting appears to increase by birth order (after the first son) and to be a function of the mothers' share of household food (Jayachandran and Pande 2013). The issue of feeding hierarchies has posed significant ethical problems for organisations providing feeding programmes and famine relief, where the policy is to allocate rations to feed all members of a household. When these are then allocated within the household according to the culturally determined feeding hierarchy, the elderly and females may receive less.

Many females are also subjected to physical and sexual abuse by males. There is strong evidence that this increases after a disaster, for at least two reasons. First, some men may feel inadequate and despondent and take out these feelings on women and children. Second, many people may be displaced and may not be living among their normal neighbours. This exposes females to men who are less constrained by peer pressure and the risk of being identified.

(3): Self-protection in relation to known hazards: in what ways does culture affect the capacity and willingness to make adequate self-protection?

One of the key debates in disaster and vulnerability analysis concerns the degree to which people are forced to live in dangerous places because of their need to follow livelihoods (Cannon 2008a, IFRC *et al.* 2014). It is certainly true that many people cannot afford to live in safe places or houses, or that they prefer to take risks in relation to hazards in order to make savings that increase their income – for instance living on dangerous slopes in cities to allow them to live nearer to income-earning opportunities (Cannon 2008b). But it is also evident that people who have choice do choose to live in dangerous places, including for example Florida (with hurricane risk) or California (earthquakes and wildfires). The (future) dangers are traded for the (immediate) benefits.

Given that many hundreds of millions of people live in locations that are affected by known hazards, what steps do they take to protect themselves (at individual and household level) in the home and workplace? Clearly there is another very significant causal connection with the income derived from the livelihood. If the household has a low income, self-protection in relation to hazards is unlikely to be possible. Such self-protection is very hazard-specific: what constitutes a safe house for an earthquake may not be effective for a flood and vice versa. Paradoxically, poor people with flimsy homes may be less likely to be hurt by an earthquake than richer people who have a substantial structure that collapses (as was relevant for some richer people in the Latur, India event in 1993). But beyond these economic factors there are significant problems (and potential advantages) related to beliefs and attitudes. An example of a cultural constraint on self-protection happens in Assam (northeast India) where migrants from other parts of the country have settled in the flood-prone Brahmaputra valley. While the indigenous people practice building their houses on stilts, the incomers (who look down on the local people) would not do this as it would have been 'against their culture', and would also mean adopting the practice of the 'inferior' locals (ICIMOD 2009, also see Schipper in this volume).

House construction is very significantly connected with culture, in that building materials and methods in some parts of the world have arisen in relation to known hazards (IFRC *et al.* 2014, chapter 5). There may also be a danger in romanticising some assertions of cultural superiority. The claimed earthquake-proof characteristics of Inca buildings are one example, where it is difficult to know if this was a deliberate reason for the construction method or an accidental outcome. The major surviving structures that are supposedly earthquake-resistant are elite or ceremonial buildings and do not represent what ordinary homes were like and their performance in earthquakes.

Changes in culture can lead to increased danger, especially the adoption of the idea in construction materials and methods that modern is better and traditional is primitive and backward. This has led to many millions of homes being constructed using concrete and poor reinforcement that is extremely dangerous in earthquakes (IFRC *et al.* 2014). However, we also know that many traditional techniques are very dangerous, especially mud (*adobe*) constructions of ordinary homes in many parts of the world (a significant factor for example in deaths in the Bam earthquake in Iran in 2003).

Another cultural aspect is also highly relevant in relation to construction safety. For example, perception of risk arising from beliefs and attitudes about natural hazards are always located within day-to-day priorities as well as future possible dangers. There are clear cases of beliefs about earthquake risk that lead construction companies to ignore seismic safety, for example because the risk is not taken seriously for the near future. While construction companies have been blamed for disastrous collapses (often linked to corruption in the use of building materials and bribes of inspectors, as for instance in the Izmit event of 1999 in north Turkey), it is doubtful that the buildings' residents would have welcomed the extra cost of earthquake resistant construction. When people build their own houses (as happens in much of the developing world) it is very unlikely that they would ensure it was safe from hazards. In the Wenchuan earthquake of 2008 in China, the scandal of the collapse of many schools that killed thousands of children was in many cases a result of corruption and bad construction. But the vast majority of deaths were of people in their own ordinary village homes. Had they been encouraged to build safely or to retrofit, it is doubtful that many would have done so: people's perception of risk leads to tragic behaviours when the culture of 'it will not happen to us' is invoked (as it is by many people in all parts of the world). This is also part of the culture clash between people and organisations: different priorities and people's willingness culturally to avoid self-protection. How this works out in terms of organisations attempting to fill the gap is covered in (4).

(4): Social protection in relation to anticipated hazards: what aspects of culture affect the type, quality and quantity of disaster preparedness?

With this component of vulnerability, we move into the behaviour of organisations and institutions that attempt to provide social protection in relation to hazards. Social protection in this context involves the ways that different interventions are made with the intention to reduce people's vulnerability to expected hazards. These can be provided by institutions at the local level such as extended family, kin group or religious institution, although these tend to be more about response (after a disaster) rather than being involved in preventive measures. At a higher scale, social protection organisations include NGOs, local and national government and international agencies (including the World Bank, regional 'development' banks and World Food Programme).

Social protection can include flood prevention measures (often 'hardware' installations such as dykes and channelling), upstream reforestation, seismic building codes and inspections of construction, retrofitting for earthquake or cyclone risk, storm and cyclone warning systems and evacuation planning, and preparedness for emergencies and post-hazard assistance. The type of activities proposed by different organisations reflects their own culture and capacities, without necessarily assessing if what they will do is the right thing, and organisations will carry out DRR activities that fit with their own scale and goals. Arrow D represents the external organisational culture that arrives with the intervention of outsiders who claim to be doing DRR. Ramalingam (2013, p. 16) argues:

> Aid actors are characterized by the 'mental models' that reflect their understanding of the world and which they use to learn about the world, frame decisions, share knowledge, and guide objectives and actions. These mental models change over time and can provide impetus for institutional change; equally, rigid mental models may reinforce particular forms of skills and knowledge, and be underpinned by certain kinds of institutions.

Most often, because it is difficult to change power systems that are largely responsible for people being vulnerable ((5) and (1)), organisations have 'mental models' (cultures) that enable them to operate without attempting to make fundamental changes.

Social protection measures are typically required when self-protection is missing, either because people are too poor to implement safe houses or locations for homes or they are 'culturally unwilling' to spend on disaster preparedness. In addition, certain types of vulnerability reduction are impossible at the household and local level and must be carried out by higher level organisations: causes of floods may be upstream and remote, warning systems require centralised scientific and planning approaches, flood measures may be large scale and legal frameworks for construction safety must be national and 'top-down'.

The type and amount of social protection that is available is determined mainly by political and economic factors in the causal relationship (Arrow E) that links governance (5) and social protection. These factors will be dealt with under the assessment of (5). But culture is also relevant to social protection in a number of ways, mainly regarding the organisational culture of the different institutions who are (or not) involved (Arrow D). Why and how organisations get involved in social protection is influenced by their perception of risk (including their beliefs and attitudes towards hazards and their self-identification with their potential to have an effect) and by their values (for instance, do they focus on particular groups of people, including some and excluding others, or claiming impartiality and universality?)

A major cultural factor is that some organisations tend to identify their position in relation to predetermined beliefs, attitudes and values that are internal to the organisations and that may not fit with the needs of the people they are claiming to support. Sometimes this can be characterised as 'the law of the instrument': someone who only has a hammer perceives every problem as being a nail. The culture of most DRR organisations is to believe that the serious hazards people face are the most important problem, and that people must share this priority and divert from dealing with daily needs and risks, leading to a mismatch of priorities and goals that hampers the effectiveness of DRR. This in turn can lead to a circularity of cultural justification in which the organisations exhibits an 'unwillingness to fail' and so finds ways to enable its activities to be evaluated positively

(Mansuri and Rao 2013, Ramalingam 2013). The dominant culture of aid leads to puzzlement, with the result that organisations ask 'Why don't they want what we know they need?' (Ramalingam, 2013 p. 91). The internal culture of the organisations enables a self-justifying coexistence of those organisations, its local partners, and the funding agency (IFRC *et al.* 2014, chapter 7).

Availability of funding for particular types of activities is another key factor in enabling organisations' cultural framing of DRR. When funding is only available for certain types of activities (for example, disaster preparedness or climate change adaptation – and now increasingly 'resilience'), then that organisations' culture makes it fit into their remit and adapts to the new agenda. DRR NGOs suddenly find that resilience is what they should have been doing all along. While almost all of the organisations engaged in these activities claim a 'moral' agenda, this can be distorted or diverted by the need to get funding:

> [. . .] while aid agencies want to do good, they also want to do well, and so the fundraising imperative can start to dominate the moral one. And so morals are usually articulated only in narrow, institutionally determined ways, and supported by a culture of 'institutional bystanders', such that, when things go in directions that might conflict with these high level values and morals, they are not brought back into the fold.
>
> (Ramalingam 2013, p. 87)

Another factor involves the belief that some organisations have in technological interventions for DRR. Often this is driven by their business model, but in many cases there is also a cultural certainty that a scientific approach is what is needed, and that a technical fix ('tech-fix') is appropriate. Such approaches have been undermined in recent years by the failure of some 'hardware' flood prevention measures in Europe and the USA and by the 'cultural' factors that undermined safety in the Fukushima reactors when they were damaged by the earthquake and tsunami in 2011 (McCurry 2012). Many technical interventions are based on a faith in the technology while ignoring the social relations (including political and economic factors) that determine how technology actually fits into social systems. This faith can even be considered analogous to religious beliefs and can involve ignoring evidence that the technological approach can fail (IFRC *et al.* 2014, Box 2.2).

Technical fixes that require large investments and civil engineering projects also often find favour with governments, donors and banks. Politicians potentially benefit from the kickbacks under such arrangements, corporations from valuable contracts, and banks (including the World Bank and other international 'development' banks) benefit from this type of business and from having large value loans to fewer clients. The large majority of disaster resilience and climate adaptation-related projects in developing countries follow this model and take very little account of how they will integrate with local people's needs and priorities. The culture of tech-fix organisations fails to understand that problems that have predominantly political, economic, social and cultural causes are not guaranteed to be fixed by proposed solutions that do not address the causes (and which can often reinforce the very power relations that are part of the problem as well).

People's culture can also affect what organisations can and cannot do for DRR. Social protection therefore either has to embed itself in existing cultures, or find ways that can support processes of change in those cultures that enable people to accept disaster preparedness. One of the key constraints on organisational effectiveness can arise from cultural and religious beliefs that lead people to oppose outside interventions, or to have behaviours that do not promote safety. For example, Protestant sects in Central America oppose disaster preparedness because it deprives god of control over the effects of nature (Schipper 2010). In Bangladesh, gender power related to religion and wider culture means that females cannot leave their home when there is a cyclone warning until they have permission from a senior male relative (Cannon 2002). This means that they (and the children in their charge) are much more vulnerable. In many countries, the culture does not allow women to learn to swim, again leading to higher risks in floods and cyclones. In a related religious context, a senior seismic engineer in a Muslim North African country has encountered a local religious leader who opposed retrofitting for earthquakes on the grounds that if a disaster happens it is god's will (personal communication). Fortunately such views are not universal, and in Dhaka, Bangladesh (which faces a major earthquake in a very dangerous context of unconsolidated ground of the delta sedimentary material) a number of construction companies are claiming that their new residential buildings (aimed at middle and upper class buyers) are resistant up to Richter scale 7 (an internet search for real estate developers will find their promotional materials).

(5) Governance and power relations: what is the significance of culture?

Governance is used here as shorthand for power relations: governance is 'the way power is exercised in the management of a country's economic and social resources for development' (World Bank 1992). It is about all aspects of power, including not only governments but also elites (including politicians who, although out of office, are often still very powerful), domestic and foreign corporations, other non-government actors who control resources, and international organisations (development organisations, banks, International Monetary Fund (IMF) etc.). In much of the world, governments are less powerful than corporations, and in some cases they have less control than international organisations such as the IMF. It is therefore important to understand all types of power relations that may affect vulnerability.

Section (5) relates to other parts of vulnerability in two different ways. The first (Arrow E) involves how power relations affect the amount and type of social protection available to people. People's vulnerability can be higher or lower depending on the culture of the institutions of power. For example, many corporations favour technical interventions for disaster preparedness and follow a belief in top-down interventions with an attitude towards people that is dismissive or alienating. Government approaches frequently follow similar beliefs, and also often hold a social class perspective that amounts to a culture of exclusion of ordinary people or particular groups (based on ethnicity, for example) or females. This leads to situations in which people's vulnerability is increased, as for instance in the USA in relation to Hurricane Katrina, Myanmar/Burma in Cyclone Nargis and China in relation to a number of disasters and health emergencies.

Governments may also increase vulnerability by preventing civil society from organising highlights of environmental and hazard risks and reducing the freedom of the media to report on risks. Although this has the appearance of being a political factor, it is in some cases directly linked to culture, as for example in China where the need to preserve social order dominates all other factors in what is or is not tolerated. Similar appeals to culture and national historical legacy have been invoked in the USA, Thailand, India and Russia in order to oppose people who argue for greater safety. The role of the media as a factor was highlighted (Drèze and Sen 1989) in the context of famine in China by comparison with India: the former preventing media reporting and covering up disasters, while India was claimed to have avoided major famines because of the ability of the media to monitor and report on problems.

Arrow F links with how power relations establish the pattern of ownership and control over resources (such as land, minerals, energy sources, forests and fisheries), how income and welfare payments are allocated, tax revenues collected and used, and the legal systems within which these are established. The strength of people's livelihoods (1) depends on the availability of assets that can be used and these are often unequally distributed because of the power systems that allocate them or fail to redistribute them (for example as with many failed land reforms in India, Bangladesh and Pakistan since their independence). This connection (Arrow F) between (5) and (1), and the degree of support through governance to social protection (the link between (5) and (4) through Arrow E), are arguably the most powerful determining processes in the vulnerability of most people. To a significant extent they are subject to the elite cultures of governments, corporations, dominant classes and the organisational cultures of the agencies that attempt to provide social protection for when self-protection and power relations are inadequate for reducing vulnerability. In a range of countries, elite culture argues that people are poor because they are lazy, incompetent, stupid or otherwise deserving of their position in society. This framing of privilege is not a good basis for disaster preparedness, and unless challenged is likely to maintain not only the increasing disparities in income and wealth that are happening in many countries (leading to impoverishment and insecure livelihoods) but also high levels of vulnerability. With climate change leading the damage to the livelihoods of billions of rural poor people, and increased frequency and/or severity of extreme climate events, the most crucial change that is needed for better disaster risk reduction is a fundamental change to the elite and global culture of inequality, resource theft and the self-justification of corporate and elite power that they deserve the inordinate share that they get.

References

Ariely, D., 2009. *Predictably Irrational: The Hidden Forces that Shape Our Decisions*. New York: HarperCollins.

Blaikie, P., Cannon, T., Davis, I. and Wisner, B., 1994. *At Risk: Natural Hazards, People's Vulnerability and Disasters*. London: Routledge.

Cannon, T., 2002. Gender and climate hazards in Bangladesh. *Gender and Development*, 10 (2), 45–50.

Cannon, T., 2008a. Reducing People's Vulnerability to Natural Hazards: Communities and Resilience. *WIDER* Research Paper 34. Helsinki: United Nations University [online]. Available from: www.wider.unu.edu/publications/working-papers/research-papers/2008/en_GB/rp2008–34 [accessed 21 October 2014].

Cannon, T., 2008b. Vulnerability, 'Innocent' Disasters and the Imperative of Cultural Understanding. *Disaster Prevention and Management*, 17 (3), 350–57.

Carney, D., n.d. Sustainable Livelihood Approaches: Progress and Possibilities for Change. London: Department for International Development (DFID) [online]. Available from: www.eldis.org/vfile/upload/1/document/0812/SLA_Progress.pdf [accessed 21 October 2014].

Crabtree, A., 2013 . Questioning psychosocial resilience after flooding and the consequences for disaster risk reduction. *Social Indicators Research*, 113, 711–28.

Donovan, K., Suryanto, A. and Utami, P., 2012. Mapping cultural vulnerability in volcanic regions: The practical application of social volcanology at Mt Merapi, Indonesia. *Environmental Hazards*, 11 (4), 303–23.

Drèze, J. and Sen, A. 1989. *Hunger and Public Action*. Oxford: Clarendon Press.

Dynes, R. R., 2005. The Lisbon Earthquake of 1755: The First Modern Disaster. In: T. E. D. Braun and J. B. Radner, eds, *The Lisbon Earthquake of 1755: Representations and Reactions*. Oxford: Voltaire Foundation, 34–49.

Hewitt, K., ed., 1983. *Interpretations of Calamity from the Viewpoint of Human Ecology*, Allen & Unwin.

ICIMOD (International Centre for Integrated Mountain Development), 2009. *Local Responses to Too Much and Too Little Water in the Greater Himalayan Region*. Kathmandu: ICIMOD.

IFRC (International Federation of Red Cross and Red Crescent Societies), n.d. Vulnerability and Capacity Assessment. *IFRC* [online]. Available from: www.ifrc.org/en/what-we-do/disaster-management/preparing-for-disaster/disaster-preparedness-tools1/ [accessed 21 October 2014].

IFRC (International Federation of Red Cross and Red Crescent Societies), Cannon, T., Schipper, L., Bankoff, G. and Krüger F., eds, 2014. World Disasters Report 2014: Focus on Culture and Disasters, Geneva. *IFRC* [online]. Available from: www.ifrc.org/publications-and-reports/world-disasters-report/.

Jayachandran, S. and Pande, R., (2013). Choice not genes: Probable cause for the India–Africa child height gap. *Economic and Political Weekly*, 48 (34).

Mansuri, G. and Rao, V., 2013. Localizing Development: Does Participation Work? *World Bank* [online]. Available from: http://econ.worldbank.org/WBSITE/EXTERNAL/EXTDEC/EXTRESEARCH/0%2C%2CcontentMDK:23147785~pagePK:64165401~piPK:64165026~theSitePK:469382%2C00.html [accessed 21 October 2014].

McCurry, J., 2012. Fukushima reactor meltdown was a man-made disaster, says official report. *The Guardian*, 5 July 2012 [online]. Available from: www.theguardian.com/environment/2012/jul/05/fukushima-meltdown-manmade-disaster [accessed 21 October 2014].

Oliver-Smith, A., 1986. *The Martyred City: Death and Rebirth in the Andes.* Albuquerque: The University of New Mexico Press.

Oliver-Smith, A. and Hoffman, S. M., eds, 1999. *The Angry Earth: Disaster in Anthropological Perspective.* New York: Routledge.

Oliver-Smith, A. and Hoffman S. M., eds, 2002. *Catastrophe & Culture: The Anthropology of Disasters.* Oxford: James Curry.

Paul, B. K. and Nadiruzzaman, M., 2013. Religious interpretations for the causes of the 2004 Indian ocean. *Asian Profile,* 41 (1), 68–77.

Ramalingam, B., 2013. *Aid on the Edge of Chaos: Rethinking International Cooperation in a Complex World.* Oxford: Oxford University Press.

Schipper, E. L. F., 2010. Religion as an integral part of determining and reducing climate change and disaster risk: An agenda for research. In M. Voss, ed., *Der Klimawandel. Sozialwissenschaftliche Perspektiven.* Wiesbaden: VS Verlag, 377–93.

Scoones, I., 1998. Sustainable Rural Livelihoods: A Framework For Analysis. *IDS Working Paper 72.* Brighton, UK: Institute of Development Studies (IDS) [online]. Available from: www.ids.ac.uk/files/dmfile/Wp72.pdf [accessed 21 October 2014].

Torry, W., 1979. Anthropological studies in hazardous environments: Past trends and new horizons. *Current Anthropology,* 20(3), 517–41.

Wisner, B., Blaikie, P., Cannon, T. and Davis, I., 2004. *At Risk. Natural Hazards, People's Vulnerability and Disaster.* 2nd ed. Routledge, London.

Wijkman, A. and Timberlake, L., 1984. *Natural Disaster. Acts of God or Acts of Man?* London: Earthscan.

World Bank, 1992. *Governance and Development.* Washington, DC: World Bank.

World Bank, 2014. *World Development Report 2015: Mind and Society.* Washington, DC: World Bank.

PART 2

Cultural linkages to vulnerability

6

CULTURES AND CONTRA-CULTURES

Social divisions and behavioural origins
of vulnerabilities to disaster risk

James Lewis

Culture in the context of risk and disaster

Our personal understandings of behaviour, human nature and of our own cultures
might commence with childhood experiences of nursery rhymes and fairy tales
(cf. Grimm and Grimm 1812, Sendak 2011); communicating of hazards, and sometimes of disasters (ZiF 2012, p. 38), starts at an early age. These often disturbing
stories recount the behaviours of wolves, rats, ogres, witches, robbers and thieves
– the wicked and the bad – and those of children, maidens, fairies and piglets of
the happy and the good – perhaps with wayward machines somewhere in between.
Are we, as products of these early formulators, as much aware of our own cultures
as we may be of those of others?

Culture in the context of this article's consideration is understood as the sum
of individual and collective behaviour, practice, thoughts and tales. Assuming that
there are several dominant cultures (however they may be distinguished) in every
culture there can exist subcultures, which coincide with some of the dominant
culture's norms and practices and differ with others.

Often assumed as historic and permanent, some national, sub-national, regional
or ethnic cultures may be more resilient than others against hazards by displays of
traditionally advantageous values and techniques. In reality, however, cultures may
move, shift, amalgamate and divide in response, for example, to pressures and
constraints resulting from power seeking opportunism and pursuit of social status,
these last arising often as reflections of personal or group insecurities. Cultures,
therefore, are not static, nor are shifts and divisions advantageous to all with one
sector behaving towards another in ways that can either increase or decrease social
vulnerability to stress, adversity and catastrophe.

Discussions of culture and human nature have traditionally dwelt on interactions
between people and their environments, with environmental extremes often being

of less concern than habitat and ever present risks of the failure of livelihoods. However, in the early 1970s, these interrelations between culture and (natural) environments were debated intensely in the academic (cf. Douglas and Wildavsky 1982, Douglas 1992) as well as in more popular scientific communities, as is reflected by, for example, Bronowski (1973, pp. 19–24): by imagination, reason, emotional subtlety and toughness, humankind came to fit most environments – not only by acceptance of them, but by adapting to them as expressions of cultural evolution.

Individuals, groups and social organisations have adapted to environments by behavioural adjustments, with behaviour expressing culture. Behavioural change has become cultural change by evolution, the content of evolution being the elaboration of new behaviour (Bronowski 1973, pp. 40–41). Considering adaptation to environments as being a form of behaviour by which culture is expressed, culture becomes a communally preferred and learned behaviour to be adopted by an entire society (Bronowski 1973, p. 48).

To assume, however, that cultures remain homogenous throughout evolutionary change and adaptation would not be realistic; other factors may affect behavioural responses to hazards and risk, such as fear for example (Lewis, Kelman and Lewis 2011). In reality, and as bi-products of evolution, change and experience, cultures will shift and divide, creating subcultures, or 'mono-cultures' of a single sect or belief, in which more recent or nascent minority cultures come to be defined and expressed – inclusive of changes for the not so good as well as changes for the good.

Cultures shift and change, they modify and grow, or they diminish, perhaps temporarily, but generally for the longer term. For example 'working class culture' shifts and changes according to its composition at any one time and, by comparison, 'mono-cultures' are considered to find shifts and changes more difficult. This has been, in my view, aptly put into words in the media, based on a widespread popular opinion and demonstrated by the following quote from a newspaper article: 'In a mono-culture [. . .] they don't have the energy – they haven't come from a culture where they've got work, they think there's a more limited range of things they can aspire to' (Vasagar 2011). Alternatively but succinctly: 'Cultures change when mind-sets change' (Siddique 2011). To further illustrate this, later in this chapter examples are described of changing cultures in Nepal and Japan.

The impermanence of cultures may be demonstrated suddenly; for example, some shifts of culture can be triggered by political change. The following highlights the consequence of the end of colonial Tanganyika and commencement of an independent national government for Tanzania:

> [. . .] in independent Africa, the struggle for power instantly assumed an extremely fierce and ruthless character. All at once, in the blink of an eye, a new ruling class arises – a bureaucratic bourgeoisie that creates nothing, produces nothing, but merely governs the society and reaps the benefits [. . .] once decades, even centuries, were needed for a new social class to emerge, and here all it took was several days. The French [. . .] called the

phenomenon *la politique du ventre* (politics of the belly), so closely was a political appointment connected with huge material gains.

(Kapuscinski 2002, p. 36)

Of more current concern, climate change is already known to be the cause of shifts in ecosystems that are placing demands for rapid adaptation upon indigenous cultures. For example, glacial retreat in Colombia (Baptiste 2011, Walsh 2011) and in Nepal (Chetri 2010) is causing reduced river flows and consequent depletion of water supplies that in turn impact upon animal husbandry, food provision and supplies of drinking water: seasonal weather has become both less predictable and more violent in its causes of similar consequences.

By such changes, cultures are obliged to adapt and evolve, to shift, migrate, divide and amalgamate as they have in the past and, given freedom without constraint or oppression, will continue to do so for the future. Such adaptations, however, will not always be in unison, nor be coordinated to conform to a pre-arranged schedule, but will behave as experience and need demand (Tsioumani 2009).

Vulnerabilities to hazards may ensue from this shifting matrix. Communities may feel the need to relocate and to re-establish themselves. Among individuals, those who perform most successfully might exercise power, either by employment or exploitation, and thereby becoming more powerfully dominant and impacting upon others, either socially, commercially or politically, in ways that may come to be a cause of vulnerability for those others.

Culture and vulnerability to hazards

Over time, the threat of environmental hazards exposes ways in which we interact and behave with one another and, by experience of their impact, ways in which our cultures emerge, respond and adjust. But, while cultural evolution is not static, neither does it necessarily reflect unity, coordination, synchronicity or common purpose. By negligence or intent, evolution is obliged to contain subcultures of difference, separateness or self-interest. Such separation may not always display purposes to the common good, intent or direction, raising doubts about principle, integrity or ethnic values that a once dominant culture may have regarded itself to have pre-empted or controlled.

As well as the positive and the 'good' although not always progressive, nor even benign and desirable – culture also encompasses the questionable, the less favourable and the negative. In consequence, behaviour, the expression of culture and perhaps of its denial influences ethics, politics, governance and governments, whether of dictatorships, democracies or tribal entities. Some behaviours are so ingrained, repetitive and powerful that they have become 'contra-cultures'; that is, less benign subcultures working against those aspects of common culture usually regarded as beneficial and 'good'.

Some individual, group or corporate behaviours have become a root cause, not only of behavioural responses to hazards (once a major concern, cf. White 1974), but also of behavioural causes of the vulnerability of others to those same and similar hazards. Some cultures, evident now, have been formed by the behavioural influence of others over long periods of time. From Nepal, Adhikary Nripal provides an example:

> In my studies of a few villages of Nepal, I found that vulnerability is a political issue. Culture is a response to politics. Before we talk about preparedness or coping mechanisms that cultures have adopted in different temporal and spatial settings [. . .] it is necessary to address the root causes of vulnerabilities [. . .]. In the Dhadhing district of Nepal, ethnic groups like Chepangs live in very steep lands [. . . and] studies have shown that they are very adaptive to those settings. But it is the policies of institutionalised discrimination [. . .] over about eight hundred years, which has forced them to live in steep slopes. They have adapted [. . .] to those harsh conditions [. . .] but that does not mean they are completely safe. Food insecurity and poverty still haunt them and they are the first victims of [. . .] landslides. They are living in the edge of vulnerability, not because of their cultural inclination but because of institutionalised discrimination against them.
>
> (Nripal 2008, quoted in Lewis 2008b)

Nripal's 'policies of institutionalised discrimination' are an example, first, of how vulnerability can be brought about by processes, often long term, over which those who are rendered vulnerable may have no control and, second, that within a national culture there may be powerful and separate subcultures, possibly at appointed or self-appointed higher social levels and of different and not always benign behaviours. Such behaviours may not be to the advantage of others, being a subcultural 'contra-culture' working against the interests of the majority of others, in the past, at the time or in the future.

What emerges is that most people's vulnerability is not of their own making but is caused by the actions of other people (Lewis 1999, p. 5), those who are more powerfully self-interested. Most people's vulnerability emanates from policies that are to their disadvantage, established by others in their own interest and which represent less honourable aspects of human nature that have become inescapably integrated within cultures and within societies comprising those cultures.

As further demonstration of these processes, large numbers of people are caused to be at risk by pernicious political, commercial and social realities that result in, for example: discrimination and displacement; impoverishment by others' self-seeking expenditure; denial of access to resources; and corrupt siphoning of public money that would be otherwise spent to the public good (Lewis and Kelman 2012). Often prevailing as 'unavoidable' or 'inevitable' political contexts, these behaviours cause vulnerability to others as consequences of often corrupt, political and commercial processes that have become 'contra-cultures', subcultures working to favour the few but in opposition to the interests of the many.

Corruption and mismanagement

Frequently and popularly described as 'cultures' within themselves, some behavioural causes of vulnerability are riding rough-shod over Disaster Risk Reduction (DRR), which is rendered meagre by comparison. For example 'tax-haven culture' (Blond 2007, White 2000) or 'offshore culture' (Preston 2011) are now exposed in their global scale (Shaxson 2011). Politicians and commercial operators, privately and corruptly, have siphoned collectively enormous amounts of money, much of it from development funding and very often from their own disaster-prone countries into private bank accounts in the countries that were the origin of the aid (Ndikumana and Boyce 2011). Although this activity has emerged as probable evidence of why some countries have remained 'less developed' and why poverty has prevailed as the principal cause of vulnerability and disaster losses, developed countries in Europe are also responsible for the illicit equivalent of billions of US dollars flowing into and out of its emerging economies (GFI 2014).

One analysis has stated that between 2000 and 2008, the cost to developing countries of crime, corruption and trade mispricing (trade as a vehicle of monetary transfer) was approximately US$6.5 trillion (GFI 2011); a subsequent report from the United Nations Development Programme (UNDP) indicating that a significant share, US$197 billion, has accrued from those countries categorised as LDC (Least Developed Countries; UNDP 2011). This latest list is headed by Bangladesh and includes numerous other recognisably disaster-prone, less developed countries.

An examination of capital sent from thirty-three African countries between 1970 and 2008 has concluded that over that thirty-eight-year period, 'capital flight' amounted to US$735 billion, a sum roughly equal to 80 percent of the combined GDP of those countries in 2008. The period over which this study has been applied indicates that the sum involved is 'not a transitory product of unusual circumstances but rather an outcome of persistent underlying causes' (Ndikumana and Boyce 2011, p. 46). A previous study by the same authors concluded that this sum was 'from assets belonging to a narrow, relatively wealthy stratum of populations while, in consequence, public external debts are born by the people through their governments' (Ndikumana and Boyce 2008, quoted in Shaxson 2011, p. 158). Similar procedures making use of offshore tax havens have operated on behalf of the rich and at the cost of the poor within and from many countries (Shaxson 2011). This aggressive milking of development funding means, obviously, that there will be less development or that development projects that take place will be less extensive or of lesser quality in their arrangement or construction.

A vivid example of how corruption increases the vulnerability of people is seen in the building industry in earthquake zones. Studies of culture in building construction have very rarely included the negative influences of corrupt management (Lewis 2003). Evolutionary traditional forms of construction have recently been largely displaced by modern reinforced concrete construction worldwide, contributing to the depletion and erosion of traditional skills (Oliver 1978, p. 10, Fathy 1986, Karababa and Guthrie 2007). Reduced construction quality and inappropriate materials may mean, among other things, that some buildings will be less able to withstand earthquakes and other hazardous extremes, thus adding to the vulnerability faced by the people who live in them. Earthquake losses, of destroyed buildings and consequent deaths and injuries, may be a reflection of inadequacies in building construction as well as of earthquake energy, but building failures also originate from behavioural content in procuration, design, execution, maintenance, administration and management (Johnson 2011, Perez and Johnson 2011). For example, corruption is known to have been responsible for illicit omissions or substitutions of materials and the bribing of construction inspectors to look the other way – or not to look at all – with a strong linkage having been recognised between corruption in construction and earthquake damage and casualties (Ambraseys and Bilham 2011, Lewis 2005, 2008a).

Overall, and by its hierarchy of bribery and graft, corruption for the benefit of the few means continued and exacerbated poverty, the established cause of vulnerability to disasters of all kinds, for the many. These consequences are then worsened by simultaneous breakdown and malfunction of hospitals, clinics and health care generally (Ndikumana and Boyce 2011, pp. 74–83).

In potentially hazardous contexts, fear may be a behavioural influence on decision making (Lewis, Kelman and Lewis 2011). Powerful subcultures may exercise their own behaviours, some not far removed from corruption; others may be, by senior level management systems, so remote from the realities of industrial, commercial or governmental processes. The Fukushima Daiichi nuclear installation in Japan has provided pre-eminent examples of managerial malpractice, since confirmed as aspects of Japanese national culture.

The 11 March 2011 Richter 8.9 earthquake and subsequent tsunami, followed by industrial nuclear explosions and consequent fallout, have conveyed lessons for all countries that have developed or are aspiring to similar technologies. Japan as a nation has been revered as a highly developed culture, renowned for its disciplined responses to earthquake risk. Within the same culture, however, has emerged an acknowledged failure to respond to tsunami risk for highly vulnerable coastal settlements, including the nuclear power installations built after government information had convinced Japanese people of their safety (Leonard 2011, McCurry 2011a, Onishi 2011).

Caused by an undersea earthquake in 2011, the 14-metre tsunami that followed it minutes later overwhelmed protective sea walls, swamped 561 square kilometres of land, destroyed entire communities and caused 14,500 deaths with a further 11,450 people reported missing. There were 130,000 homeless evacuees (ADRC 2011, figures as of May). Whereas earthquake resistant construction protected city buildings against collapse, coastal sea walls had not been designed to withstand a tsunami of such height and power (*The Economist* 2011), nor had nuclear installations been designed to withstand the overtopping of sea walls or their structural failure in severe earthquake. The result was the flooding of the Fukushima nuclear power plant, causing its crucial cooling system to fail. Consequent overheating caused explosions of the power plants that, in turn, created the biggest nuclear disaster since that of Chernobyl in 1986.

Degrees and levels of structural resistance are preconditioned both by precedent and by cost, a balancing of which being a matter of research, debate and judgement. But, whereas precedent and cost are matters of fact and information, judgement is behavioural and may be influenced by other behavioural inputs. Further, once achieved, most structures rely upon regular inspection and essential maintenance for safe continuation of their function, the management of these aspects being also subject to not always benign behavioural influences.

In the case of the Fukushima Daiichi nuclear installation, it was reported that, after only three days of inspection, government regulators approved a ten-year extension for Reactor 1, the oldest on the site, despite warnings of structural cracks that made standby generators vulnerable to corrosion from rain and seawater penetration (McCurry 2011b, Tabuchi, Onishi and Belson 2011). These generators were required to supply standby power for essential cooling of the reactors in the event of failure and so prevent dangerous overheating.

Shortly after the granting of this extension, the power company admitted shortcomings in the inspection procedure and of maintenance management of all six reactors but, only two weeks later, the 11 March earthquake and tsunami set off the Fukushima nuclear crisis (Onishi and Glanz 2011).

The productive life of existing power plants had allegedly been extended due to public protest against new nuclear installations, but in some cases they may have been extended beyond their technological or managerial capacity. The following serious shortcomings within managerial processes were exposed:

- 'Cosy ties' existed between the ministry and the power company, former ministry officials taking 'lucrative jobs' within the company, and 'unhealthy ties' existed between power plant operators and the regulators who oversaw them: 'Expert panels like the one that recommended the extension [of time] are drawn mainly from academia to backstop bureaucratic decision-making and rarely challenge the agencies that hire them' (Onishi and Glanz 2011).
- Cover-up by the power company of information concerning large cracks and falsification of inspection records to save on repair costs (Onishi and Glanz 2011).
- Weak inspection and poor effectiveness of Japan's Nuclear and Industrial Safety Agency, having been appointed by the Ministry of Economy, Trade and Industry – the ministry responsible for the development of Japan's nuclear industry (Onishi and Glanz 2011).

In a country from which the word originates, '*tsunami*' first appeared in government guidelines for nuclear installations only in 2006, with high dependency upon anecdote and 'living memory'. Precedent included recollection and documentation, but with inadequate historical research. As a result, Japanese sources stated that their 'nuclear establishment' largely disregarded potentially destructive tsunamis. The Fukushima Daiichi plant, first commissioned in 1971, constructed breakwaters and sea walls designed to resist typhoons, not tsunamis, *and* without responding to non-binding advice issued in 2002 that raised the then projected tsunami wave height: its argument being that there was no precedent (Onishi and Glanz 2011).

Sea walls were constructed higher than the highest tsunamis on record but, with what is described as a typically Japanese deterministic rather than a probabilistic approach that would have taken unknowns into account, Japanese engineers continued to work to what they believed were maximum recorded earthquakes, whether for reasons that were '[. . .] cultural, historical or simply financial [. . .]' (Onishi and Glanz 2011).

While tsunami science in Japan slowly advanced, however, construction design remained comparatively static, with information on sources of potential overtopping of Fukushima's anti-tsunami safeguards not being recognised, for example, by a source similar to the 1993 R7.8 earthquake on Japan's western coast. Furthermore, research made two years after the completion of Fukushima Daiichi revealed that, in the year 869, the Jogan Sanriku earthquake caused a tsunami that reached a mile inland in an area immediately north of where the Fukushima plant now stands (ADRC 2011, Onishi and Glanz 2011).

In Japan, superior technology is demonstrably achievable, but managerial status may be decreed on the basis of a person's age. Therefore, inherent respect from junior employees can inhibit questioning of experience in decision making. For some new and probably imported technologies and industries, what appear to have been questionable are decision making processes in design, application, maintenance and management.

In July 2012, the required report to the Japanese Parliament by the Fukushima Nuclear Accident Independent Investigation Commission stated that the Fukushima nuclear accident had been '[. . .] rooted in government–industry collusion and the worst conformist conventions of Japanese culture'. The report also warned that the Fukushima plant may have been damaged by the earthquake before the arrival of the tsunami. It went on to blame a 'tepid response' to the crisis of collusion between the company, the government and regulators who had all 'betrayed the nation's right to safety from nuclear accidents' and manipulated a 'cosy relationship with regulators to take the teeth out of regulations' (Tabuchi 2012).

The same report criticised 'a culture in Japan that suppresses dissent and outside opinion', and expressed the view that the causes of the nuclear accident 'are to be found in the ingrained conventions of the Japanese culture: our reflexive obedience; our reluctance to question authority; our devotion to "sticking with the program"; our groupism; and our insularity'. But by the creation of an independent investigation commission and avoidance of a reliance on bureaucratic self-examination that could be clouded by self-defence, the report itself, which contained hints on how Japanese society needs to change, was a break with Japanese precedent (Tabuchi 2012).

Catastrophic failure of modern technology brought about by a massive natural hazard has caused the entirety of an ancient and revered culture to re-examine itself. Subsequent cultural changes eventually will be assisted by, and during, the long aftermath of sudden national shock. Behavioural inadequacies in Japan do not make Japan unique; other countries are not excluded from the possibility of similar experiences.

Coming back to the Nepal example mentioned above, historical political change there caused established cultures to adjust over long periods of time but, that cultures change for the better, requires a preferably positive subsequent adjustment given necessary freedom without intrusion by ill-informed administrations, arbitrary regulation or the 'contra-cultures' of rampant corruption, misappropriation and cover-up (Lewis 2005).

Expressions of need are behavioural, as are measures taken in response to need. Policies and decision making, themselves behavioural processes, involve other behaviours such as inquiry, conversation, discussion and debate. Technological design in response to identified need, inclusive of earthquake resistance, is also a behavioural process, itself involving multiple layers of decision making necessary for the physical outcome. The resulting structure may respond well to the expressed need or, because of aspects of decision making, may be inadequate and may fail. Where this occurs, failure will be assumed to be of technology, but will be as much to do with behavioural decision making as with external forces.

Disaster risk reduction (DRR)

Identified as 'drivers of vulnerability' (UN/ISDR 2009), poor urban governance, increasing urbanisation, vulnerable rural livelihoods, climate change and declining ecosystems can all, in some degree, be interpreted as physical manifestations of behavioural decision making. Without exploration of the origins of those 'drivers', no modification of them can be proposed, the implication of the UN document being of 'given' and immutable prevailing conditions. Reduction of vulnerability could, however, ensue from identification of causative pressures and behaviours and a reduction of disaster incidence and risk could ultimately be achieved.

Currently, disaster risk reduction makes little or no acknowledgment of behaviours. That greed and self-indulgence serve to exacerbate the inequality (Dorling 2010), deprivation and poverty at the root of other people's vulnerability is being disregarded as another unfortunate reality. These behaviours nonetheless prevail within and beyond most concepts of governance and operate powerfully and pervasively *inter alia* contrary to the objectives of disaster reduction at domestic, group, corporate and institutional levels.

Power and its manifestation is itself an aspect of culture carried by belief systems (Hewitt 2008) and is as much a part of 'culture' as is say, resilience, *laissez-faire* or corruption – or of their denial. Corruption itself succeeds either through cultural absorption and consequent acceptance or by other cultures of oppression and fear.

Managerial malpractice succeeds in secrecy and *laissez-faire*, but negative behaviour can be made to diminish by greater transparency in governance.

As an aspect of DRR, what can community resilience (cf. Lewis 2013) achieve against pervasive corrupt criminality entrenched within the same communities? What might be called 'culture' by some may be 'corruption' to others, corrupt behaviour recognised as a problem only if and when its consequences of impoverishment and subsequent subservient poverty, destitution, injury and death are identified and realised.

As it is, the entire concept of DRR is required perpetually to trade on well-meaning innate optimism, positive drive and honourable 'common good' for the safer future of humankind. What is not expressed is that governments, societies and communities – whether rural or urban – include the negative as well as the positive, the corrupt as well as the ethical and the powerfully active as well as the inactive, inert and the weak (Cannon 2007). Good governance, for the initiating or permitting of favourable systems and procedures, has also to prevent and inhibit those other systems that, while advantageous to the powerful few, result in widespread deprivation for the many (Lewis and Kelman 2012).

People bereaved, displaced, disabled or disadvantaged by catastrophe – those represented by disaster statistics – are not usually the source of the causes of their vulnerability. Decisions made, and those not made, in the board rooms, staterooms, ministerial departments, parliaments and committee rooms may be at the root of people's disadvantagement, impoverishment, poverty and consequent vulnerability.

Whether or not there is yet a prevailing DRR culture or, if DRR can yet be recognised as an expression of prevailing cultures, seemingly unavoidable 'contra-cultures' are similarly active at the same time that, as yet, DRR has not recognised, identified or aimed to prevent. Meanwhile, the causes of vulnerability and of disaster risk creation (Lewis and Kelman 2012) proceed apace and the long, ongoing, causative histories of vulnerabilities to natural hazards, and the disasters that ensue, continue to demonstrate these incongruities and impracticalities.

As DRR becomes internationalised across all countries, governments, laws and religions, diverse cultures and their consequent behaviours will apply in the achievement of this globally proclaimed ethos. But disaster risk reduction has not yet recognised behavioural or cultural issues and the massive negative presence of behavioural 'contra-cultures'. So far, it has welcomed the positive, but has ignored the interactions of the negative in the social realities of the contexts it seeks to serve.

For DRR to succeed, examination of the *status quo* is required for exposure of the real causes of disasters in governance, technological failure, social and economic conditions, and of the underlying attitudes and behaviours by which vulnerability and risk are perpetrated, perpetuated and exacerbated. Without such action, 'contra-cultures' will remain hidden and continue their socially destructive processes.

Examination of the *status quo* would itself be a positive behavioural response, commenced by analyses of disaster aftermath and with retrospection of management and decision making, in much the same thorough and in depth way as has been undertaken by Japan. A project under the auspices of the International Council for Science (ICSU) entitled 'Forensic Investigations of Disasters' (FORIN 2011) is being established as part of a more demanding and penetrating approach to understanding the causes of disasters, the results of which would potentially become the basis of evidence-based decision making and increased accountability for responsible policy making in disaster risk reduction.

Those childhood stories – early conveyors and formulators of our cultures – could serve us well in the real world at large where behaviours abound of the epitomes and counterparts of wolves, rats, ogres, witches, robbers, thieves – and wayward technologies. The question should be asked: if cultures can be so influenced, what are the future implications for disaster risk reduction of some current corporate and individual behaviours?

References

Ambraseys, N. and Bilham, R., 2011. Corruption kills. Comment. *Nature*, 469, 153–5.

ADRC, 2011. Great East Japan earthquake: Preliminary observations. *Asian Disaster Reduction Centre and the International Recovery Platform* [online]. Available from: www.adrc.asia/documents/disaster_info/2011March11_EastJapan_EarthquakeReport_%20final.pdf [accessed November 2011].

Baptiste, B., 2011. Climate change is affecting traditional knowledge in Colombia (Interview with Constanza Vieira). *Latin America and Caribbean Communication Agency* [online]. Available from: www.alcnoticias.net/interior.php?lang=688&codigo=20203&PHPSESSID=e7189abf9e060637b233798a61d9c9ef [accessed November 2011].

Blond, P., 2007. Conservatives must reject capitalism [online]. Available from: www.thefirstpost.co.uk/8968,news-comment,news-politics,conservatives-must-reject-capitalism [accessed February 2011].

Bronowski, J., 1973. *The Ascent of Man*. London: BBC Books.

Cannon, T., 2007. Reducing people's vulnerability to natural hazards: Communities and resilience. Paper presented at the WIDER Conference on *Fragile States-Fragile Groups: Tackling Economic and Social Vulnerability*. Finland: Helsinki, June 2007 [online]. Available from: www.wider.unu.edu/publications/working-papers/research-papers/2008/en_GB/rp2008-34/ [accessed 31 December 2010].

Chetri, M. B. P., 2010. *Effects of Climate Change in Nepal and Beyond*. Paper presented to the *Annual Conference of the Disaster Management Institute of Southern Africa*, Jeffrey's Bay, South Africa, 8–10 September, 2010.

Dorling, D., 2010. *Injustice: Why Social Inequality Persists*. Bristol: Policy Press.

Douglas, M. and Wildavsky, A. B., 1982. *Risk and Culture: An Essay on the Selection of Technical and Environmental Dangers*. Berkeley, CA: University of California Press.

Douglas, M., 1992. *Risk and Blame: Essays in Cultural Theory*. London, New York: Routledge.

The Economist, 2011. An earthquake in Japan: When the earth wobbled. *The Economist* [online], 11 March. Available from: www.economist.com/blogs/asiaview/2011/03/earthquake_japan_0&fsrc=nwl [accessed September 2011].

Fathy, H., 1986. *Natural Energy and Vernacular Architecture: Principles and Examples with Reference to Hot Arid Climates*. Chicago, IL: University of Chicago Press.

FORIN, 2011. Forensic Investigations of Disasters (FORIN) report. *PreventionWeb* [online]. Available from: www.preventionweb.net/english/professional/publications/v.php?id= 25016 [accessed November 2013].

GFI, 2011. Illicit financial flows from developing countries: 2000–2009. *Global Financial Integrity* [online]. Washington, DC. Available from: http://iff-update.gfip.org/ [accessed January 2011].

GFI, 2014. US$68.9 billion flowed illegally into or out of emerging EU economies in 2011. *Global Financial Integrity* [online]. Washington, DC. Available from: www.gfintegrity. org/content/view/676/70/ [accessed February 2014].

Grimm, J. and Grimm, W., 1812. *Children's and Household Tales (Kinder und Hausmärchen)*. Berlin and Göttingen.

Hewitt, K., 2008. *Personal Communication*, 3 October.

Johnson, C. G., 2011. *State Failing to Enforce Seismic Rules for Schools*. SFGate [online]. San Francisco, CA. Available from: www.sfgate.com/cgi-bin/article.cgi?f=/c/a/2011/04/ 08/MN5J1IPS52.DTL [accessed April 2011].

Kapuscinski, R., 2002. *The Shadow of the Sun: My African Life*. London: Penguin.

Karababa, F. S. and Guthrie, P. E., 2007. Vulnerability reduction through local seismic culture. *Technology and Society Magazine, IEEE*, 26 (3), 30–41.

Leonard, H. B., 2011. Preliminary observations on the Japanese 3/11 earthquake and tsunami. *Harvard Kennedy School, Programme on Crisis Leadership* [online]. Available from: www. ash.harvard.edu/extension/ash/docs/earthquake.pdf [accessed 24 October 2012].

Lewis, J., 1999. *Development in Disaster-Prone Places*. London: IT Publications.

Lewis, J., 2003. Housing construction in earthquake-prone places: Perspectives, priorities and projections for development. *Australian Journal of Disaster Management*, 18 (2), 35–44.

Lewis, J., 2005. Earthquake Destruction: Corruption on the Fault Line. In: Transparency International, ed., *Global Corruption Report 2005: Corruption in Construction and Post-Conflict Reconstruction* [online]. Berlin: Pluto Press, 23–30. Available from: www. transparency.org/whatwedo/pub/global_corruption_report_2005_corruption_in_construction_ and_post_conflict [accessed November 2011].

Lewis, J., 2008a. The Worm in the Bud: Corruption, Construction and Catastrophe. In: L. Bosher, ed., *Hazards and the Built Environment: Attaining Built-In Resilience*. London: Taylor & Francis, 238–63.

Lewis, J., 2008b. *The Creation of Cultures of Risk: Political and Commercial Decisions as Causes of Vulnerability for Others. An Anthology* [online]. Available from: www.islandvulnerability. org/docs/lewis2008risk.pdf [accessed November 2011].

Lewis, J., 2013. Some realities of resilience: A case study of Wittenberge. *Disaster Prevention and Management*, 22 (1), 48–62.

Lewis, J. and Kelman, I., 2012. The good, the bad and the ugly: Disaster risk reduction (DRR) versus disaster risk creation (DRC). *PLOS Currents Disasters* [online]. Available from: http://currents.plos.org/disasters/article/the-good-the-bad-and-the-ugly-disaster-risk-reduction-drr-versus-disaster-risk-creation-drc/ [accessed February 2014].

Lewis, J., Kelman, I. and Lewis, S. A. V., 2011. Is 'fear itself' the only thing we have to fear? Explorations of psychology in perceptions of the vulnerability of others. *The Australasian Journal of Disaster and Trauma Studies* [online], 2011–13, 89–103. Available from: http://trauma.massey.ac.nz/issues/2011–3/AJDTS_2011–3_03_Lewis.pdf [accessed December 2011].

McCurry, J., 2011a. Fukushima Daiichi nuclear power plant operator 'ignored tsunami warning'. *The Guardian*, 29 November [online]. Available from: www.guardian.co.uk/ world/2011/nov/29/fukushima-daiichi-operator-tsunami-warning [accessed November 2011].

McCurry, J., 2011b. Japan nuclear firm admits missing safety checks at disaster-hit plant. *The Guardian*, 22 March [online]. Available from: www.guardian.co.uk/world/2011/mar/22/japan-nuclear-power-plant-checks-missed?INTCMP=SRCH [accessed November 2011].

Ndikumana, L. and Boyce, J., 2008. New estimates of capital flight from sub-Saharan African countries: Linkages with external borrowing and policy options. *Political Economy Research Institute, University of Massachusetts*. Amherst, MA [online]. Available from: www.peri.umass.edu/fileadmin/pdf/working_papers/working_papers_151–200/WP166.pdf [accessed February 2014].

Ndikumana, L. and Boyce, J., 2011. *Africa's Odious Debts: How Foreign Loans and Capital Flight Bled a Continent*. London: Zed Books.

Nripal, A., 2008. Culture and risk: Understanding the socio-cultural settings that influence risk from natural hazards. *ICIMOD/SIDA 2008* [online]. Moderator: Ken Hewitt. Theme 1: The Role of Socio-Cultural Settings in Influencing Peoples' Capacities to Deal with Risk from Natural Hazards and to Adopt or Reject Modern Safety Measures. Available from: www.mtnforum.org/sites/default/files/forum/files/participants-contributions-carthreads1–2.pdf [accessed November 2011].

Oliver, P., ed., 1978. *Shelter and Society*. London: Barrie & Jenkins Ltd.

Onishi, N., 2011. Safety myth left Japan ripe for nuclear crisis. *The New York Times*, 24 June [online]. Available from: www.nytimes.com/2011/06/25/world/asia/25myth.html?scp=4&sq=Norimitsu%20Onishi&st=cse [accessed September 2011].

Onishi, N. and Glanz, J., 2011. Japanese rules for nuclear plants relied on old science. *The New York Times*, 26 March [online]. Available from: www.nytimes.com/2011/03/27/world/asia/27nuke.html?sq=No [accessed September 2011].

Perez, E. and Johnson, C. G., 2011. Audit: Sloppy oversight increases risk of unsafe school buildings. *California Watch* [online]. Available from: http://californiawatch.org/daily report/audit-sloppy-oversight-increases-risk-unsafe-school-buildings-13964 [accessed 3 August 2012].

Preston, P., 2011. Peter Preston's counsel after reading Treasure Islands: Abandon hope (Book promotion) [online]. Available from: http://treasureislands.org/peter-prestons-counsel-after-reading-treasure-islands-abandon-hope/ [accessed September 2011].

Sendak, M., 2011. *Bumble-Ardy*. New York: HarperCollins (Emma Brockes' Review. *The Guardian*, 3 November 2011).

Shaxson, N., 2011. *Treasure Islands: Tax Havens and the Men Who Stole the World*. London: Bodley Head.

Siddique, M., 2011. Thought for the day. *Radio, BBC 4*, 6 December [online]. Available from: www.bbc.co.uk/podcasts/series/thought [accessed December 2011].

Tabuchi, H., 2012. Inquiry declares Fukushima crisis a man-made disaster. *The New York Times*, 5 July [online]. Available from: www.nytimes.com/2012/07/06/world/asia/fukushima-nuclear-crisis-a-man-made-disaster-report-says.html?_r=1 [accessed 2 August 2012].

Tabuchi, H., Onishi, N. and Belson, K., 2011. Japan extended reactors life, despite warning. *The New York Times*, 21 March [online]. Available from: www.nytimes.com/2011/03/22/world/asia/22nuclear.html?_r=2 [accessed November 2011].

Tsioumani, E., 2009. Traditional knowledge and climate change: From Anchorage to Copenhagen. *UNU-IAS Traditional Knowledge Initiative* [online]. Available from: www.unutki.org/default.php?doc_id=170 [accessed September 2011].

UNDP, 2011. Illicit financial flows from the least developed countries: 1990–2008. *UNDP*. New York [online]. Available from: http://content.undp.org [accessed May 2011].

UN/ISDR, 2009. The United Nations 2009 global assessment report on disaster risk reduction [online]. *United Nations International Strategy for Disaster Reduction Secretariat.* Available from: www.preventionweb.net/english/hyogo/gar/report/index.php?id=1130 [accessed 28 November 2010].

Vasagar, J., 2011. Education chief identifies white working-class pupils as big challenge (quoting Liz Sidwell, Schools Commissioner for England). *The Guardian*, 23 September [online]. Available from: www.guardian.co.uk/education/2011/sep/23/education-chief-white-working-class-challenge?INTCMP=SRCH [accessed September 2011].

Walsh, D., 2011. Climate change takes a toll on cultures. *The New York Times*, 27 September (quoting Brigitte Baptiste, Alexander von Humboldt Biological Resources Institute, Ministry for the Environment, Columbia) [online]. Available from: http://green.blogs.nytimes.com/2011/09/27/climate-change-takes-a-toll-on-cultures/ [accessed September 2011].

White, G. F., ed., 1974. *Natural Hazards: Local, National Global.* New York, London, Toronto: Oxford University Press.

White, C., 2000. Monaco: Principality under pressure. *Euromoney.* London [online]. Available from: www.euromoney.com/Article/1004044/Monaco-Principality-under-pressure.html [accessed February 2011].

ZiF, 2012. The culture and perception of risk. *ZiF-Mitteilungen* 1/2012. Zentrum für Interdisziplinäre Forschung. Center for Interdisciplinary Research. Universität Bielefeld.

7

THE CULTURAL SENSE OF DISASTERS

Practices and singularities in the context of HIV/AIDS

Klaus Geiselhart, Fabian Schlatter, Benedikt Orlowski and Fred Krüger

Introduction: defining disasters is a cultural act

Defining a disaster is a cultural act: naming a disaster first starts with recognising societal conditions that are evaluated to be disastrous. Plainly, this can only be done in reference to values and norms. In a second stage, conditions tagged as disastrous become traced to either a hazardous event or to social circumstances and dealings that have reduced people's resilience, rendered them vulnerable, or hampered their appropriate response to threats. It is thus intelligible that in different cultures, depending on particular perceptions and values, disasters can be substantially different things. Disaster definitions, for example, are created by organisations, institutions or individuals with regard to their own practical assignments. But, then, how can 'a culture' be identified, a situation be narrowed down to a 'disaster', and appropriate disaster risk reduction (DRR) measures be initiated?

In a postmodern and postcolonial context, or after the cultural turn in the social sciences generally, it has become obvious that it is crucial to envision a notion of culture beyond essentialisms or static conceptualisations. Only by this means can culture become a useful and applicable concept for social research and DRR. From this point of view, culture can neither be defined for a certain time or space nor, strictly speaking, in respect to social groups.

It is easy to state that cultures are fluid, unstable, and in permanent transition. However, if we employ 'culture' in empirical research and in DRR practice we are in danger of falling into the trap of essentialisation. What kind of social unit has which specific culture? We are thus confronted by two challenges: how can cultures in a given society be identified and acknowledged in their multiplicity and fluidity, and how can such a broad-based notion be translated into DRR practice? For the purpose of this chapter, let us assume that cultures can best be defined by fields of practice. These are certain contexts where practices are institutionalised in order to render life and social actions more predictable, or reliable.

Nonetheless, we have to acknowledge that certain common values and sets of knowledge are involved in collaborative practical dealings. At least an implicit agreement and mutual understanding are needed for the sake of accomplishing certain tasks. Cultures can then be identified by fields of practice involving conventionalised dealings, agendas, guidelines, explicitly codified tasks and targets.

Within these fields, however, individuals sometimes act beyond common conduct. Especially those situations or settings that are subsequently termed 'disasters' occur as singular incidents where unanticipated moments emerge. Such situations, which are deemed complex, hazardous or as constituting an emergency, as well as processes interpreted as slow-onset crises, are characterised by institutional breakdowns and by a multiplicity of singular situations or occurrences. These singularities lead to shifting assemblages of actors and create proceedings that distinctly differ from conventionalised practices. As expressions of ongoing or momentary doing, it is thus worthwhile to include such singularities in the assessment of disaster processes. This may lead to a way forward in the analysis of linkages between cultures and disasters.

In this chapter, we focus both on institutional fields of practice and on individual dealings within the setting of the HIV and AIDS pandemic in Botswana. We argue that by looking at an 'arena of singularities', which is often neglected in DRR-related research and intervention, a better understanding of the 'cultural' dimensions of disasters can be fostered.

The findings presented here are based on empirical studies conducted during different field campaigns between 2004 and 2013 in Botswana. This research was part of two projects funded by the Deutsche Forschungsgemeinschaft (German Research Foundation) on the social effects of HIV/AIDS and its associated intervention schemes; the projects also included geographical and anthropological research about traditional healers.

In the following, after a more precise explication of our basic terms, we will illustrate the diverse conceptualisations of HIV/AIDS as a disaster in Botswana. We will look at different fields of practice and lay open their internal rationalities and contradictions. Such fields include biomedical intervention, support groups and traditional practices of healing, among others. These different spheres clash, unite or coexist in the arena of singularities when individuals navigate through different institutions. We will identify different aspects of those singularities and assess them with regard to their possible influence on future HIV/AIDS intervention schemes.

Culture: fields of practice and the arena of singularities

The 'cultural turn' in the social sciences effectively eroded an essentialist notion of culture. Cultures can no longer be regarded as a fixed set of norms, values and practices that are confined and obligatory to a certain community or geographical area. There has been a shift towards conceptualising culture as the process of signification and production of meaning (Rutherford and Bhabha 1990, p. 210,

Hall 1992). In the course of this cultural turn, notions of culture have become increasingly restricted to semiotic and discourse effects. However, this has been contested by the argument that such understanding disregards the restraints of lived-in worlds. Processes of identity building, even in a postmodern liquid life (Bauman 1996), can never be completely independent from given social conditions and their materiality (Jackson 2000). In the wake of this call for a 'rematerialisation' of social geography, new conceptualisations such as the actor-network theory were then employed. Notions of 'assemblages' (Anderson and McFarlane 2011) or 'techno-logical zones' (Barry 2006) were developed in order to conceptualise the normative and standardising effects of technologies and processes of societal organisation. To secure their functioning, organisations, as well as informal institutions, demand certain qualifications of every individual who wants to become active within them, thus forming zones of qualification (Barry 2006). These can be understood as fields of practice, with compulsory sets of knowledge required of their associates. Further-more, these zones of qualification imply normative and value-loaded endeavours that are firmly connected to the associates' societal assignments. Organisations find it necessary to agree on programmes or agendas and thus create value-loaded sets of knowledge, procedures and attitudes.

Accompanying these concepts, approaches to praxis (cf. Dewey 1929, Bourdieu 1977, Giddens 1984, Schatzki 2003) help to analyse how knowledge is incorporated in institutionalised practices. According to Dewey (1929), people have a basic need for certainty. This causes them to construct explanations about the world in order to make life predictable and themselves capable of action. Knowledge plays a crucial role in this process. It derives from a collective historical process: based on the experiences of many people, explicit assumptions about the world and of how to best deal with it are deduced. People negotiate these assumptions and try to make the outcomes of their own experiences mandatory. Knowledge is thus rooted in praxis and becomes realised in conventionalised practices in which it proves its significance.

This applies not only to explicit knowledge but, more importantly, to partially conscious knowledge. People's actions are not entirely deliberate; instead, people skilfully perform according to how they have been socialised in their roles and professions. Practical knowledge (Giddens 1984), the habitus (Bourdieu 1977), experience (Dewey 2005 [1934]) or institutional logic may determine action more than the deliberation of a single person. 'Culture' therefore also entails a perform-ative dimension. We thus see culture as a reification of lived and assumingly valid knowledge sets that are reproduced and permanently contested in interaction. Culture is thus neither static nor essentialist.

Soja (1999) and Bhabha (1996) have conceptualised 'third space' as the sphere of the lived experience of individuals, as a space where people establish their individuality 'in between' conventionalised cultures. As regulations can never cover all the eventualities of social life, individuals, no matter which position they obtain in an institutionalised setting, are challenged by the demands of organisational structures as well as by their individuality and thus need to act creatively (Berk and Galvan 2009). When entering different fields of practice, individuals encounter different cultures within their everyday lives. People might feel torn or disrupted when in between settings, as each setting carries its own norms, values and knowledge systems (e.g., people's religious and ethical values might be challenged by their professional conduct when an employer demands them to act in terms of their company's strategy). Under these circumstances, individuals might then develop new and individualised practices beyond institutionalised routines or solutions. Taking this pragmatist perspective further, we conceptualise an arena where an individual's solutions and explanations appear singularly in specific situations and moments. It is an arena where, despite the normative order in which these situations or moments are embedded, the trajectories of individuals bring up singular solutions. The 'arena of singularities' may potentially influence the broader societal discourse about disasters and, consequently, drive future DDR implementations and research.

Fields of practice and their disasters: a disease in Botswana

In the late twentieth century, Botswana began to experience a severe health crisis. Death was omnipresent. Almost everybody knew somebody who had recently died or who was currently in decline and enduring weight loss, coughing, darkening of the skin, failure to control one's bowels and other symptoms. Frequent funerals, with processions making their way from the homes of victims to local cemeteries, made the crisis obvious. As funerals are traditionally organised as family gatherings that last more than a week, work and social life sometimes came to a near standstill. Traditional healers dealt with the crisis in terms of *boswagadi* (Tabalaka 2007) or *meila* (Rakelmann 2005), conditions arising or diseases caused when people 'have had sexual intercourse with a widow or widower before the time of mourning has lapsed' (Mmualefhe 2007, p. 9). However, traditional medicine had no cure for those diseases.

The discovery of HIV in the 1980s laid the foundation for interpreting Botswana's health crisis as the effect of a pathogen. The disease was named AIDS (Acquired Immune Deficiency Syndrome) and thus described as a biomedical condition. Infection rates in southern Africa were found to be extremely high. There were concerns that the regional epidemic might cross borders and become a pandemic that would also affect the Global North. A global alliance was established to combat HIV and AIDS and a massive intervention scheme was implemented as a development initiative. Modern and foreign expertise intruded into local contexts and existing social structures were regarded as being incapable of countering the disease because their health related knowledge was seen as being inadequate. Obviously, 'modern' knowledge was also not capable of providing a cure, yet in the early intervention phase expectations were high that pharmacology would soon develop effective medication. However, initial interventions largely confined their focus to education, research and monitoring.

Intervention knowledge in Botswana has not simply replaced alternative explan-ations, although intervention schemes have become quite effective since the early 2000s with the introduction of new medication regimes. At present, different perceptions exist as to what constitutes this disaster. The omnipresence of people suffering is of course dreadful, and the sheer number of those who are affected calls for naming the current situation a disaster. Beyond this, however, there are widely varying opinions about the 'true' nature of the disease and how deeply it has eroded basic structures of society. Different fields of practice and the internal logic of each may serve to illustrate this: governmental response is based on categories that see seroprevalence, multiple deaths and socioeconomic impact as the major aspects of the HIV and AIDS crisis that need to be addressed. The self-help move-ment, which evolved in the course of the growing crisis, emphasises stigma and discrimination as the most disastrous effects on people affected by the disease. In turn, traditional healers highlight the erosion of society's spiritual basis, which was triggered not directly by modern medication and treatment schemes but mainly by the marginalisation of traditional explanations that has occurred in the wake of modern intervention, as the most devastating consequence of HIV and AIDS.

The governmental response to the HIV/AIDS crisis

Governmental response to HIV and AIDS in Botswana can be deemed successful, at least if measured by the achievement of its own targets or by international standards. For our analysis it is not useful, and indeed not necessary, to contest these achievements that have undoubtedly saved many thousands of lives. We argue, however, that this form of intervention, due to its biomedical success, is very powerful and excludes other perceptions of the crisis. This in turn has hampered a more thorough handling of the crisis and has caused multiple frictions in society.

The legitimation of governmental intervention

On the level of national and international institutions, the HIV/AIDS pandemic in southern Africa is monitored with the help of statistical surveys, mostly based on the collection and analysis of demographic and clinical data. Although reliability of the data used in HIV epidemiology can be questioned – for example, in 2007 UNAIDS reduced its global HIV prevalence prognosis by six million people (UNAIDS 2007) – the outcomes of such surveys are an important base for navi-gating and evaluating intervention practice (Halperin and Post 2004). Measuring and quantifying data enables the ordering and illustration of complex developments. The numbers behind the HIV/AIDS pandemic show a devastating picture and serve to legitimate the government's intervention in Botswana.

Owing to the high rate of infected people, mortality in Botswana has soared since the early 1990s, when the pandemic was detected and its devastating extent first estimated. The average life expectancy decreased from 64 years in 1990 to under 50 years in 2002. For well over a decade now, HIV seroprevalence among adults in Botswana has lingered above 20 percent (Figure 4). At the moment, Botswana's 23 percent seroprevalence is the second-highest in the world (UNAIDS 2013a). These numbers, of course, only give a vague notion of the qualitative dimension of the crisis that is affecting all areas of life. Illness and dying have become an everyday experience. Yet a numerical description of the past and present conditions of this pandemic serves as an important element for broad governmental intervention. Botswana has been struck hard by HIV/AIDS, but it has also had a stable economy and government; thus it was a promising partner for international donors to cooperatively implement a broad HIV/AIDS intervention scheme in 2004, the first of its kind in Africa (Ramiah and Reich 2006). Notable effects of mitigation measures were not able to be observed until the implementation of the governmental HIV/AIDS therapy programme 'Masa', which takes its name from a Setswana word meaning 'a new dawn'. This programme includes a free and lifelong antiretroviral therapy for HIV-infected citizens meeting the World Health Organization's (WHO) clinical criteria. In the wake of the treatment programme, which proved highly efficient in reducing deaths directly related to AIDS, life expectancy began to rise again (Geiselhart and Krüger 2007). Although there is still no cure for an HIV infection, decreased mortality and the recovery of patients under medication helped to mitigate the societal impact of the crisis and also increased quality of life and wellbeing of affected persons and their families (Krüger and Samimi 2010).

The introduction of the national treatment scheme 'Masa' was undoubtedly based on the perception held by the Botswana Government, health organisations, medical personnel and international donors that the HIV/AIDS crisis was a biomedical disaster that needed immediate and massive biomedical intervention. From this perspective, focus was clearly best given to counselling as well as medication in order to reduce viral loads and treat HIV-induced opportunistic diseases (WHO 2006). Statistical figures notably confirmed the treatment scheme's success.

The measurable success of Botswana's antiretroviral therapy (ART) programme legitimises its governmental intervention but suggests two important developments. First, because the number of AIDS related deaths is declining and life expectancy is rising, more people are *living* with HIV in Botswana. In 2011, more than 300,000 people had to cope with HIV in their everyday lives (Figure 4). Second, the great impact of ART on life expectancy and other health indicators seems to justify a decidedly biomedical conceptualisation of the threat (NACA/ACHAP 2008).

The modern constitution of the intervention against HIV/AIDS

A theoretical perspective is useful to understand the rationalities behind the governmental intervention practice. Bruno Latour (1993) identifies the differentiation between the natural and social sciences as characterising a specific understanding of the world that he calls the 'modern constitution'. He explains how a view of nature and, on the other side, a view of culture historically developed to become

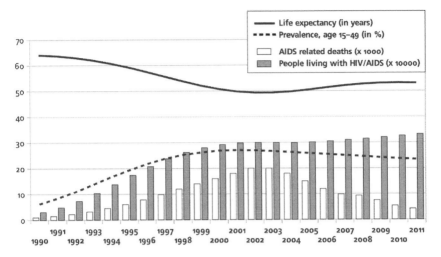

FIGURE 4 Selected HIV/AIDS indicators for Botswana

Source: data aggregated from UNAIDS: www.aidsinfoonline.org, 11 May 2013; UNAIDS 2013b. Note: Other UNAIDS statistics suggest that life expectancy at birth fell below 40 years in the early 2000s.

two separated realms of scientific inquiry. In terms of the modern worldview, both of these realms are constantly concerned with what Latour calls a process of 'purification' in order to clear their object of investigation from artefacts of the opposite scientific approach. This separation, based on the underlying distinction between 'nature' and 'culture', has become extremely effective in producing research findings applicable in, for instance, disaster medicine or other emergency intervention. Modern medicine analyses the body by means of the natural sciences and in terms of separable chemical or mechanical processes. These processes can be isolated in clinical experiments in order to explore the causality of their functioning (Foucault 1975 [1963]). This biomedical conceptualisation separates processes, pathogenesis and transmission from the causes that facilitate a disease's spread within society, a separation that confirms the distinction between a causally determined nature and a cultured society of people with habits, behaviours and attitudes.

Against this theoretical background, we argue that in Botswana, we find a 'modern constitution' regarding the intervention against the HIV and AIDS crisis. It is manifested in a widely executed separation between biomedical and social interventions (Figure 5). On one hand, there is an application of modern medicine and a structural expansion of the modern health system. On the other hand, with respect to social intervention, we find a monitoring of social indicators that are regarded as representing the negative effects of HIV/AIDS. Furthermore, factors that are under suspicion of increasing infection rates are monitored. Social intervention is confined to biomedical logic too because the modern explanation of HIV/AIDS is where all the governmental and institutional consideration starts from. The strict hierarchy in each realm of the national approach to HIV and AIDS

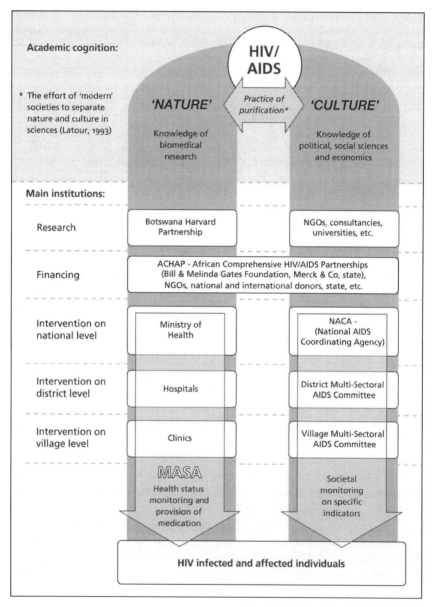

FIGURE 5 The structure of national response against HIV and AIDS in Botswana interpreted against the theoretical backdrop of Latour's concept of modernity

implies a tendency of the needs of people directly affected by HIV and AIDS becoming successively faded out on higher levels of the intervention.

Biomedicine: the Masa programme

Botswana was the first country in Sub-Saharan Africa to implement a national antiretroviral therapy (ART) scheme. Its countrywide implementation was facilitated by a fairly well established pre-existing health infrastructure, with clinics and health posts able to provide primary health care free of charge in every major village. The Ministry of Health's aims have been very clear: its first priorities have been to expand the number of clinics and outreach stations, as well as increase the number of patients receiving ART, according to high ranking programme officers in the Ministry of Health (personal communication, 5 April 2012). Their strand of argumentation stresses the achievements of the ART programme, which are measured quantitatively, as illustrated in Figure 5. On these grounds, Masa is regarded as successful. There is, however, evidence that some aspects which lie beyond effects that can be measured quantitatively are neglected.

Adherence to ART, which requires the regular intake of medication in combination with the adoption of preventive strategies, is internationally seen as one of the most important tools in the management of HIV. Non-adherence, in contrast, is threatening in two ways: intake interruptions may lead to treatment failure associated with, for example, an increase of viral load and the recurrence of symptoms, as well as to an increased chance of transmission. Furthermore, the virus may develop drug-resistant quasi-species; if so, the medication regime of affected patients has to be changed to the next 'line' of drugs, often with more side effects and higher medication costs. There are concerns that, in the long term, HIV could develop resistance against many available lines of antivirals, thus rendering ART schemes useless and boosting the number of new infections due to an increased chance of transmission. Non-adherence thus threatens the entire intervention programme.

In contrast to its internationally acknowledged importance, the issue of adherence is only marginally being addressed within the Masa programme, despite a variety of relevant measures already put in place: first, health care providers are being taught in adherence counselling; second, adherence is monitored through pill counts (counting the pills a patient has remaining after a given time and thus checking whether the prescribed number of pills have been taken); third, patients who do not follow their ARV prescriptions are supposed to be accompanied by health care assistants who counsel them to stay or get back on medication. Adherence monitoring is sometimes combined with blood tests where T cells expressing CD-4 are counted, as the number of these so-called CD-4 cells functions as a proxy that can help determine the condition of a person's immune system and the stage of the disease.

Determining adherence from pill counts might be an adequate method in resource-poor settings (McMahon *et al.* 2011), yet this sort of technical assessment does not sufficiently take into account how people have to cope with medication intake on a daily basis. Due to financial restrictions and a lack of human resources, ongoing support in particular is largely unavailable. Doctors often do not even speak Setswana and, in general, are sometimes not receptive to how their patients include their medication regimes into daily life nor do they know what problems patients are facing in order to adhere to medication regimes. This does not necessarily mean that health care providers do not take an interest in patients' lives. In Botswana, however, healthcare facilities are dramatically understaffed and providers overburdened, a condition directly related to the fact that social aspects of living with the disease, and the prolonged medication associated with it, are not prioritised.

Societal monitoring: addressing socioeconomic facts

Forming a second aspect of the governmental response to HIV/AIDS, findings from a wide variety of research and monitoring studies have found their way into discourse of the HIV and AIDS pandemic. A wide number of economic (IMF 2001, BIDPA 2000), educational (Abt 2002) and behavioural studies (Hope 2003) have documented that HIV/AIDS in Botswana has struck all societal functions and concerns. By these means, HIV/AIDS has been displayed as a complex multifaceted pandemic and this has given legitimation and also prioritisation to decision makers and donors for implementing and widening the Masa programme to a multi-sectoral approach (MoH 1997).

IMF (2001) and BIDPA (2000) are two organisations that have assessed the negative economic effects of the pandemic. According to their findings, without the economic impact of the disease, Botswana's annual GDP increase would have been 3–4 percent higher in the 1990s. Labour productivity in particular was drastically reduced during the period assessed because the disease mostly affected the working-age population. Additionally, government expenditures rose due to higher treatment costs. The pandemic's impact on education was also assessed. Due to higher disease-related absence rates of teachers and students, education standards from prior years could not be maintained (Abt 2002). Such assessments strongly supported the implementation of the ART programme because of its predicted positive micro- and macroeconomic effects (NACA n.d.).

Research in this vein has evaluated a whole set of economic and social consequences of the disease (for instance, economic impact or educational setbacks) and found them to be disastrous. However, all these studies deduce their findings from the biomedical causal chain of infection–illness–death and thus mainly have had an interest in factors that monitor the pandemic's progression (such as sick leave, death toll, etc.) or that are seen as influential with regard to infection rates (HIV-related knowledge, for example). In 2000, NACA (National AIDS Coordinating Agency) was established to pool all information about the progress of the societal effects of the epidemic in Botswana; NACA mainly gathers its information from the grassroots level, but does not implement any support schemes itself. The second strand of the state's intervention is basically a monitoring system with no means of implementing programmes and little intention to explore the realities of life of the people affected by HIV and AIDS (Figure 5). In fact, whereas such everyday realities are taken account of on the grassroots level, they are successively faded out with each step up the organisational hierarchy. Self-help movements can be seen as a response to this institutional gap.

Self-help movement: stigmatisation and participation

Before 2000, when government response was still relatively slack and largely incapable of adequately dealing with the pandemic due to lack of information and resources, a large number of self-help initiatives were active in Botswana. Many of these received some degree of financial support from the government or external sources. Activists disclosed their infection status, travelled to South Africa and the United States, and 'imported' knowledge about how to best support PLWHA (people living with HIV/AIDS). Yet since 2002, the number of those groups has decreased considerably. A set of factors has led to this development, a major one being the government's focus on a new intervention scheme that is designed to take a mix of biomedical and socioeconomic aspects into account but that has found little need for self-help initiatives.

With regard to biomedical rationalities, support groups aim at raising awareness about the threats of multiple concurrent sexual partnerships and reinfection. They inform about a broad set of issues such as safe male circumcision, the relationship between tuberculosis and HIV and adherence requirements, and they also bolster treatment literacy. Self-help initiatives primarily understand themselves as forums where people affected and infected can socialise and share personal experiences in order to improve their wellbeing and self-esteem, receive psychosocial support and, to a certain degree, foster recovery from AIDS related illnesses. For many people affected, support groups are the only discrimination-free places they experience; they feel they do not need to hide their HIV positive status and that they can talk freely about their daily problems. Stigma and discrimination, the shock of being diagnosed as HIV positive, familial and social relationships, problems at work or the challenges of having to organise monthly visits to the clinic are important topics that can be openly discussed. All these topics also relate to treatment adherence, as they touch on barriers for regular medication intake. Not only do stigma and discrimination take place in the broader societal setting but also inside the family or among friends. Stigma is thus the main topic that support groups address publicly. The groups are in favour of an integrative society where PLWHA are regarded as full-value members and where their experience of 'being positive' is seen as an expertise that can help to combat the disease.

The self-help initiatives actively promote 'accepting oneself', a life attitude and coping strategy to acknowledge and incorporate the fact that a person is HIV positive. Learning to 'accept oneself' is seen as a key to 'living positively' with the infection. It is usually a long process of experiences and therefore needs ongoing supportive counselling, which the state's intervention scheme does not provide. 'Accepting oneself' goes along with addressing various psychosocial burdens such as self-doubt, questions of guilt and anxiety about the future, as well as restructuring one's familial relations and circle of friends (Geiselhart 2009, p. 69).

After the introduction of the ART scheme, for many PLWHA, stigmatisation and social exclusion is now the major threat to their wellbeing. From the support groups' point of view, stigma and discrimination form what the disaster of HIV and AIDS is all about. Many support groups had effective programmes in place to mitigate the hardship of everyday life. However, due to lack of funding, most groups have since disbanded. In order to safeguard the national treatment scheme Masa, and also to bundle resources, the Botswana Government is suspicious of all activities where it cannot wield influence or exercise direct control. The Government fears counterproductive effects that might erode the ART programme's achievements. There is, therefore, very little governmental support for initiatives whose targets reach beyond Masa's core concept, cater for an ongoing psychosocial support for PLWHA or attempt to reduce stigmatisation and discrimination.

The marginalisation of dingaka

Traditional healing is another field of practice worth exploring. In this section, we refer to traditional medicine as those healing techniques that point to indigenous spiritual conceptions, as well as to tribal 'culture' in a broader sense. To honour the specific quality of this realm and emphasise that the term 'traditional' is a distinctive category, set aside from modern perceptions, we will use the Setswana word *ngaka* (doctor, healer; plural: *dingaka*). The field of dingaka is highly individualised. Most healers have received little formal education and are often illiterate (Staugard 1989, Andrae-Marobela *et al.* 2010, p. 26). They often undergo a yearlong 'apprenticeship' and are instructed by experienced healers, with their techniques varying according to different schools of training. Dingaka claim to gain their knowledge about herbs, treatment and other subjects directly from the spiritual world or from dreams, and their training is to a large extent based on techniques supporting their ability to receive that intuition. A few healers even claim to have been taught exclusively by natural spirits. In general, dingaka use oracles for diagnosis. The technique most often practised is to 'cast the bones'. Four pieces of bone, wood or ivory are thrown on the floor and interpreted with regard to the way they come to rest (van Binsbergen 2003). Dingaka understand illness not only as a bodily problem but also caused by impaired social relations, distraction from the spiritual world or natural forces. They discuss social relations as well as current societal developments with their clients. New information is therefore permanently integrated into their practices. The field of healing is thus, in principle, very flexible and 'under constant influence from religious and cosmopolitan medical sources' (Ingstad 1990, p. 28).

As mentioned above, dingaka have described certain conditions and health related symptoms as *boswagadi* or *meila*. When physicians were introducing the new diagnosis of HIV and AIDS to exactly the same symptoms and accused healers as being ignorant, they disregarded the notion that the dingakas' perceptions were not irrational at all: a ngaka 'is not seeing a new thing at all. He sees that old problem of boswagadi' (Tabalaka 2007, p. 69). One can safely assume that dingaka had been concerned with the noticeable impairments in patients long before the modern intervention and, even if they had no cure for it, their explanation of the disease in terms of local perceptions made sense of the phenomenon. In order to avoid practices that might have negative effects on patients, it is thus important to look at the way dingaka perceive modern knowledge about HIV and AIDS and whether they integrate it into their techniques.

Regarding HIV and AIDS, healers are often accused by governmental and other public institutions of contributing to the spread of the virus rather than helping to tackle the pandemic. In the course of the modern intervention, indigenous knowledge has often been rendered inadequate, untrue or mythical:

> There was a need for cultural change, given the fact that AIDS threatens to wipe out the population of Botswana [. . .]. We are in a war situation where we have to make many tough decisions and forget about some of our cultural beliefs in order to save millions of lives.
>
> (Tabalaka 2007, p. 63)

With the onset of the pandemic, modern biomedical knowledge was promoted as being superior and traditions such as *seantlo*, a custom whereby a widow is married to a brother of her former husband in order to secure her livelihood, were evaluated as supporting the spread of HIV.

As outlined above, the modern health system in Botswana became quite effective in combating AIDS. Alternative treatment practices and medicines were seen by many as outdated, and medical professionals generally assumed that people would gradually refrain from 'traditional' health practices. However, modern medicine repeatedly displayed its negligence in paying attention to patients' own experiences and perceptions of illness; the fact that physicians sometimes tend to ignore that their medical explanations are not comprehensible to all of their patients became increasingly evident. In addition, the mere prescription of medication gives little assistance for coping with diseases mentally, nor does it give explanations with regard to spiritual worldviews. It is here where dingaka often perform much better, and in Botswana, 'traditional' health care has therefore remained an important health system in its own right (Staugard 1989, p. 18, Andrae-Marobela *et al.* 2010).

Health anthropology has shown that even symbolic actions such as healing rituals can have effects on physiology (Kirmayer 2004). The WHO (2002, p. 11) acknowledges the growing influence of alternative medicine worldwide, not only in peripheral regions where modern health care is not available. Practitioners of such medicine adopt healing techniques from all over the world. Stollberg (2002) attests a globalisation and localisation of alternative medicine. The WHO therefore suggests an integration of alternative healing services into public health systems. Botswana, however, is one of the few countries in southern Africa that, according to our own research, still has not established a legal cooperation to this effect (cf. WHO 2002, p. 17). In the beginning of the 1990s, some attempts to install a collaboration between public health actors and dingaka were begun, but local healers still feel patronised by governmental bodies (WHO 1990, Chipfakacha 1994). In a field study we carried out in Botswana, John P. Setilo, president of the Baitseanape ba Setso mo Botswana ('experts in culture', an umbrella organisation for traditional healing practitioners) and president of the Botswana Dingaka Association, stated:

> We are working closely with the Ministry of Health. That I would admit. But we are working closely with them, not as equal partners really, because at the end of the day all they do is to run workshops for us, for purposes of wanting us to help solve their problems. They concentrate maybe on HIV/AIDS. Nowadays there is diarrhoea. So they will workshop us on diarrhoea. And that's it.
>
> (Interview, 18 August 2012)

Traditional healing practices have largely been disconnected from everyday modern lifestyles, but are not at all confined to oblivion. We came across cases where people would openly discredit dingaka but consult them secretly when modern therapy would not provide a desired effect. Some also sought treatment associated with witchcraft. Setilo confirmed this was true even in the case of leading politicians:

> Our leaders will pretend that they don't believe in traditional medicine or whatever [. . .]. There is a lot of hypocrisy. Look at it this way, elections are coming in 2014 and a good number of us will make a lot of money. Because the very same people who will be saying traditional medicine is rubbish, will come running to us because they want *muti* [traditional medication] [in order] to be re-elected [. . .] There's a lot of hypocrisy.
>
> (Interview, 18 August 2012 [annotation by authors])

Healers have to choose whether they will be content with the small fees they receive for healing consultations or whether they will practice witchcraft, which is well paid but ostracised or even illegal.

In the course of AIDS related interventions based on biomedical notions, both of disease and of what constitutes a health disaster, little has been done to make rationales compatible to existing beliefs and actions. In fact, traditional know-ledge has often been contested and opposed. It is not possible to simply extinguish certain customs as they are integrated into a complex system of understand-ings combined with practices made conventional. While AIDS related intervention schemes were unaware of or ignored existing rationales, the biomedical conceptual-isation of the disease was ignored by some traditional healers because it appeared as a foreign entity. There was no serious attempt for mutual understanding and learning. Indigenous knowledge systems 'are dynamic systems embedded in chang-ing social, economic and environmental relations' (Andrae-Marobela *et al.* 2010, p. 4), though one could add that this is true only as long as their knowledge is not rendered inadequate, but appreciated. The character of the HIV and AIDS disaster from the perspective of dingaka lies in the fact that, aside from the immense loss of lives, they themselves have become marginalised. In the course of the implementation of modern medicine, healers have widely lost their status as authorities.

Ruptures and unexpected alliances in the arena of singularities

Above, we have analysed different fields of practice and related perceptions regarding a current health crisis in Botswana. Those perceptions can be viewed as somewhat stable and continuous because they are, in one way or the other, institutionalised and organised. Next, we will look at single events and their trajectories that make these fields of practice fluid.

The arena of singularities

According to a pragmatist view, praxis is a continuum that is permanently in flux, flow and transformation. Praxis is an endlessly diverse sphere of social interaction that roots in the past and heads forward to the future by permanently passing an infinite number of singular moments. The distinction between universality and singularity has long been a philosophical concern. One of the key thinkers of the philosophical tradition of pragmatism, Charles S. Pierce, suggests that only perceptions of moments are singular (Berwing 2012). In our everyday practice, we automatically distinguish what is universal from what is singular by what Mead (1934, p. 186) calls 'importation of the social process'. We permanently ask ourselves what others would think and thus try to identify what is commonly accepted or intersubjectively valid in our community. In doing so, we operate with symbols of signification and try to grasp societal discourse. This process is an elementary component of socialisation and the development of self-identity for each and every individual. It propels the emergence and repetition of conven-tionalised practices and, on the societal level, the effort to make society reliable through organisations and informal institutions.

Only if we define what 'normal' is can we grasp what is 'exceptional'. Once we have identified common practices, we can also determine if something is unusual or extraordinary due to the present situational constellation. By recognising what is singular, we can gain individualised experiences and thus can integrate new meaning into our personal systems of signification and, furthermore, argue for these meanings to be recognised in societal discourse (Berwing 2012).

Singularities appear when common practices fail, produce unwanted effects or need to be altered. Individuals who are involved in such situations often experience emotional effects, but this might never be recognised on the societal level. Assessing the arena of singularities is thus regarding actual life in order to assess what is not considered within conventionalised perceptions. Such assessment is about identifying ruptures between practices and lived-in worlds. Furthermore, we have to examine whether alliances have been formed with the potential of influencing common practices. Taking singularities into account is about gaining an integrative understanding that bridges the gaps between the differing worldviews of distinct fields of practice.

Ruptures and alliances: Botho, dingaka and HIV-effective traditional medicine

Regarding HIV/AIDS, in order to understand how individuals navigate through different fields of practice, it is necessary to look closely at the daily lives and routines of PLWHA. Ruptures in these routines emerge where people encounter different sets of knowledge, values and beliefs with diverging explanations. One example is stigmatisation, a major factor that hampers adherence to antiviral drug regimens. Different solutions for HIV/AIDS related problems may compete against each other. It is apparent that the parallel existence of the modern and the traditional health systems (Staugard 1989, p. 18), whereby the consultation of the latter is mostly kept a secret, offers contradictory explanations. Individuals often feel torn between these different belief systems and negative side effects from incompatible treatment schemes may occur. Research from other regional contexts also shows that there are negative effects on health when different health systems do not communicate with each other (Barcan 2010, p. 141). Patients do not necessarily get advice as to where they can probably get helpful alternative treatment.

In the arena of singularities, we sometimes find activities and perspectives that have the potential to bridge the rift between institutionalised fields of practice. To illustrate this, we highlight three examples of different 'alliances' of ideas, concepts and actors: a) the moral codex of 'Botho', b) the Botswana Dingaka Association (BDA), and c) the identification of herbs with potential value for HIV/AIDS treatment.

In the Botswana traditional moral codex of Botho, individuals can only be themselves because of others. This fundamental notion of the relationship between individuals and their societal environment has far-reaching implications for consultancy and treatment approaches. According to most modern assertions, illness is a matter of the body, and information about it should thus be given confidentially only to the affected individual. However, according to Botho, illness is a concern of the whole community when a person is sick. 'When one is affected, we are all affected. Even more interestingly, when a traditional doctor is called, the divination is never individualized' (Mmualefhe 2007, p. 9). The conditions to stay healthy need to be provided for all individuals by means of a strong and supportive community. Botho emphasises that despair reduces resilience, thus distrust, envy and hostility make people sick. Therefore, it seems impossible for someone to feel happy when another is in sorrow. In this assertion, the signifying character of the HIV/AIDS disaster is stigmatisation and discrimination of people who are HIV positive. Botho can bring 'healing where there is no cure' (Mmualefhe 2007, p. 13). Mmualefhe offers a theological perspective that holds the potential of finding a common basis for the appreciation of support groups and dingaka in the context of HIV and AIDS.

With its specific claims, its individual chairpersons and its historical development, the BDA can be seen as an alliance situated in the arena of singularities. This alliance of interests is capable of something beyond the reach of individual persons: making oneself heard. The BDA argues for an integration of traditional health services into Botswana's health legislation. This would give the BDA authority to coordinate and set standards for ngaka practices. Currently, as traditional healing is highly individualised, even some practices that would seem to foster HIV transmission have survived. On 4 November 2011, the journal the *Voice* titled an article 'Muti Ritual Left My Baby HIV+', drawing on a case where a child was allegedly infected in the course of a spiritual healing procedure. Setilo (chairperson of the BDA) is aware of the fact that such outdated techniques are still practised and surmises: 'If there was legislation this wouldn't be happening. You see? These things would not be happening' (Interview, 18 August 2012).

As we write, the BDA is involved in negotiations with representatives of the government. The progress of its requests likely depends on the negotiation skills of its chairpersons, who are more formally educated than the majority of dingaka. On the one hand, this causes problems of legitimation with respect to other healers who do not understand the chairpersons' political strategies. It has already led to the secession of some alternative healer associations from the BDA. On the other hand, the fact that some dingaka who are capable of leading political negotiations have taken the initiative to participate in governmental processes is a historical chance for traditional healing to become more appreciated in contemporary Botswana.

Research done on plants used in traditional medicine has confirmed that many of them are indeed effective (Staugard 1989). In the course of recent surveying (Andrae-Marobela et al. 2010) researchers came across rumours that some healers were able to heal HIV and AIDS. A local woman claimed that in her dreams her ancestors had told her a recipe for a medicine capable of healing AIDS (Keoreng 2013). Taking this claim seriously and investigating it further, scientists found that some herbs used in traditional medicine do significantly inhibit HIV-1 replication (Leteane et al. 2012). This exploration and discovery were made possible from an encounter between people from completely different societal realms. Andrae-Marobela, a highly trained biochemist, met with a poor uneducated woman from a local community on a mutual basis of understanding and the two took each other seriously. These findings are surprising in that Leteane *et al.* (2012) speak of them as carrying the potential of reversing the 'laboratory-to-clinics' process to 'clinics-to-laboratory'. 'We report here an example where traditional medical knowledge led to the identification of an extract that might potentially be useful in the treatment/management of HIV/AIDS' (Leteane *et al.* 2012, p. 55).

It is interesting to note that these overlapping fields of practice, or the situations created through everyday action, may lead to new solutions to tackle threats and crises. However, this is sometimes met with scepticism. Political reaction to the use of traditional plants is one example of this. Botswana Vice President Kedikilwe argues:

> We are aware and are concerned about reports that some traditional herbs have been shown to cure AIDS or even reduce chances of HIV infection. As a country, it would be unfortunate if we rushed to such options, as they have potential to take us where we were some 28 years back.
>
> (Kedikilwe 2013)

From the perspective of the national HIV/AIDS intervention, everything that diverges from the current practice of biomedical intervention might lead to lower acceptance of the ARV treatment scheme and must thus appear as threatening the success of the national public health response.

Conclusion: identifying fields of practice and acknowledging the arena of singularities

Our work has departed from the critique of modern intervention logic and their tendency towards cultural existentialism. Top-down interventions are often criticised for not adequately recognising experiences and rationalities that take place at the grassroots level, which often results in misdirected implementations.

Instead, we propose an assessment of fields of practice as structural elements that order societies. These might be global intervention strategies, national policies, economic or technological networks, religious or moral regimes, social undertakings or indigenous techniques and procedures. These strategies are planned, and activities all carried out, by organisations, institutions or societal groups. Their logic of justification might generally be convincing, and their legitimisation seemingly easy, because negative effects mostly occur only in the arena of singularities, which is so easily overlooked. It is here where the rationales and policies of such strategies are confronted with the lived-in worlds of individuals. But it is also here where such confrontations and contestations are often hidden from view, neglected or deemed irrelevant precisely because they are dismissed as singular, non-representative occurrences.

No matter whether we analyse these fields of practice as practice-arrangement bundles (Schatzki 2003), fields of habitus (Bourdieu 1977) or fields of experience based on knowledge and practices (Dewey 1929, Berk and Galvan 2009), we find them to be major determiners of the perceptions of those who work within these fields. We analyse conventionalised and institutionalised practices as structuring social life, making it reliable and stable.

Furthermore, when focusing on human agency, we have to recognise that human action is indeterminate. Assemblage approaches (Anderson and McFarlane 2011), in their reference to Deleuze and Guattari (1987) in particular, emphasise the rhizome-like nature of the social, which results in infinite multiplicity and diversity within social life. This leads us into the arena of singularities: we need an eye for the unique and manifold situations that occur while conventionalised practices are being performed.

The arena of singularities accommodates creativity, especially when difficulties, ruptures or uncertainties emerge and certain practices in specific situations are rendered inadequate. Under such circumstances, individuals cope with situations via singular experiences: if such situations demarcate ruptures in societal dealings with threats, challenges or crises, the experiences of involved individuals are singular. Even if these experiences are had on a broad quantitative basis, they primarily will be distinct and unconnected. Individual people will initially regard these experiences as something personal due to the very concreteness of their immediate life-situation. The absence of a common explanation of the main characteristics of such experiences averts individuals from seeing themselves as a part of a collective process. But alliances might be deployed to give voice to these experiences with the intention of bringing about a change in established orders. For many PLWHA, for instance, experiencing rejection when being discriminated against and blaming themselves for having contracted the virus had been private problems until the self-help movement addressed discrimination and stigmatisation as issues that needed to be dealt with on a societal level.

We suggest that the extraordinary conditions of disastrous situations particularly need the acknowledgment of singularities because the conditions that are experienced as disastrous demand creativity from the individuals involved, which in turn brings up a multiplicity of singular ideas. Our approach may better reveal those practices that develop from below and may allow for an assessment of people's potential to generate solutions. It is not necessary to agree with local perceptions or to understand local knowledge. But if it is observable that people have developed practices that foster their own coping with those aspects of a crisis that they themselves evaluate as being disastrous, then we should accept and recognise these practices.

The potential that lies in the arena of singularities is still largely hidden because individual solutions are often deemed irrelevant to a broader view of disaster mitigation. This in turn is simply because different worldviews are discounted, neglected or ignored, the interconnectedness of singularities underestimated, and the forming of alliances (if detected) either not taken seriously or deemed a threat to hegemonial practices and interests. As we hope to show by our example of the HIV/AIDS pandemic in Botswana, events in the arena of singularities point the way forward to a reinterpretation and reorganisation of social issues concerning disaster and risk reduction and management. Future DRR policies will determine whether these ideas will gain influence, but empirical research might certainly help to reveal and evaluate what is happening in the arena of singularities and advocate the integration of these findings into DRR.

References

Abt Associates South Africa Inc., 2002. The impact of HIV/AIDS on education in Botswana. *Abt Associates South Africa Inc* [online]. Available from: www.undp.org.bw/docs/bots_education_final.pdf [accessed 11 November 2013].

Anderson, B. and McFarlane, C., 2011. Assemblage and geography. *Area.* 43 (2), 124–7.

Andrae-Marobela, K., Ngwenya, B. N., Monyatsi, K. N., Okatch, A., Masizana, A. and Muzila, M., 2010. *Documentation and Promotion of Indigenous Knowledge-Based Solutions for Botswana – An Ethnosurvey.* Gaborone: Centre for Scientific Research, Indigenous Knowledge and Innovation (CesRIKI).

Barcan, R., 2010. Spiritual boundary work: How spiritual healers and medical clairvoyants negotiate the sacred. In: E. B. Coleman and K. White, eds, *Medicine, Religion, and the Body.* Leiden, Netherlands: Brill, 129–46.

Barry, A., 2006. Technological Zones. *European Journal of Social Theory,* 9 (2), 239–53.

Bauman, Z., 1996. From pilgrim to tourist – or a short history of identity. In: S. Hall and P. Gay, eds, *Questions of Cultural Identity.* London: Sage, 18–36.

Berk, G. and Galvan, D., 2009. How people experience and change institutions: A field guide to creative syncretism. *Theory and Society,* 38 (6), 543–80.

Berwing, S., 2012. Jenseits des Sprachkäfigs. Potenziale der Peirce'schen Semiotik für eine Foucault'sche Kulturgeographie. *Berichte zur deutschen Landeskunde,* 86 (1), 67–81.

Bhabha, H. K., 1996. Culture's in-between. In: S. Hall and P. Gay, eds, *Questions of Cultural Identity.* London: Sage, 53–60.

BIDPA (Botswana Institute for Development Policy Analysis), 2000. *Macroeconomic Impacts of the HIV/AIDS Epidemic in Botswana.* Gaborone, Botswana: BIDPA.

Binsbergen van, W., 2003. The translation of Southern African Sangoma divination towards a global format, and the validity of the knowledge it produces. Paper read at the symposium *World views, Science and Us,* Brussels, Centre Leo Apostel, Free University Brussels, Belgium. Available from: www.shikanda.net/general/paper_brussels_BIS.pdf [accessed 10 June 2003].

Bourdieu, P., 1977. *Outline of a Theory of Practice.* Cambridge: Cambridge University Press.

Chipfakacha, V., 1994. The role of culture in primary health care. *South African Medical Journal,* 84 (12), 860–2.

Deleuze, G. and Guattari, F., 1987. *A Thousand Plateaus. Capitalism and Schizophrenia.* Minneapolis, MN: University of Minnesota Press.

Dewey, J., 1929. *The Quest for Certainty. A Study of the Relation of Knowledge and Action.* London: Georg Allen & Unwin.

Dewey, J., 2005. *Art as Experience.* London: Penguin Publishing Group.

FAO/WHO, 2002. Living well with HIV/AIDS. A manual on nutritional care and support for people living with HIV/AIDS. *FAO/WHO* [online]. Available from: ftp://ftp.fao.org/docrep/fao/005/y4168E/y4168E00.pdf [accessed 4 February 2013].

Foucault, M., 1975. *The Birth of the Clinic. An Archaeology of Medical Perception.* New York: Vintage Books.

Geiselhart, K., 2009. *Stigma and Discrimination: Social Encounters, Identity and Space. A Concept Derived from HIV and AIDS Related Research in the High Prevalence Country Botswana.* Thesis (PhD). Available from: http://nbn-resolving.de/urn:nbn:de:0168-ssoar-290930 [accessed 10 April 2014].

Geiselhart, K. and Krüger, F., 2007. Die HIV/AIDS-Krise—Botswanas strategische Antwort als Vorbild? *Geographische Rundschau*, 59 (2), 54–61.

Giddens, A., 1984. *The Constitution of Society.* Cambridge: Polity Press.

Hall, S., 1992. The question of cultural identity. In: S. Hall, D. Held and T. McGrew, eds, *Modernity and its Futures.* Oxford: The Open University, 273–327.

Halperin, D. and Post, G., 2004. Global HIV prevalence: The good news might even be better. *The Lancet*, 364 (9439), 1035–6.

Hedberg, I. and Staugard, F., 1989. *Traditional Medicine in Botswana: Traditional Medicinal Plants.* Gaborone, Botswana: Ipeleng Publishers.

Hope, R., 2003. Promoting behavior change in Botswana: An assessment of the peer education HIV/AIDS prevention program at the workplace. *Journal of Health Communication: International Perspectives*, 8 (3), 267–81.

IMF, 2001. *The Macroeconomic Impact of HIV/AIDS in Botswana.* Washington: International Monetary Fund, Working Paper WP/01/08.

Ingstad, B., 1990. The cultural construction of AIDS and its consequences for prevention in Botswana. *Medical Anthropology Quarterly*, 4 (1), 28–40.

Jackson, P., 2000. Rematerializing social and cultural geography. *Social and Cultural Geography*, 1 (1), 9–14.

Kedikilwe P., 2013. Opening Remarks by his Honour the Vice President Dr. Ponatshego Kedilkwe at the National Aids Council. *National AIDS Council* [online]. Available from: www.gov.bw/Global/NACA%20Ministry/wana/OPENING%20REMARKS%20BY %20VP%20NAC%20%20-%2025%20April%202013.pdf [accessed 17 October 2013].

Keoreng, E., 2013. Palapye woman claims to have HIV/AIDS herb. *The Patriot*, 8 April 2013 [online]. Available from: http://thepatriot.co.bw/palapye-woman-claims-to-have-hivaids-herb/ [accessed 23 September 2013].

Kirmayer, L. J., 2004. The cultural diversity of healing: Meaning, metaphor and mechanism. *British Medial Bulletin*, 69, 33–48.

Krüger, F. and Samimi, C., 2010. Entwicklung und Umwelt im Südlichen Afrika – Chancen und Herausforderungen des Gesellschaftlichen und Ökologischen Wandels. *Geographische Rundschau.* 62 (6), 4–10.

Latour, B., 1993. *We Have Never Been Modern.* Cambridge: Harvard University Press.

Leteane, M. M., Ngwenya, B. N., Muzila, M., Namushe, A., Mwinga, J., Musonda, R., Moyo, S., Mengestu, Y. B., Abegaz, B. M. and Andrae-Marobela, K., 2012. Old plants newly discovered: Cassia sieberiana DC and Cassia abbreviata Oliv. Oliv. root extracts inhibit in vitro HIV-1c replication in peripheral blood mononuclear cells (PBMCs) by different modes of action. *Journal of Ethnopharmacology*, 141 (1), 48–56.

McMahon, J., Jordan, M., Kelley, K., Bertagnolio, S., Hong, S., Wanke, C., Lewin, S., Elliott, J., 2011. Pharmacy adherence measures to assess adherence to antiretroviral therapy: Review of the literature and implications for treatment monitoring, *Clinical Infectious Diseases*, 52 (4), 493–506.

Mead, G. H., 1934. *Mind, Self, & Society. From the Standpoint of a Social Behaviorist.* Chicago, IL: The University of Chicago Press.

Mmualefhe, D. O., 2007. Botho and HIV&AIDS: A theological reflection. In: J. B. R. Gaie and S. K. Mmolai, eds, *The Concept of Botho and HIV&AIDS in Botswana.* Eldoret, Kenya: Zapf Chancery Publishers Africa: 1–27.

MoH, 1997. *Botswana HIV and AIDS Second Medium Perm Plan MTP II 1997–2002.* Gaborone, Botswana: Ministry of Health.

NACA, n.d. *Botswana National Strategic Framework for HIV/AIDS 2003–2009.* Gaborone: NACA.

NACA/ACHAP, 2008. *HIV/AIDS in Botswana: Estimated Trends and Implications based on Surveillance and Modelling.* Gaborone: NACA/ACHAP.

Rakelmann, G. A., 2005. Process of integration of AIDS into daily life in Botswana: From a foreign to a local disease. *Curare,* 28 (2/3), 153–68.

Ramiah, I. and Reich, M., 2006. Building effective public-partnerships: Experience and lessons from the African Comprehensive HIV/AIDS Partnerships (ACHAP). *Social Science & Medicine,* 63 (2), 397–408.

Rutherford, J. and Bhabha, H., 1990. The third space – Interview with Homi Bhabha. In: J. Rutherford, ed., *Identity: Community, Culture, Difference.* London: Lawrence and Wishart, 207–21.

Schatzki, T. R., 2003. A new societist social ontology. *Philosophy of the Social Sciences,* 33 (2), 174–202.

Soja, E., 1999. Thirdspace: Expanding the scope of the geographical imagination. In: D. Massey, J. Allen and P. Sarre, eds, *Human Geography Today.* Malden: Blackwell, 295–322.

Staugard, F., 1989. *Traditional Medicine in a Transitional Society. Botswana Moving Towards the Year 2000.* Broadhurst, Botswana: Ipelegeng Publishers.

Stollberg, G., 2002. Heterodoxe Medizin, Weltgesellschaft und Glokalisierung: Asiatische Medizinformen in Westeuropa. In: G. Brünner and E. Gülich, eds, *Krankheit verstehen.* Bielefeld: Aisthesis, 143–58.

Tabalaka, A. B., 2007. The significance of cultural and religious understanding in the fight against HIV&AIDS in Botswana. In: Gaie, J. B. R. and Mmolai, S. K., eds, *The Concept of Botho and HIV&AIDS in Botswana.* Eldoret, Kenya: Zapf Chancery Publishers Africa: 61–70.

UNAIDS, 2007. *AIDS Epidemic Update.* Geneva: UNAIDS.

UNAIDS, 2013a. *Global Report – UNAIDS Report on the Global AIDS Epidemic 2013.* Geneva: UNAIDS.

UNAIDS, 2013b. *Epidemiological Status. World Overview. UNAIDS* [online]. Available from: www.unaids.org/en/dataanalysis/datatools/aidsinfo/ [accessed 14 May 2013].

WHO, 1990. *Report of the Consultation on AIDS and Traditional Medicine: Prospects for Involving Traditional Health Practitioners.* Francistown, Botswana: Traditional Medicine Programme & Global Programme on AIDS.

WHO, 2002. WHO Traditional Medicine Strategy 2002–2005. *World Health Organization* [online]. Available from: http://whqlibdoc.who.int/hq/2002/who_edm_trm_2002.1.pdf [accessed 22 April 2014].

WHO, 2006. *From Access to Adherence: The Challenges of Antiretroviral Treatment.* Geneva: WHO.

8

RELIGION AND BELIEF SYSTEMS

Drivers of vulnerability, entry points for resilience building?

E. Lisa F. Schipper

Introduction: why do we need to prod at belief systems?

> *Without taking into account [. . .] socio-cultural beliefs, it is difficult to understand the manner in which the poor perceive and respond to natural hazards and disasters.*
>
> Hutton and Haque 2003, p. 417

The mainstream discourse around both disasters and climate change acknowledges the importance of social vulnerability in determining risk, and encourages the use of social vulnerability assessment as a crucial component of understanding risk. Yet religion and traditional beliefs are an important sub-set of influential social drivers that continue to be overlooked in these methodological and conceptual approaches, because they are either too poorly understood by scientists who deal with vulnerability to hazards, or because they are sensitive topics. This chapter argues for the importance of considering religion, sociocultural traditions and belief systems in assessing vulnerability to natural hazards and identifies some of the issues that need further exploration.

Vulnerability is a crucial factor influencing how natural hazards affect the world. The academic debate about what it means may have calmed a bit, but there is a clear disparity in views about how it can be characterised (Füssel 2009, Hinkel 2011). To distinguish between social and natural systems, vulnerability is often divided into different 'types', such as social, economic, physical or biological. This is helpful for those who seek to quantify the information, but ultimately breaking down vulnerability in this way reduces its complexity. It is precisely because of this complexity that vulnerability is difficult to understand and address, but it is also in this complexity that the richness of vulnerability is found. One of the first dimensions of vulnerability to be excluded from vulnerability assessments is belief systems – a broad category including traditional beliefs and religion – because they are considered too sensitive or taboo to characterise and address from an external perspective. Yet, as this chapter argues, to omit belief systems from a study of social vulnerability may be to ignore one of the most important drivers of vulnerability to natural hazards.

Belief systems and how they affect vulnerability is probably one of the least well understood dimensions of risk. The term 'belief systems' refers to any set of ideas that stem from spirituality, mysticism or faith in divinity. They are often referred to as religion, but also include beliefs that are expressed through superstitions, mythology and folktales. Typically, this refers to shared principles – i.e., not just one individual's personal beliefs. Often, they are centred on some form of the divine, but also include philosophies, ideologies and world views that have sociocultural significance. The important role of religion in influencing perceptions of hazards, as well as attitudes about reducing risk, has been explored elsewhere (Schipper 2010). This chapter aims to take that discussion one step further towards identifying some key issues that need exploration to better understand how risk reduction can become part of existing belief systems.

Despite anecdotes about how people's beliefs inspire them to behave in thought-provoking ways before, during and after a disaster, they are often left as that – just anecdotes (Bhalla 2014). Nevertheless, there is growing consensus among academics and practitioners that sociocultural and behavioural dimensions need to be integrated into actions to address vulnerability to climate change and natural hazards to maximise the potential for long term success. Doing this requires a thorough understanding of how people reason and what leads them to behave in certain ways, which is not typically part of projects or programmes designed to address climate change or disaster impacts. This requires drawing on fields such as anthropology, psychology, sociology and behavioural economics to develop a transdisciplinary understanding of how culture and behaviour influence social vulnerability.

Understanding belief systems is now more crucial than ever. The need for transformation rather than simply adaptation to climate change has become a popular way of describing the extensive societal changes needed to live in a warmer world (Pelling 2010). A large portion of the changes that are necessary have to do with attitudes and behaviour. This is true not only for actions to reduce greenhouse gas emissions – as is well recognised – but also to adapt to the consequences of climate change (Inderberg *et al.* 2015). Despite this, there is little discussion in the climate change adaptation policy and practice arena of how to transform societies' perceptions of risk in a way that will facilitate adjustment to new climate conditions. Partly this results from a lack of awareness among the decision making actors, and partly it is driven by the framework of short-term projects and funding within which they function. However, it also reflects unwillingness to influence traditional beliefs, social norms and other sociocultural aspects that determine perceptions.

Culture is the key

Climate change and disaster risk reduction experts have recognised that adaptation and vulnerability reduction are often context-specific. This is not only because the dynamics of change and risk are affected by local ecosystem dynamics and geomorphology, but also because local norms, customs and belief systems shape people's worldviews, particularly in places where external influence is limited. According to Hewitt (2008, ix), 'what the people living at risk know and do about natural hazards and disaster risks is mediated by a range of factors including social conditions (such as age, gender, wealth, ethnicity) and cultural settings (language, beliefs, traditions, customs)'. Responses to hazards are in part determined by an understanding of the reason for the hazards and how to respond to them; for some this means more prayers, for others it means having rain gutters and seasonally clearing debris. The belief systems that guide how individuals and societies live play a large part in determining their perspectives on risk, influencing whether they decide to make changes that will minimise current and future risk. Culture, as expressed in, for instance, belief systems, influences people's risk because it influences their exposure and perceptions of what a hazard represents. Furthermore, it shapes the norms by which we define the acceptability of risk. What to some may be too risky may to others be part of their daily challenges. Thus, culture also influences people's risk indirectly by influencing their livelihood choices, settlement locations, social networks and time availability for preparedness activities, including education. Culture may be one of the most important characteristics for certain societies, where culture alone may decide whether people will be adversely affected by climate change or disaster risk.

The topic of culture and responses to environmental change is not new. A rich set of literature can be found on issues that relate to spirituality and nature, and how contemporary value systems tend to deemphasise the human–nature connection (Hulme 2009). Additional literature examines perceptions specifically, as well as our relation to risk (cf. Douglas and Wildavsky 1980, Beck 1992, Gaillard 2007). Yet another set of scholars have examined values in society (cf. Rokeach 1973, O'Brien 2009). More specifically related to religion, work has been done on the role of faith in the recovery process following a disaster (cf. Davis and Wall 1992, Massey and Sutton 2007, Merli 2005, Schlehe 2010), religious explanations of nature (cf. Peterson 2001, Orr 2003), and the role of religion in influencing positions on environment and climate change policy (cf. Kintisch 2006, Hulme 2009). Despite this work, the conclusions mostly do not offer guidance to policy and decision makers about how to take sociocultural systems into account when assessing vulnerability and designing policy, projects and programmes on disaster risk reduction and adaptation to climate change.

Civilisations have survived disasters, but they have also fallen in the face of adverse environmental conditions, not able to cope with the challenges presented to them (Adger and Brooks 2003). Those who have managed to get by have learned and taught lessons about how to minimise exposure and sensitivity to such risks. However, in many places people continue to be adversely impacted by natural hazards on a recurring basis, showing few signs of any learning process. So if people have always dealt with risk, but do not appear to have adequate capacity to adapt, what can we infer about the process of acquiring and adopting such knowledge into existing sociocultural fabric? Since culture plays a significant role in shaping how knowledge and understanding of risk is applied and interpreted, how can it allow for flexibility in behaviour and perceptions?

Importantly, culture is not a static characteristic – it too evolves over time. Attitudes that appear seemingly impenetrable at one point have been shown to shift as new needs arise. In Assam in northeast India, caste-based roles in society have been abandoned to allow for groups who were previously banned from certain livelihood activities to undertake them as a response to the loss of land for agriculture caused by recurring floods (ICIMOD 2009). In the same area, people from a higher caste are starting to build houses on stilts to protect themselves, even though this sort of construction is the tradition of a lower caste and has previously been considered taboo for the higher group (ICIMOD 2009). Evidence of such cultural shifts suggests that it may be possible to identify the triggers for these changes in attitude that shift cultural norms and recreate the circumstances that enable these triggers to have an effect.

Schipper (2010) explores examples of how religious belief influences perceptions and attitudes as well as behaviour and response to hazards. The different ways in which these play out are summarised in Tables 1 and 2. These provide an overview of the types of situations that can be encountered on the ground (Table 1), and the way in which people respond to them based on different perspectives (Table 2). The examples indicate that there are reasons that people respond the way that they do – the beliefs provide a logic that guides their actions and thoughts. In other words, their actions are justified by the way in which their beliefs help them understand the world. But the tables also show that the same perspective has many different types of responses. For example, even when people are in agreement that a hazard is 'caused' by God, this does not result in the same reaction. For those with a fatalistic attitude, it means that nothing can be done, or even that the 'suffering' that is brought is an important part of life on Earth. For others, it is a call to demonstrate creativity, strength and general ability to take action.

Although religion and religious leaders are often the source of these attitudes, the way in which religious texts or traditions are interpreted and taught will shift over time and according to the religious leader. For example, liberation theology, promoted by the Catholic Church during the 1970s, took hold particularly strongly in Central America because it resonated with the social struggles going on during the time. In El Salvador and elsewhere the Catholic Church empowered the poor to take action against their oppressors (Allen 2000), but it went so far that the Vatican felt that the 'church from below' became out of control. Consequently, the Vatican distanced itself from the movement in the 1980s, although it persisted among the Jesuits in Central America. During the civil war in El Salvador, the US Government sponsored Evangelical Protestant missionaries who were tasked with recruiting poor people to reduce the influence of liberation theology, which was by then deeply engrained in the minds and attitudes of the guerrillas (Haggarty 1988).

The rise of the Evangelical Church in Central America demonstrates how new ideas can be introduced through missionaries. But the goal of integrating risk

TABLE 1 Different religious approaches to perceptions/attitudes and behaviour/responses

Linkages/ *Possible approaches*			

Beliefs and perceptions/attitudes

• Beliefs determine attitudes about hazards: cause, reason, magnitude, location, adverse consequences	• Beliefs determine attitudes about risk: cause, degree of danger, people at risk	• Beliefs determine attitudes about disaster: cause, magnitude, impact, location, people affected	• Beliefs determine attitudes about responding to risk: spiritual consequences, effectiveness of responding

Beliefs and behaviour/response

• Beliefs require behaviour that increases vulnerability to hazards: e.g., requiring certain attire that restricts swimming during floods, requiring prayer during dry periods that takes time away from finding alternative income	• Beliefs include activities that directly address environmental degradation and factors that increase risk	• Beliefs implicitly or explicitly discourage/ encourage anticipatory behaviour to reduce vulnerability to hazards	• Beliefs implicitly or explicitly discourage/ encourage reactive behaviour to respond to impacts

Source: table based on Schipper 2010.

reduction is not to change existing belief systems, but rather to allow them to exist alongside them. So, what are the conditions that would enable a cultural shift towards a risk reduction society? In order to answer this question, characterising how belief systems evolve over time is crucial. Some influences such as wars, deadly pandemics or development of transformative technology have influence not only in the locations where they are taking place, but also much further, while others are internal, such as changes in local political structures or the arrival of missionaries. The next section focuses on questions to further explore this change.

Key issues for further exploration

One of the most urgent questions resulting from research on culture and risk is how to overcome the cultural barriers (Schipper 2010). This has important implications for disaster risk reduction efforts around the world, whether they are initiated by large international agencies or regional governments; without a good

TABLE 2 Different perspectives of causes of hazard and risk and attitudes about responding

Perspective	Cause	Response (1)	Response (2)	Response (3)	Comments
Hazards and disasters cannot be controlled	• Fatalistic: God punishes bad behaviour by sending hazards and disasters	• Good behaviour	• Do nothing; fate cannot be changed	• God is testing humans; vulnerability to hazards should be reduced to avoid disasters	• For some, there is no difference between hazards and disasters, because the causal linkage between hazards and disasters is decided by god. • Another view is that suffering is not caused by god, it is a consequence of human actions. Good and moral behaviour will eliminate suffering.
People are victims	• God	• Do nothing; suffering is necessary	• Do nothing; fate cannot be changed	• Pray to avoid losses and loss of life	• People are helpless victims. Humans must experience suffering to appreciate the difficulties in life and value positive situations. For some, good comes out of suffering. • This attitude also reflects the view that hazards cannot be controlled, and consequently disasters cannot be controlled.
Disasters are not natural	• Hazards are natural but disasters are a consequence of high vulnerability determined by social, political and economic factors	• Reduce vulnerability to hazards	• Reduce factors that cause hazards (greenhouse gas emissions for climate change; soil erosion for landslides and floods, etc.)	• Build infrastructural defences	• Religious beliefs often focus on hazards or disasters, but rarely consider vulnerability to them as a cause of disasters. Vulnerability does not appear frequently in religious discourses, which focus more on capacity to overcome difficulties than on reasons underlying difficulties.

Source: table based on Schipper 2010.

understanding of the cultural landscape, interventions to reduce disaster risk can totally fail. For instance, without understanding that women stayed at home rather than escape floods in Bangladesh during the cyclone in 1991 because their culture dictates that they must be accompanied by a male relative to leave their houses (Bradshaw and Fordham 2013), interventions might focus on attempting to educate women about where to go, which will not help if they are still not allowed to leave their houses. At the same time, how can such actions incorporate cultural aspects effectively? Should one intervene from the outside when faced with customs that increase disaster risk – and if so, how? In some cases, people are unaware that their behaviour is increasing their vulnerability to hazards, but should they be told – and if so, how?

Evidence indicates that people sometimes continue with their adverse behaviours knowingly. This may be because they are bound to a tradition through their culture. Is it up to an outsider to intervene? One of the most germane examples is where traditional ceremonies themselves are considered to be coping strategies, such as in Pratt's (2002) example from Kenya where community response to drought in Kenya includes praying. In this example, the act of praying often involves other actions, which bring together the community and consequently make them better prepared mentally and sometimes physically to deal with hazard risk (Pratt 2002). Thus, while some may think prayer is an irrational form of fatalistic behaviour, it simultaneously provides other types of benefits that are unrelated to beliefs, which may also reduce sensitivity and exposure to hazards and create a sense of community that is fundamental during an emergency.

As one of many potential stressors that influence how vulnerable people are to natural hazards and climate change, belief systems have a role both in motivating understandings of hazards as well as in driving interest in taking action to reduce sensitivity and exposure to them. Those working on adaptation to climate change and disaster risk reduction rarely have the background necessary to understand the role of belief systems in determining or reducing risk. In order to incorporate these understandings in policy and decision making, it will be useful to apply established analytical frameworks, such as those used in anthropology and other fields, to identify the role that culture plays in determining attitudes and perceptions, as well as behaviour and response. But new approaches for analysis may also need to be developed.

Schipper (2010) identified several possible entry points for studying religion and risk, shown in Table 3. This chapter focuses on understanding what the triggers are that allow for parallel belief systems also known as syncretism. In this case, it will be important to examine all of these dimensions to get a complete sense of the role that religion plays. Perhaps there is only a limited role for religion in supporting development in a given location, for example, case A-1, in which case it may be possible to present this as the place in which risk reduction can be promoted. It may not be as practical to present risk reduction as the solution to overcoming disasters emotionally, if this is presently the role occupied by a belief system.

TABLE 3 Entry points for examining religion in the context of disaster risk reduction and adaptation to climate change

Focus on capacity	Focus on reducing risk	Focus on responding to disasters
A-1 Role of religion in supporting development	**B-1** Role of religion in influencing policy on environment and climate change (positively)	**C-1** Role of religion in helping people to emotionally overcome disaster (mental health)
A-2 Role of religion in encouraging social capital (organisation) for coping during difficult times	**B-2** Role of religion in raising vulnerability to hazards	**C-2** Role of religious institutions in supporting disaster relief and recovery processes
A-3 Role of religion in influencing preventive and reactive responses to disaster risk and climate change	**B-3** Role of religion in reducing vulnerability to hazards	**C-3** Role of religion in influencing relief and recovery processes (rebuilding, planning)

Source: table based on Schipper 2010.

The overarching question to address is: how can risk reduction be made compatible with existing belief systems? Several sub-questions help us to answer this question. Two focus on theoretical and methodological approaches, and two on lessons and strategy:

1   What theoretical frameworks for understanding culture and behaviour can be found in the disciplines of anthropology, behavioural economics, psychology, sociology and other fields?
2   What methods have been used in scholarly examinations of sociocultural drivers of vulnerability to climate change and natural hazards?
3   What can be learned from societies with parallel belief systems, such as dual religions? How could this be replicated intentionally through project work?
4   How can a better understanding of sociocultural drivers be introduced into adaptation policy and practice?

Thomalla et al. (2015) identify a number of additional questions that are important to consider around the assumptions we hold as external researchers. These include asking how external interventions can incorporate cultural aspects effectively. Further, they ask whether one should 'intervene' from the outside when faced with behaviour that increases disaster risk – and if so, how? They point to the fact that people are many times unaware that their behaviour is increasing their vulnerability to hazards, but wonder whether they should be told – and if so – how? (Thomalla et al., 2015).

Conclusion: the way forward

Relatively rapid evolution of belief systems are possible, on their own, or when triggered by something external. The example from Assam shows that floods can be such a trigger. External factors can also contribute, e.g., elections that change political perspectives and offer different groups of people new opportunities. Trying to understand the way in which risk reduction can be consciously absorbed into sociocultural structures will require looking beyond belief systems to understand the forces that matter in shaping attitudes. However, it is in places where religion strongly dictates actions and attitudes that this research will be the most necessary, because it is also these communities that are the most vulnerable.

This research is not attempting to 'brainwash' anyone into a new set of beliefs centred on risk reduction. The purpose is not to be missionaries of a better way of viewing the world. Rather, this is a proposal for a way towards understanding how risk reduction can be introduced so that they resonate with existing beliefs. This 'educating' of risk reduction is already taking place, but usually does not take into account that it may fall on deaf ears when it is not presented in a way that allows for beliefs – which may be contradicting – to exist in the same space. Belief systems must be recognised by policy and project planners as institutions, which must be part of the process to reduce vulnerability. In this way, this important dimension of vulnerability will hopefully not be left out as we move towards more climate change and a greater need for risk reduction.

References

Adger, N. and Brooks, N., 2003. Does global environmental change cause vulnerability to disaster? In: M. Pelling, ed., *Natural Disasters and Development in a Globalising World*. London: Routledge.

Allen Jr., J. L., 2000. Key principle of liberation theology. *National Catholic Reporter* [online]. Available from: www.natcath.org/NCR_Online/archives/060200/060200i.htm [accessed 30 July 2014].

Beck, U., 1992. *Risk Society: Towards a New Modernity*. London: Sage Publications.

Bhalla, N., 2014. Lower-caste people get less aid when disaster strikes. *Reuters Newsreport*, 29 January 2014 [online]. Available from: http://in.reuters.com/article/2014/01/29/india-low-caste-dalits-idINDEEA0S03M20140129 [accessed 30 July 2014].

Bradshaw, S. and Fordham, M., 2013. *Women, Girls and Disasters*. Review for DFID, UK.

Davis, I. and Wall, M., 1992. *Christian Perspectives on Disaster Management: A Training Manual*. Teddington, UK: Interchurch Relief and Development Alliance and Tearfund.

Douglas, M. and Wildavsky, A., 1980. *Risk and Culture*. Berkeley, CA: University of California Press.

Füssel, H. M., 2009. Review and quantitative analysis of indices of climate change exposure, adaptive capacity, sensitivity, and impacts. *Background Note to the World Development Report 2010*. Washington, DC: World Bank.

Gaillard, JC, 2007. Alternative paradigms of volcanic risk perception: The case of Mt. Pinatubo in the Philippines. *Journal of Volcanology and Geothermal Research*, 172, 315–28.

Haggarty, R. A., ed., 1988. *El Salvador: A Country Study*. Washington, DC: Federal Research Division, Library of Congress.

Hewitt, K., 2008. Moderators Comments during E-Conference: Culture and Risk: Understanding the Socio-Cultural Settings that Determine Risk from Natural Hazards. 22 September and 8 October 2008. Kathmandu: ICIMOD and Mountain Forum.

Hinkel, J., 2011. Indicators of vulnerability and adaptive capacity: Towards a clarification of the science–policy interface. *Global Environmental Change*, 21, 198–208.

Hulme, M., 2009. *Why We Disagree About Climate Change*. Cambridge: Cambridge University Press.

Hutton, D. and Haque, C. E., 2003. Patterns of coping and adaptation among erosion-induced displacees in Bangladesh: Implications for hazard analysis and mitigation. *Natural Hazards*, 29 (3), 405–21.

ICIMOD (International Centre for Integrated Mountain Development), 2009. Local Responses to Too Much and Too Little Water in the Greater Himalayan Region. Kathmandu: ICIMOD.

Inderberg, T. H., Eriksen, S., O'Brien, K. and Sygna, L., eds, 2015. *Climate Change, Adaptation and Development: Transforming Paradigms and Practices*. London: Routledge.

Kintisch, E., 2006. Evangelicals, scientists reach common ground on climate change. *Science*, 311 (5764), 1082.

Massey, K. and Sutton, J., 2007. Faith community's role in responding to disasters. *Southern Medical Journal*, 100 (9), 944–5.

Merli, C., 2005. Religious interpretations of tsunami in Satun Province, Southern Thailand: Reflections on ethnographic and visual materials. *Svensk Religionshistorisk Årsskrift*, 14, 154–81.

O'Brien, K. L., 2009. Do values subjectively define the limits of climate change adaptation? In: W. N. Adger, I. Lorenzoni and K. L. O'Brien, eds, *Adapting to Climate Change: Thresholds, Values, Governance*. Cambridge: Cambridge University Press: 164–80.

Orr, M., 2003. Environmental decline and the rise of religion. *Zygon*, (38) 4, 895–910.

Pelling, M., 2010. *Adaptation to Climate Change: From Resilience to Transformation*. London: Routledge.

Peterson, A., 2001. *Being Human: Ethics, Environment, and Our Place in the World*. Berkeley, CA: University of California Press.

Pratt, C., 2002. Traditional early warning systems and coping strategies for drought among pastoralist communities. Northeastern Province, Kenya. Working Paper No. 8. Medford: Tufts University, Feinstein International Famine Centre, Fletcher School of Law and Diplomacy.

Rokeach, M., 1973. *The Nature of Human Values*. New York: Free Press.

Schipper, E. L. F., 2010. Religion as an integral part of determining and reducing climate change and disaster risk: An agenda for research. In: M. Voss, ed., *Climate Change: The Social Science Perspective*. Wiesbaden: VS-Verlag, 377–93.

Schlehe, J., 2010. Anthropology of religion: Disasters and the representations of tradition. *Religion*, 40, 112–20.

Thomalla, F., Smith, R. and Schipper, E. L. F., 2015. Cultural aspects of risk to environmental changes and hazards: A review of perspectives. In: M. Companion, ed., *Disasters' Impact on Livelihoods and Cultural Survival: Losses, Opportunities, and Mitigation*. Boca Raton, FL: CRC Press.

9

THE DEEP ROOTS OF NIGHTMARES

Andrew Crabtree

Webs of relations

Can there be floods even after the Kosi Barrage has been built upstream? [. . .] Can man win against Mother Kosi? [. . .] See what happens when you try to dam the Kosi!

A bolt of thunder crackled and lightning strikes one of the fields. The world is drowning in a liquid darkness [. . .] It is the end of the world [. . .] Disaster! Catastrophe! Hapless helpless villagers attempt to pacify Mother Kosi by beating their drums and cymbals and singing songs of appeasement. Young men set about cutting lathes and bamboos to construct makeshift shelters. Shrill voices emerge from fearful throats to the accompaniment of drums and cymbals: 'O Mother Kosi, I fall at your feet! I will offer flowers to you-o-o-u-u' [. . .]

From 'Old Story, New Moral' by Phanishwarnath
Renu (2010)

Renu's short story (see textbox 'Old Story, New Moral') turns into a political satire about blame, responsibility and political gain. Fifty years after being written, it was reprinted in *The Hindu* at the start of the 2008 Kosi flood (Bihar, India), calling the flood an 'annual ritual' out of which politicians gain much mileage. The story here suggests that politicians should lose much mileage, but Nitish Kumar was nonetheless returned to power as Chief Minister of Bihar in a landslide victory.

The research presented in this chapter has its origins in epidemiological studies on the psychosocial consequences of flooding disasters in developing countries. These show that mental health problems can be severe and pervasive, 'even' a 10 percent incidence rate of a disorder translates into large numbers, especially given the lack of mental health facilities in developing countries (Crabtree 2012). While such studies are important in providing an overview of 'outcomes' at any one time and in pointing to unmet needs, they are weak in terms of causal explanations for two main reasons. First, disasters are a meeting of hazard and vulnerability (Wisner *et al.* 2004). Epidemiological studies tend not to examine the vulnerability context as a whole but are limited to a few predicative factors (usually exposure, age and gender). Second, they do not reveal any secondary stressors that may influence outcomes (Crabtree 2012). An additional limitation to these studies is that they aim to establish the extent of particular disorders but they do not try to explain why certain mental health problems do not arise. The research presented here examines these issues. It is based on fieldwork carried out in the village of Rajni in Bihar, India, eighteen months after the 2008 Kosi River flooding.

Oliver-Smith succinctly states that 'disasters are best conceptualised in terms of the web of relations that link society (the organisation of relations among individuals and groups), environment (the linkages with the physical world in which people and groups are both constituted and constituting), and culture (the values, norms, beliefs, attitudes, and knowledge that pertain to that organisation and those relations)' (Oliver-Smith 1999, pp. 28–9). This chapter examines these relations. In doing so, it draws upon Pierre Bourdieu's work, which provides a rich set of conceptual tools to explicate just such relations. Thus the next section briefly outlines Bourdieu's ideas. Thereafter, the chapter turns to the 2008 Kosi River flood, outlining the vulnerability context for the event before presenting the specifics of the actual case. The final section presents conclusions. One main conclusion is that psychosocial interventions of various kinds, while important, will only provide partial resolution to a vulnerability issue. In a Bihari context, dealing with psychosocial problems requires fundamental sociocultural changes, land reforms in particular.

OLD STORY, NEW MORAL

Depression in the Bay of Bengal causes a cyclone. Snow on some Himalayan peak melts. Rain-heavy clouds begin to gather and unleash their fury over the dense jungles of the Terai. The plains await the impending catastrophe, tense and still.

The cattle grazing by the banks of the Kari-Kosi River sniffed something in the air and shied away in startled panic. An old cow ran helter-skelter with her tail in the air. The old cowherd looked closely at the water in the river. He cupped a handful and sniffed it. It smelt of the earth and was reddish-ochre. Reddish water meant water from the hills – did it mean floodwaters were soon to be upon them?

The young cowherd laughed at his fears. But the shivering in the bodies of the animals increased. They stood in a herd beside the riverbank staring at the swirling waters and began showing increasing signs of panic. And not one of them – not even the newest-born calf – put his mouth to the grass to graze.

Fields of tender wheat, corn and jute spread on either side of the tributaries of the Kari-Kosi – the Panar, Bakra, Lohandra and Mahanadi rivers – as though someone has painted the land a rich dark green with a coarse thick brush. Mango groves and courtyards echo with the sweet songs of Madhushravani. And in the air there is the heady smell of the red, fuchsia and yellow veils belonging to the brides fluttering in the breeze.

The Easterly wind sings and dances and whirls inside a bamboo grove. And keeping tune with it, countless spirits and demonesses swing from the branches of the trees chuckling and whooping with fearsome delight.

In the patter of the rain a piteous indistinct cry emerges from the darkness to shiver through the village – He-e-e-e-o-o-o-o-o!

The spirits of the bamboo grove run towards the fields with countless flickering fireflies studded in their veils. Mothers clutch scared children to their bosom. Someone standing in a field far away beside the river bank again cries out for help. He-e-e-e-o-o-o-o-o!

Why is the goddess of the fields crying in the middle of the night? Is disaster about to strike?

The cry for help becomes fainter and in its place there is a roar, a terrifying thunderous angry growl: Gooo-o-o-o.

The growl seems to be coming closer. The people from the northern-most villages shout in one voice: 'Floods! O God, help!'

'The waters of the Bakra River are sloshing past its eastern and western banks. There is water up till the waist beside the boundary of my fields.'

'Have pity on us, Queen Kosi!'

Can there be floods even after the Kosi Barrage has been built upstream? . . . Can man win against Mother Kosi? . . . See what happens when you try to dam the Kosi!

A bolt of thunder crackled and lightning strikes one of the fields. The world is drowning in a liquid darkness . . . It is the end of the world . . . Disaster! Catastrophe! Hapless helpless villagers attempt to pacify Mother Kosi by beating their drums and cymbals and singing songs of appeasement. Young men set about cutting lathes and bamboos to construct makeshift shelters. Shrill voices emerge from fearful throats to the accompaniment of drums and cymbals: 'O Mother Kosi, I fall at your feet! I will offer flowers to you-o-o-u-u . . .'

And dancing to the tune of these words, Kosi Maiya comes traipsing into the village and within the blink of an eye, fields and granaries, homes and barns and trees – everything begins to dance to this tune: Ta-ta-thaiya, ta-ta-thaiya, Dhin-tak-dhinna, Chhamak-kat-chham!

Now there are no drums and cymbals, no songs and pleas; only clamour and confusion!

'Mother! Help! I'm gone! . . . Father! Help! . . . Watch it! . . . Be careful! . . . There, there . . . Hey Sugni! . . . Ramlalva! . . . O Dear Lord Mahadev! . . . I am drowning! Help! . . . The water has reached the chest in the courtyard!

. . . Not here, not here! This thatch is not strong enough! . . . Snake! Snake! . . . Where is the rope? . . . Here, pass the sickle . . . Ta-ta-thaiya, ta-ta-thaiya, Dance, Dance, Kosi Maiya, Dhin-tak-dhinna, Chhamak-kat-chham!'

In the patchy light of early morning the old vulture sitting on the plume of the tallest palm in the village saw: for miles all round there is nothing but swirling, lapping ochre waters, and in the middle occasional islands of what were once villages and in these villages one or two houses and sitting on their roofs a huddle of people. And there, in the distance, the corpse of a buffalo – and beyond that drowned fields showing the drooping plumes of wilted maize.

The clouds are gathering once again. The wind is picking up once more. Mercy! Have mercy on us!

*

The dream of this area's lifelong Servant of the People – the candidate defeated in the last elections – has finally come true! Mother Kosi has given him yet another opportunity to 'serve' the people. Hail to thee, Mother! May you live long! This time, God willing, he will not rest till he has defeated his opponent. He is busy trying to establish telephone contact with the District Magistrate and various Ministers of State: 'Hello! Hello!'

The regional correspondent of a major Delhi-based newspaper has come across a major breaking news story after years, but: 'What? You don't have telegram forms? . . . Trrrrinnngggg.'

'Hello, I am the party leader Sharma speaking. About 50 villages are completely drowned. No, sir, not boats, villages . . . villages are drowned. The DM must be informed, sir. . . . The MLA? . . . But sir, he is from the opposition party! . . .'

The party worker caught hold of the newspaper reporter at the post office counter, took him to his camp and said, 'Write . . . let it be remembered that such a flood has never struck before . . .'

'But 10 years ago . . .'

'Who remembers what happened 10 years ago? Now, write, as soon as I got the news I reached the flood-affected areas in the middle of the night . . . Mine should be the very first response in print.'

The reporter uses his journalistic horse sense: 'But the MLA has already given his statement – to the First Press of India – on the telephone itself.'

Sharma's face fell . . . God has granted him this opportunity to serve the people after so long and the opponent gets first shot at making a statement to the press? The enemy gets to be first? At the time of the Chinese invasion too he had got left behind in making speeches and collecting funds. And this time, again?

'Look here, how many flood-affected villages did I say? 50? Make it 250
... If more villages are affected we will get more 'relief' for this district. I can
do anything for the good of my people and my district. And if it so pleases
God, by tomorrow 200 villages can get submerged.'

Early next morning the reporter sent another urgent dispatch: 'Last night
another 250 villages were submerged due to the breaking of the Bardaha Dam.'
And Sharma was working the telephone lines again: 'Hello! Hello Patna . . .'

The traders and moneylenders of Rampur *kasbah* were quick to understand
that such an opportunity for 'auspicious gain' does not come every day. At
the time of the Chinese invasion they had missed the bus. This time while all
the fuss was going on about the drought God had sent them the floods.

'Brothers! Brothers! This evening. There will be a grand assembly. Of the
people of Rampur. At the local town hall. To set up a committee. For the flood-
affected. Brothers! . . .'

'It's come! The relief wagon has come! . . .'

'The Minster for Irrigation is coming!'

'Donate generously, Brother, give rice-clothes-money . . .'

'Long Live Freedom!'

The boys of the two schools of Rampur *kasbah* – the Middle and Higher
Secondary – took out processions, singing songs and collecting old clothes.
By evening they had split into two groups. The quarrel flared from arguments
to abuses and knives and bamboos being taken out to beat each other.

However, the Servant of the People, *Sharmaji*, was elected undisputed to
the post of Chairman of the Relief Committee.

All the famous leaders of the major political parties have descended on
Rampur *kasbah* with their entourage of 'workers'. Each has set up his own
camp. A group of government doctors and nurses has arrived. Government
officers are occupying every room in the Dak Bungalows. A Coordination
Meeting has been convened. Each political party has proposed the name of
a representative for the Vigilance Committee. Almost every party is split into
two groups: the 'official' group and the 'dissident' group. In every camp, a
half-buried discontent smoulders.

'. . . Tomorrow the Chief Minister will make a "flying inspection".'

'. . . The Union Minister for Food and Civil Supplies is also flying down.'

'. . . The Minister for River Valley Development has issued a statement.'

'. . . And the relief is on its way. Ten trucks laden with rice-flour-oil-cloth-
kerosene-matches-sago are on their way.'

'Brothers! This evening. There will be a public gathering. In the Municipal
Grounds. Where the strictest discontent will be expressed. Against the
present government's relief efforts. And the arbitrary setting up of the Relief
Committee . . .'

*

Helpless, starving and thirsty people surrounded by water on all sides – marooned atop thatches, trees and mounds – saw the boats arrive.

The nearest boat has a flag – a Congress flag!

The one behind has one too – but it is a different colour!

'. . . Long Live Mahatma Gandhi!'

'Huh? Why Mahatma Gandhi? What has he to do with anything?'

'Don't get so excited, or the branch will break.'

The boats draw closer. The Servant of the People himself rides the first one. 'Brothers, even though you did not vote for me in the last election, I contacted the Chief Minister, the Food Supplies Minister, the Irrigation Minister as soon as I heard of your plight . . .'

The workers of the opposition party ride the second boat. In one voice, they oppose Sharmaji's claims: 'You are using the government boat and government aid to wrongfully campaign for the Party . . .'

Sharmaji's boat sprinkles the submerged villages with speeches and goes away. The people in the boat directly behind it oppose every word.

'A bigger boat is on its way.'

'Brother, is there only a boat coming or does it carry something? The children are faint with hunger. My daughter is about to die . . .'

Two dozen boats plied the waters till the evening, gathering people. At night the relief officer told the Vigilance Committee in plain words: 'The boats must not carry the flag of any political party . . . Nothing should be given without taking a thumbprint or signature . . . We simply cannot provide *bidis* . . . It is unethical to either praise or criticise any political party while distributing relief supplies. Those who continue to indulge in such practices will henceforth not be given any responsibilities by the Committee.'

There is no work yet for the doctors and nurses. They are busy playing 'indoor' and 'outdoor' games – 'Game-ball' . . . 'Two spade' . . . 'Hey Miss Bannerji' . . . 'No trump'.

Arrangements have been made for the flood-affected villagers on the high ground near the railway bridge – under the trees, beside the *haat*, in the school compound. And also for people to stay in villages where the floodwaters have not entered but which are surrounded by high waters. Boats ply the waters regularly, carrying rations to and fro. Several groups of doctors and nurses have been deployed to run health centres at such places.

The waters are gradually receding. The ferrymen are beginning to feel cooped up in the camps. These free-spirited people hear the news of the receding waters and become restless. They are creatures of the water. They can stay for months in mud and water. . . . Their thumbs have blackened with all the thumb imprints they have been giving. . . . No, no . . . That dog-faced officer was trying to entice my Sugni, did you know? . . . A bunch of thieves, that's what they are!

A major newspaper from Delhi published an expose: 'The engineers of the Public Works Department have shown remarkable short-sightedness in building several small dams on minor rivers and their tributaries, such as the old distributary channels of the Kosi. That is why those villages which were never affected by the floods are submerged under water this year. Inept government servants . . .'

Another daily newspaper placed the blame squarely on the officials of a neighbouring country and said, 'In building a barrage close to the border of our state, the neighbouring kingdom has stopped the drainage outflow of all the small rivers in northern Bihar. A flood such as this would never have happened had the concerned officials bothered to consult our experts before building the barrage.'

The local rag declared the flood a 'manmade' calamity and announced: 'It isn't the neighbouring state, but the oarsmen of the neighbouring country who have drowned us!'

Eventually, rats were found to be responsible for the breaking of the Bardaha Dam. Rats had dug up countless tunnels through its foundation, making it weak and porous – in less than a year!

*

A flock of vultures is soaring in the skies. Countless dark wings – like black clouds hovering overhead. And on the earth bloated carcasses of dead animals. And destroyed crops and rotting plants in the fields. The stink! Oh, that terrible malodorous smell!

Straggling groups of people – heads bent, clutching babies, herding their pitifully few cocks, hens, goat and sheep, some on wagons and carts, others on their shoulders – are returning to villages where not a trace of their huts remains nor even a pinch of grain in their fields. But their feet are racing homewards. After nearly 30–35 days of hellish exile, a flood of love has overwhelmed their hearts – love for their homeless villages and mud-filled fields.

The gods have come once again to live in these spirits buried under government relief, debt and assistance. For days they have fought amongst themselves for survival and pleaded with government officials for relief supplies. Greed made them fight, steal, covet . . . Satan had come to roost in their hearts.

In the month of Ashwin the sun appears without fail to awaken the earth. Green tufts of grass glimmer in the drying mud.

A skein of wild geese circles above, crying 'paink-paink'. At this moment everything seems dear and sweet – even the kites, crows and vultures. Are those cranes in the water-filled ditches or Koka flowers? Branches of the Haarsinghar are heavy with sweet-smelling flowers. The lilt of welcome, of new arrival, is in the air. The mother is coming! That beggar woman? Is she the Goddess Annapurna, the provider, the fertile Mother Goddess?

The Ashtami moon laughs amid the clear serene waters of the Kari-Kosi.

Eventually, rats were found to be responsible for the breaking of the Bardaha Dam. Rats had dug up countless tunnels through its foundation, making it weak and porous – in less than a year!

Source: 'Old Story, New Moral' (original title in Hindi 'Purani Kahani, Naya Paath'), short story by Phanishwarnath Renu (2010), translated from the Hindi by Dr. Rakhshanda Jalil.

Bourdieu's conceptual framework

In 'Distinction', Bourdieu (1984) sums up his ideas in the following schematic formula: [(*habitus*) (*capital*)] + *field* = *practice*. This formula, I suggest, can be a way of explaining the sociocultural context of a disaster. I shall take each of Bourdieu's concepts in turn.

Habitus is a cultural theory of action (Swartz 1997) and a 'structured and structuring structure' (Bourdieu 1990, p. 170). It is a way of connecting structure with agency and agency with the re-creation of structure. Bourdieu uses the concept to reject both the determinist structuralist tradition of Durkheim or Lévi-Strauss, who give little importance to agency, and the individualism of Sartre and rational choice theory, which give little importance to social structures. Habitus is in part ingrained; it is 'the strategy-generating principle enabling agents to cope with unforeseen and ever changing situations' (Bourdieu and Wacquant 1992, p. 18). At the individual level, societal structures, regularities or principles provide the context of individual agency and affect our dispositions, tendencies, propensities and inclinations – our way of being and doing. Yet we are not overly socialised; there is also the creative agent who plays a role in structuring structures. In our present case, the caste system provides the overall structure within which action takes place, yet it does not determine everything. Indeed, during the first phases of the flood, villagers discarded the caste system and helped one another and doing so was one of the strongest mitigators of harm (for similar observations, cf. Oliver-Smith 1986, Bonanno *et al.* 2010).

Symbolic systems are, for Bourdieu, one part of the structuring structures that are in turn structured by structuring agents. A habitus includes beliefs, ideas, values and norms usually associated with culture as enumerated by Oliver-Smith. Some of these, *doxa*, are taken for granted and are unquestioned basic assumptions about the world. The importance Bourdieu gives to agency underlines the fact that culture is not just passive, something we 'have/possess', but is dynamic, being constantly created and re-created.

Societies are dynamic because of the aims people and groups aspire to, the strategies they undertake in order to lead the lives they value, and the desire to increase capital. Bourdieu (Bourdieu and Wacquant 1992) identifies four main forms of capital: economic (finances, land, property), cultural (such as skills and education), symbolic (prestige, honour, caste status) and social (networking, patronage). Those who have power, who have accumulated various kinds of capital, often exercise symbolic violence. This involves denying recognition to others. The basic insight here is that humans are social beings and thus lack of respect from others can be devastating (Bourdieu 2000). Bourdieu's list of forms of capital is not exhaustive. Despite their importance, environmental issues are absent from Bourdieu's work. Taken from ecological economics, the term 'natural capital', the stock of natural ecosystems, can be employed to make up this deficit (Brand 2009). In the current case, cultivatable land may, for example, be considered natural capital. All forms of capital can be accumulated (or lost) over time, passed on from one generation to the next, or, within limits, exchanged. Power is related to the possession, or lack, of different forms of capital.

Bourdieu's notion of field was not present in his earlier work, undertaken in Algeria in the 1960s. Later, he would define the term and see fields such as the educational field and the economic field as becoming increasingly distinct in modern societies. A field's structure is determined by the force between the players within it, which is dependent on the capital each player possesses and the relevance of that capital to a particular field. Power is thus seen as relational, and fields are sites of struggle, domination and subordination. Fields consist not only of individuals but also of institutions, organisations and markets. All fields are hierarchical and relate to the overall field of power. Bourdieu saw the field of power in France as being determined by two forms of competing capital: economic and cultural. Conflicts thus related to both economic (material) and symbolic, cultural, resources.

Schematic formulae have their limitations, and the formula above is perhaps best thought of as a heuristic device to show that the concepts of habitus, capital, field and practice are inextricably interrelated. However, the formula helps us guard against using Bourdieu's concepts in isolation – a practice which is fairly widespread (Bourdieu and Wacquant 1992). We now turn to see how Bourdieu's thinking can illuminate our understanding of the psychosocial consequences of the Kosi flood.

Root causes and structuring structures: Bihar prior to the 2008 Kosi flood

Disasters are deeply rooted (Wisner *et al.* 2004). Similarly, the caste system, with is historical roots in the third century BC, remains the dominant habitus (structuring structure) of social, cultural and economic structures in rural Bihar and pervades all forms of capital, beliefs and practices. This structuring structure has not remained static but has undergone important changes, primarily as related to colonisation and caste conflict.

Symbolic 'systems' can be paradoxical. The villagers of Rajni described themselves as *Kabir Panthis* (followers of *Sant Kabir* the truth-seeker) and *Vaishnavas*. However, while *Kabir* rejected the caste system, Hinduism and Islam in favour of a direct relationship with an incarnate God, *Vaishnavas* are Hindus who support the caste system and worship a multiplicity of gods and goddesses and thus maintain the existing structure. How the two systems came to be combined is unknown as there is a lack of written historical records, but it is a case of actors reshaping their cultures to give them their own meaning (Bahuguna, undated). The *Vaishnava* influence proved the stronger, leaving the social structure, and the symbolic violence done to the lower castes, untouched. Vegetarianism seems to be the sole remnant of *Kabir's* teaching.

It should also be stated that culture can play a double role both as a cause of psychosocial problems relating to the caste system and as a mitigator of such problems. On the one hand, Goddess Kosi can be found in the Hindu scriptures and hence is part of the Hindu system. Though stories vary, she is usually wild and untamed (Mishra 2009) yet, as related in Renu's story, there is a long tradition of trying to appease her by singing songs in her praise at times of flooding. As we shall see later, this tradition of appeasement rituals also has played a significant positive role in psychosocial interventions.

The caste system and its corresponding symbolic violence were further reinforced by British colonial rule's adoption of the *zamindar* system. The *zamindari* were mainly upper-caste Hindus with control over large areas of land. The system allowed them to collect the level of taxes they thought correct as long as they met a preset amount to be given the colonial rulers (Mathew and Moore 2011). This made the *zamindari* comparatively rich and, at the same time, their extraction of surplus meant that lower down the chain, landlords and cultivators had little incentive to invest in agriculture and growth declined. Thus the *zamindar* system is often accredited as the reason for Bihar's 'backwardness' (Sharma *et al.* 2012). When the *zamindar* system was abolished in Bihar in 1948, the main beneficiaries were the upper backward castes who were the superior tenants; benefits were not passed on to lesser tenants or landless labourers.

In Bihar, economic capital in the form of land ownership still remains skewed: 96.5 percent own just 66 percent of the land, yet a further 33 percent was owned by 5.3 percent of the landowning community (Bandyopadhyay 2009). Of those living in rural Bihar, 84 percent own less than 0.4 hectares of land per capita (Singh *et al.* 2011). Today, the agricultural sector accounts for 21 percent of Bihar's GDP (Sharma *et al.* 2012). The system of sharecropping covers approximately 35 percent of the land. Consequently, growth in the rural sector and a decrease in poverty are dependent on land reforms, the enactment of which successive governments have 'resisted' (Bandyopadhyay 2009).

In 1990, a fundamental division into two 'fields' took place: the 'caste social' and 'caste political' (Jha and Pushpendra 2012). The former refers to the traditional caste system as embodied in the scriptures. Caste political, by contrast, refers to a new configuration in which the political caste has become disjointed from the social and economic fields. In the first instance, this rupture was due to the rise to power of Lalu Prasad Yadav, who became Chief Minister of Bihar in 1990 based on support from the upper backward castes. Since that date, 'government' in Bihar has depended on a variety of caste coalitions (cf. Jha and Pushpendra 2012 for extensive details). For present purposes, the emergence of the caste political has had two important consequences. For present purposes, the emergence of the caste political has had important consequences. Lalu introduced what Jha and Pushpendra term 'governance without government'. He was concerned with undermining, both practically and symbolically, the administrative power of the upper castes both in terms of the administration and of the police. The result was a large scale increase in corruption and lawlessness (including caste armies and massacres), as well as a lack of further land reforms – Lalu's supporters who had reaped the benefits of *zamindari* abolition were not interested in further reform. Both the lawlessness prevalent at the time and Lalu's favouring of his own Yadav caste above others in his coalition enabled Nitish Kumar to come to power in 2005 under the slogan of 'justice and development'. To attain power, he formed a coalition that stretched from the upper castes to the Dalits playing, in part, the card of land reform. However, in practice, possible reforms were blocked by the upper castes in Kumar's coalition, thus perpetuating inequalities (Jha and Pushpendra 2012). Skewed ownership thus remains, and corruption, though reduced, continues to be endemic.

The *zamindari* system also played a major role in the building of a culture of embankment building in reaction to flooding and consequent crop and revenue loss. The response to any flood appeared to be 'build an embankment' and, as the *zamindari* were to benefit from flood control, they were given the responsibility of managing, supervising, overseeing, maintaining and designing the embankments (Singh 2008). Changing the flow of the Kosi in one area would lead to increased flooding in another, resulting in conflicts and the building of additional embankments to safeguard the land that had now become more flooded. The culture of embankment building has continued. In 1954, when the Bihar flood policy was introduced, there were 160 km of embankments; that figure is now more than 3,000 km. During the same period, flood-prone areas have increased to 6.89 million ha. Two basic problems remain unresolved. First, the Kosi's sediment load of some 80 million tons per annum means that its river bed will always rise, increasing the potential for flooding. Second, maintenance of the embankments is often neglected, causing extensive flooding (Mishra 2008). Following extensive floods in 1953, the Kosi Barrage was built 5 km into Nepal, under agreement between the Indian and Nepalese governments. The work was completed in 1963; nevertheless, serious flooding has occurred in 1963, 1971, 1984, 1987, 1991 and 1995 (Reddy *et al.* 2008). The 2008 breach followed this pattern.

The circular migration of lower caste men to undertake agricultural work in the Punjab or jobs in Delhi has impacted the position of women in Bihar. Traditionally, they have been subject to the symbolic violence of being restricted by Hindu customs to the household, seldom moving outside the *tola* (hamlet), and having little power in decision making (Shattuck 1999). It would be wrong to overemphasise changes, but the absence of men has meant that women have a greater say in decision making and an increase in mobility, though these are often restricted to the village level. It has also meant additional work, taking on tasks that previously carried out by men (Datta and Rustagi 2012).

All in all, the local vulnerability context can be summed up thus: although Bihar is rich in natural capital (Bandyopadhyay 2009), it remains one of the poorest regions in the world, having a multidimensional poverty rate of about 81 percent, which is similar to that of Sierra Leone or Guinea (UNDP 2010). The overall structuring structure of the region has been the caste system and a related set of caste positions, economic structures, conglomerate beliefs and symbolic systems. The majority of rural Biharis are small-scale landowners, or agricultural labourers who lack all forms of capital: economic, symbolic, social and cultural (the average literacy rate – as a basic approximation for cultural capital – for Bihar being 42 percent, the lowest in India), and are correspondingly weak in terms of power. We now turn to see how this played out in the 2008 Kosi flood.

Rajni and the psychosocial consequences of the 2008 Kosi River flooding disaster

Madhepura is one of the poorest districts in Bihar (UNDP 2009) and Rajni is a fairly representative village of the district, lying in one of the areas worst hit by the flooding. The village has approximately 11,500 inhabitants and a sex ratio of 916 females per 1,000 males. Approximately two thirds of its total workforce of 6,000 are landless agricultural labourers and a further 1,300 are cultivators. About 31.5 percent of the villagers belong to Scheduled Castes.[1] Approximately 70 percent of the villagers in Rajni are illiterate. With the exception of two *pucca* (concrete) houses, almost all the buildings in the village are *kutcha* (thatched). Rajni has no paved roads or electricity.

Methodology

The original intention of this research was to employ a mixed methods approach. Qualitative interviews were held to allow for the emergence of unexpected factors. These were the basis for quantitative interviews aimed to strengthen generalisability. No survey concerning mental health had been carried out in the area, or Bihar as a whole, prior to the disaster, thus baseline data depended on recall. A total of four focus group interviews were held in different areas of the village in order to capture potential variances in exposure, which is one of the strongest predictors of mental health outcomes following floods (Norris *et al.* 2002, Norris 2010, Crabtree 2012). A separate focus group for women from all areas of the village was held to ensure that women's voices were heard. Each meeting lasted for approximately half a day.

Semi-structured individual interviews were undertaken with the former *Mukhia* (the village head at the time of the flood), a traditional healer, a 'quack',[2] and two health workers. Further semi-structured interviews (with two men and two women) were held to permit a more in depth understanding of people's emotions and how and why they have changed over time. Finally, semi-structured interviews were made with three doctors and the hospital administrator at the nearest hospital at Murliganj. All interviewees had life threatening experiences during the 2008 flood.

The intention was to use the 'General Health Questionnaire 12' (GHQ-12), a widely used screening test for disorders, as part of the quantitative study. However, it was found that the official translation of the GHQ-12 was too literal; in particular question number twelve, concerning happiness, failed to capture the meaning of the original. Hence only information from the first eleven questions could be used. These suffice to diagnose some symptoms of depression.

Research was stopped prematurely for safety reasons, as there were armed gangs in the area. As a result, two small minority groups, Muslims and a 'scheduled' tribe, were not included in the study. These groups account for a total of 1.6 percent of the village population.

Flood phases and stressors

The Kosi River breached its embankment on 18 August 2008. Some villagers were warned of the flood the next day; however, these warnings were dismissed as the 'usual flooding'. In fact, flooding of this magnitude was simply outside the villagers' experience and beyond their imagination. Villagers disagreed as to when the flood arrived (the morning of 21 or 22 August). Nevertheless, it is probable that the threat to people's lives and loss of their livelihoods, especially through the loss of cattle, could have been significantly reduced if these warnings had been acted upon. As the villagers identified these threats as being the two major stressors from the event/disaster, it is highly likely that the negative psychosocial consequences would have been significantly reduced. This may have been the case in many areas in the region depending in part on how close people were to the breach, as few people received warnings (UNDP 2009).

Flood onset

Villagers' immediate response to flood onset reflected their habitus, their dispositions to act in certain ways to give the flood meaning and their *doxa*. Thus, some people started to build makeshift dams that later proved totally inadequate and led to guilt for having made the wrong decision. Others tried to appease Goddess Kosi by singing songs to her (as in Renu's short story) in the belief that she was punishing them for their evil deeds (this contrasts with fatalist understandings of disasters, as the villagers saw themselves as the cause). Consequently, there was an absence of anger, otherwise a common psychological response to disasters (IFRC 2009): villagers had no reason to be angry; instead, Goddess Kosi was angry with them. Goddess Kosi was part of the villagers' *doxa*, and she thus made the flood meaningful.

At the onset of the flood, one man who had a weak heart died of a heart attack and another man was swept away by the floods (had the warning been understood, it is possible that both fatalities would have been prevented). As one man stated, the situation 'was too pathetic and too horrible to describe'. Everyone in the village felt their life threatened and felt intense fear and helplessness, thus meeting criterion A for Post-Traumatic Stress Disorder (PTSD; DSM-IV[3] criteria).

The villagers of Rajni perceived flood onset as the worst time. In focus group discussions, people were asked questions concerning additional criteria for PTSD, namely intrusive recollection (B), avoidance/numbing (C), hyper-arousal (D), duration of symptoms (of over one month) (E), and functional significance (defined as clinically significant distress or impairment in social, occupational or other areas of functioning) (F). As the research was carried out eighteen months post-onset, all interviewees/participants met criterion E. The focus group interviews revealed that most people met criterion B, in the form of frequent nightmares. Some reported avoidance (C) in the form of avoiding escape routes, or hyper-arousal (D) during the monsoon season, though this problem had diminished once the embankment had been repaired.

Although re-experiencing a disaster is common, few studies ask what people specifically re-experience and why. In Rajni, nightmares of villagers predominantly, but not exclusively, concerned a loss of cattle rather than an immediate threat to villagers' lives resulting from the flood waters. Cows play a crucial role in the villagers' lives and livelihoods. Cow dung is used for fires and as fuel for cooking, as manure for the fields and as a building material for houses. But the additional deep root cause related to the symbolic value of cows is that, as the villagers are *Kabir Panthis* and therefore vegetarian, cow's milk plays a central role in their diet. Thus, intrusive recollection was focused more on people's culturally defined livelihoods and long term wellbeing than it did to the immediate threat to their lives from drowning, which would otherwise be the emphasis of PTSD criterion A.

The focus group members, the traditional healer, the 'quack' and the health workers considered two men to be 'non-functioning'. One had lost everything and 'given up' and the other, who was of nervous disposition before the flood, became even more so afterwards.

Perhaps the most important feature of the initial phase of psychosocial consequences of the flood was the temporary breakdown of the structuring structures of cultural norms and caste distinctions within Rajni by the villagers. The common threat to lives led to changes in social and symbolic capital as villagers helped each other and shared what food there was across caste distinctions and a sense of common purpose was created (Oliver-Smith 1986). This increased social cohesion was crucial for mitigating negative mental health outcomes. Several epidemiological studies have shown that social capital can reduce the risk of developing mental disorders (cf. Patrick and Patrick 1981, Suar et al. 2002, Norris et al. 2005). Given the devastation and loss of livelihoods, one might expect suicide rates to increase. However, according to the doctors at the nearest hospital in Murliganj, suicide rates in the area remained constant. According to villagers in Rajni, this was due to increased social coherence during the flood.[4]

Time away

Throughout the area affected by the flood, 15.6 percent of people remained in their villages – primarily due to worry of theft, 33.9 percent stayed in government camps, 3.8 percent in NGO camps, 18.9 percent stayed with relatives or friends and a further 24.1 percent made other arrangements on their own (UNDP 2009, p. 14). Villagers from Rajni also went to a variety of destinations depending on social capital, with relatives and friends being the first choice. Most went to camps, and some moved from one place to another. Caste was not an important factor here. A few stayed in the village throughout the period to look after cattle and prevent theft (safeguarding their livelihoods).

According to focus group interviews, several people found their lives threatened for different reasons while they were in the camps (PTSD criterion A). In one case, flood water rose to almost the top of the embankment where the camp was situated before subsiding. People also felt threatened by cobras and scorpions, which were searching for the same high ground as the villagers.

At the start of their life in the camps, the women participating in the study expressed anxiety, as being in a state of 'numbness' and feeling 'indecisive and helpless', not knowing what to do. Others said that their 'senses were not working' and that they were in grief. Throughout the period of living in the encampment, they were worried about the loss of cattle and household belongings (economic capital). Life in the camps varied: dignity (symbolic capital) and safety were important issues. One woman related that when there was no food, water or help, their status 'was even worse than beggars'. While some villagers reported being treated equally and without discrimination – which was far from a universal experience in the flooded area (Dalit Watch 2008) – other villagers complained about corruption in the distribution of relief materials and about receiving less food than those distributing it. There was also a suggestion that the local mafia was involved in the corruption – a legacy of Lalu's governance without government.

Return, loss of livelihood and the future

The structuring structures of caste, and the deeply rooted poverty and subsistence living which result from them, meant that the villagers also felt their lives threatened (PTSD criterion A) upon return to Rajni. Much economic capital, which had already been scarce, had been lost. Villagers faced food shortages, lack of purchasing power and health threats – truths that were common throughout the flooded region (UNDP 2009). In fact, the majority of deaths (two children and two women) occurred after return. Approximately 60 percent of the *katcha* houses had collapsed. Much of the *kharif* harvest had been lost and an estimated 30 percent of the land had been lost due to silting and waterlogging. Fruit trees had dried up and new pests had arrived. Most cattle had been lost, as were household goods and farming equipment. The loss of land meant that landless labourers were in an even more precarious position than before. According to individual interviews, some villagers still felt that their lives were threatened eighteen months after the onset of the flood.

Attempts to reduce the flood's negative impact on livelihoods were limited. The Bihar Government failed to deliver the three rehabilitation packages it promised to each flood victim, causing considerable anger and resentment. The first package worth 2,250 rupees (approximately 40 US dollars) plus 100 kg of food grain was delivered, but a second package of 2,090 rupees came without grain, and there was no third package. The Mahatma Gandhi National Rural Employment Guarantee Act (MGNREGA) entitles rural households to at least one hundred days of work per household per annum. However, the villagers of Rajni reported that they were cheated and were only paid for one third of the time they had worked. Illiteracy (lack of cultural capital) meant that the labourers could not check their job cards. Gaining microcredit loans from commercial banks was also problematic as bribes had to be paid to be awarded such loans. There was an increase in circular migration, and the lack of economic and symbolic capital meant that it was most commonly Dalit males who migrated, leaving their wives feeling lonely, afraid and insecure, with an increased burden of work until the men returned.

Despite its potential for becoming life threatening, loss of livelihood did not lead to symptoms of PTSD, but rather to symptoms of depression, as indicated by the eleven questions of the GHQ-12 questionnaire that could be used. Villagers said they were unable to make decisions or find initiative, had difficulty sleeping, felt depressed and unhappy, and were unable to concentrate. Some study participants took comfort in the (false) prediction that the world would end in December 2012. Symptoms were related to both actual loss of livelihood and despondency about the slim future prospect of land reform due to the inter-caste conflicts discussed earlier. This contrasts strongly with epidemiological studies which see psychological states as resulting from *past* experiences alone and do not take expectations into account.

Gender differences

After exposure to trauma, epidemiological evidence suggests that gender is a clear predictive factor in relation to mental health outcomes. However, it is far from clear as to why this is so (Kimberling et al. 2009). A 2010 review of the epidemiological literature (Bonanno et al. 2010) suggests that their subjective experience of disasters provides the best explanation for why women are particularly negatively affected by traumatic events. Other influences include women's role as caregivers, responsibility for households, experience of violence and perception of having less social support (Kimberling et al. 2009).

The general trend of women being more negatively affected by disasters/trauma than men was supported by focus group interviews in Rajni, and is supported by the only epidemiological study following the 2008 Kosi floods (Telles et al. 2009). Male interviewees maintained that women are more 'sensitive' than men (cf. Liu *et al.* 2006 for a similar explanation). However, the evidence from Bihar suggests that women face additional stressors. Again, symbolic capital is essential in explaining these distinctions. The Bihar Voluntary Health Association (BVHA 2010), which worked on psychosocial interventions during the floods, found that women and adolescent girls who were not used to going outside their *tolas* or mingling with unfamiliar men suffered as a result. Privacy was lost, especially in relation to hygiene (UNDP 2009). Research by the International Centre for Integrated Mountain Development in the Saharsa district reported that 'self-imposed starvation to avoid having to defecate is a common occurrence in flood-affected areas' (ICIMOD 2009, p. 28), as Muslim women considered it shameful to defecate in the open.

PTSD is not only related to life threatening circumstances; it can also be caused by violence to women in the form of rape or other abuse. The BVHA reported 'During the rescue work many women faced violence and misbehaviour which added to their sorrow and grief' (BVHA 2010). During field research rumours of cases of rape during evacuation or in the camps were heard. This topic was off-limits in interviews with women, but as one man said, 'I can confirm that it happened, but I refuse to tell you anything more.' After this statement, he left. This may not relate directly to women from Rajni, and it is not clear how widespread such incidences were.

Culture played an important role in determining which problems did not arise. Being unable to carry out appropriate burial rituals can have major consequences in some cultures (Rao 2006). In Rajni, I was told that this was not a problem as people were poor and, traditionally, poor people are thrown into the rivers. The suicide rate in the area generally did not change; this was due to both negative cultural attitudes towards suicide and the strong sense of social cohesion that developed. Due to cultural restraints, cases of addictive disorders did not increase. There was no suggestion of an increase in post-traumatic symptoms; this was not necessary because the religious beliefs of villagers gave the flood meaning as it was. These were not necessarily conscious decisions, but unconscious dispositions to act or not act in certain ways – a result of people's habitus.

Healing and not healing

Recent literature on disasters has drawn attention to the importance of local knowledge regarding coping methods. There is an important line of thought found within psychosocial literature that concerns itself with the 'tyranny of Western expertise', arguing in favour of local, culturally appropriate traditions (Marsella 2010). Similarly, psychosocial interventions have been attacked for their lack of cultural sensitivity and irrelevance for people who are reconstructing their lives (Summerfield 2005). These interventions are a new feature of disaster manage-ment in Bihar (BVHA 2010), and some entities call for psychosocial interventions on the basis of perceived needs (Jha and Raghavan 2008, UNDP 2009). Many of the villagers in Rajni received psychosocial help when they were in the camps.

This help included singing songs to Goddess Kosi and yoga lessons by one of India's most famous yoga gurus, Baba Ramdev, who made a brief stop to the encampments, and later yoga lessons conducted by Ramdev's followers. But 'local' and 'culturally sensitive' yoga interventions are open to question, as yoga was not previously practised in the area. Counselling, a Western form of intervention, was both given and received by the villagers; those who did not receive psychosocial help stated that they would have appreciated it. Eighteen months after the onset of the flood, villagers expressed their desire for long term psychosocial interventions.

One man from Rajni attended a psychiatrist to gain help with his sleeplessness and nightmares. He was then given sleeping tablets, which according to one of the health workers (his mother) had helped his ailments. The traditional healer, who used a combination of mantras and mustard seed, did not perceive any increased demand following the flood. His own assessment (said with a smile) was that results were mixed; sometimes healing worked and other times it did not. The village quack reported having treated four men who he described as 'senseless because of fear', with very strong heart palpitations and generalised trembling. He described them as 'depressed', for which he gave them vitamin B tablets (not a local remedy) that the quack said had helped (the symptoms suggest panic attacks on DSM-IV criteria, in which case the symptoms would disappear of their own accord). All in all, this means that far under one percent of the villagers sought help, and the rest dealt with problems themselves. Neither Western expertise nor local knowledge was of much influence. According to the interviews, there was a gender division regarding coping strategies. Women tried to cope with their problems by talking and men by restrained drinking at festivals. Yet there was clearly a desire for help.

Conclusion

Whereas epidemiological literature sees a flood as one stressor, villagers in Rajni faced at least two traumatic stressors (flood onset and loss of livelihood). Some villagers faced additional traumatic stressors, namely potential flooding of the camp, exposure to poisonous animals and, in the case of women, potential rape. Nonetheless, research evidence suggests that villagers did not suffer from full-blown PTSD. Two people were deemed to have mental health problems and nightmares were widespread, as were symptoms of depression. Four people experienced symptoms of panic attacks. Additionally, there were a large number of comparatively smaller but nonetheless important stressors, which were particularly prominent for women, such as shame, lack of privacy and loneliness. At the same time, a number of problems found elsewhere did not arise. There was no increase in addictive disorders, suicides, anger or grief due to lack of bereavement rituals. Nor was there any evidence of post-traumatic growth (Tedeschi and Calhoun 2004).

These findings, this chapter suggests, are best explained not in terms of a few predictive factors but from a Bourdieuan perspective. The psychosocial consequences of the flood had deep roots going back to the third century BC. The structuring structures of caste, the *zamindari* system, the culture of building and relying on unstable embankments (which was the immediate cause of the flood), and the region's symbolic and political systems all meant that the villagers' main concern regarding the flooding was a culturally defined loss of livelihood which led to symptoms of depression.

Perhaps the most important mitigator against psychosocial harm was the increase in social capital and the correlate reduction of symbolic violence, which arose at the start of the flood. This involved a breakdown of the caste system and the social distinctions and power relations that the system implies. Reducing the number of depressive symptoms will involve improving people's livelihoods, which is the villagers' top priority, and which requires deep, radical, sociocultural changes: land reform, reducing women's dependency, ending caste discrimination, abolishing corruption and improving health and education. I share the concern stated below:

> We are [. . .] concerned about what happens even when it is admitted that social and economic factors are the most crucial. There is often a reluctance to deal with such factors because it is politically expedient (i.e., less difficult for those in power) to address the technical factors that deal with natural hazards. Changing social and economic factors usually means altering the way power operates in society. Radical policies are often required, many facing powerful political opposition.
>
> (Wisner *et al.* 2004, p. 7)

Five and a half years later, the one-man Kosi Commission has yet to publish its report.

> *Eventually, rats were found to be responsible for the breaking of the Bardaha Dam. Rats had dug up countless tunnels through its foundation, making it weak and porous – in less than a year!*
>
> From 'Old Story, New Moral' by
> Phanishwarnath Renu (2010)

Notes

1 Castes recognised by the Indian Constitution as being traditionally disadvantaged.
2 In India, a quack is 'one who practices a form of medicinal system without qualification, training and registration from the appropriate council or authority' (Gupta 2010, p. 795). Gupta estimates that there are 1.5 million quacks working in India.
3 The fourth version of the American Psychiatric Association's Diagnostic and Statistical Manual of Mental Disorders.
4 Eighteen months later, some villagers maintained that old divisions were gone as now they were all poor; others argued that divisions and structural divisions were reappearing. The maintenance of livelihood patterns and work relations suggests the latter.

References

Bahuguna, R. P., n.d. Conflict and assimilation in medieval North Indian Bhakti: An alternative approach. *SAP-History Monograph* [online]. Available from: http://jmi.ac.in/upload/departments/history/drs/Conflict%20And%20Assimilation%20In%20Medieval%20North%20Indian%20Bhakti.pdf [accessed 04 April 2013].

Bandyopadhyay, D., 2009. Lost opportunity in Bihar. *Economic and Political Weekly*, 44 (47), 12–14.

BVHA (Bihar Voluntary Health Association), 2010. Unpublished report.

Bonanno, G. A., Brewin, C. R., Kaniasty, K. and La Greca, A. M., 2010. Weighing the costs of disaster: Consequences, risks, and resilience in individuals, families, and communities. *Psychological Science in the Public Interest*, 11 (1), 1–49.

Bourdieu, P., 1984. *Distinction: A Social Critique of the Judgment of Taste*. Cambridge, MA: Harvard University Press.

Bourdieu, P., 1990. *Logic of Practice*. Cambridge: Polity Press.

Bourdieu, P., 2000. *Pascalian Meditations*. Cambridge: Polity Press.

Bourdieu, P. and Wacquant, L. J. D., 1992. *An Invitation to Reflexive Sociology*. Chicago, IL: University of Chicago Press.

Brand, F., 2009. Critical natural capital revisited: Ecological resilience and sustainable development. *Ecological Economics*, 68 (3), 605–12.

Crabtree, A., 2012. Climate change and mental health following flood disasters in developing countries. A review of the epidemiological literature: What do we know, what is being recommended? *The Australasian Journal of Disaster and Trauma Studies*, 1, 21–30.

Dalit Watch, 2008. For a morsel of life: Report on the flood relief camps in Bihar. *Dalit Watch* [online]. Available from: www.idsn.org/uploads/media/Dalit_Watch_Report_floods_Bihar_2008.pdf [accessed 24 August 2011].

Datta, A. and Rustagi, P., 2012. *Status of Women in Bihar: Exploring Transformation in Work and Gender Relations*. New Delhi: Institute for Human Development.

Gupta, P. J., 2010. The role of quacks in the practice of proctology. *European Review for Medical and Pharmacological Sciences*, 14 (9), 795–8.

ICIMOD (International Centre for Integrated Mountain Development), 2009. Life in the shadow of embankments—Turning lost lands into assets in the khosi basin of Bihar, India. *ICIMOD* [online]. Available from: www.indiaenvironmentportal.org.in/reports-documents/life-shadow-embankments-turning-lost-lands-assets-koshi-basin-bihar-india [accessed 24 August 2011].

IFRC (International Federation of Red Cross and Red Crescent Societies), 2009. *Community based Psychosocial Support: Participant's book*. Denmark: IFRC.

Jha, M. K. and Pushpendra, 2012. Governing caste and managing conflicts. Bihar, 1990–2011. *Policies and Practices*, 46, 3–33.

Jha, M. K. and Raghavan, V., 2008. Disaster in Bihar: A report from the TISS assessment team. *Tata Institute of Social Sciences* [online]. Available from: www.mcrg.ac.in/Development/P_writing/Manish1.pdf [accessed 25 August 2011].

Kimberling, R., Mack, K. P. and Alvarz, J., 2009. Women and disasters. In: Y. Neria, S. Galea, F. H. Norris, eds, *Mental Health and Disasters*. Cambridge: Cambridge University Press.

Kumar, M., 2012. Governing flood migration and conflict in North Bihar [online]. Available from: www.academia.edu/1549952/Governing_Flood_Migration_and_Conflict_in_North_Bihar [accessed 14 November 2012].

Liu, A., Tan, H., Zhou, J., Li, S., Yang, T., Wang, J., Liu, J., Tang, X., Sun, X. and Wen, S. W., 2006. An epidemiologic study of posttraumatic stress disorder in flood victims in Hunan China. *Canadian Journal of Psychiatry*, 51 (6), 350–4.

Marsella, A. J., 2010. Ethnocultural aspects of PTSD: An overview of concepts, issues, and treatments. *Traumatology*, 16 (4), 17–26.

Mathew, S. and Moore, M., 2011. State incapacity by design: Understanding the Bihar story. *IDS Working Paper*, 366.

Mishra, D. K., 2008. The kosi and the embankment story. *Economic and Political Weekly*, 43 (46), 47–52.

Mishra, D. K., 2009. The legends of Kosi. *HiMAL South Asian* [online]. Available from: www.himalmag.com/component/content/article/468-The-legends-of-Kosi.html [accessed 04 April 2012].

Norris, F. H., Friedman, M. and Watson, P., 2002. 60,000 Disaster victims speak, Part I. An empirical review of the empirical literature, 1981–2001. *Psychiatry*, 65 (3), 207–39.

Norris, F. H., Baker, C. K., Murphy, A. D. and Kaniasty, K., 2005. Social support mobilization and deterioration after Mexico's 1999 flood: Effects of context, gender, and time. *American Journal of Community Psychology*, 36 (1/2), 15–28.

Norris, F. H., 2010. Summary and overview of disasters in developing countries. In: A. J. Marsella *et al.*, eds, *Ethnocultural Perspectives on Disasters and Trauma: Foundations, Issues, and Applications*. New York: Springer SBM, 383–94.

Oliver-Smith, A., 1986. *The Martyred City: Death and Rebirth in the Andes*. Albuquerque, NM: University of New Mexico Press.

Oliver-Smith, A., 1999. What is a disaster? Anthropological perspectives on a persistent question. In: A. Oliver-Smith and S. M. Hoffman, eds, *The Angry Earth*. London: Routledge.

Patrick, V. and Patrick, W. K., 1981. Cyclone '78 in Sri Lanka – The mental health trail British. *Journal of Psychiatry*, 138, 210–16.

Rao, K., 2006. Lessons learnt in mental health and psychosocial care in India after disasters. *International Review of Psychiatry*, 18 (6), 547–52.

Reddy, D. V., Kumar, D., Saha, D. and Mandal, M. K., 2008. The 18 August 2008 Kosi River breach: An evaluation. *Current Science*, 95 (12), 1668–9.

Renu, P., 2010. *Panchlight and Other Stories*. New Delhi: Orient Blackswan.

Sharma, A. N., Datta, A. and Ghose, J., 2012. *Development Research on Bihar, 2000–2010. A Compendium*. New Delhi: Institute for Human Development.

Shattuck, C., 1999. *Hinduism*. London: Routledge.

Singh, P., 2008. The Colonial State, *zamindars* and the politics of flood control in North Bihar (1850–1945). *Indian Economic Social History Review*, 45 (2), 239–59.

Singh, S. K., Panday, A. C. and Nathawat M. S., 2011. Rainfall variability and spatio temporal dynamics of flood inundation during the 2008 Kosi flood in Bihar State. *India Asian Journal of Earth Sciences*, 4 (1), 9–19.

Suar, D., Mandal, M. K. and Khuntia, R., 2002. Supercyclone in Orissa: An assessment of psychological status of survivors. *Journal of Traumatic Stress*, 15 (4), 313–19.

Summerfield, D., 2005. What exactly is emergency or disaster 'mental health'? *Bulletin of the World Health Organization*, 83 (1), 76.

Swartz, D., 1997. *Culture and Power. The Sociology of Bourdieu*. Chicago, IL: The University of Chicago Press.

Tedeschi, R. G. and Calhoun, L. G., 2004. Posttraumatic growth: Conceptual foundations and empirical evidence. *Psychological Inquiry*, 15 (1), 1–18.

Telles, S., Singh, N. and Joshi M., 2009. Risk of Posttraumatic Stress Disorder and depression in survivors of the floods in Bihar. *Indian Journal of Medical Science*, 63 (8), 330–4.

UNDP (United Nations Development Programme), 2009. Kosi floods 2008, How we coped! What we Need? Perception Survey on Impact and Recovery Strategies. *UNDP* [online]. Available from: www.undp.org.in/content/pub/CrisisPrevention/Kosi-Floods-2008.pdf [accessed 26 August 2011].

UNDP (United Nations Development Programme), 2010. *Human Development Report, 2010.* New York: Palgrave Macmillan.

Wisner, B., Blaikie, P., Cannon, T. and Davis, I., 2004. *At Risk: Natural Hazards, People's Vulnerability, and Disasters.* 2nd ed. London: Routledge.

PART 3

Unequal risks

Staging and reducing disaster risk

10

CELEBRITY CULTURE, ENTERTAINMENT VALUES ... AND DISASTER

David Alexander

Introduction

This chapter considers the role of celebrity in modern disasters and examines the effect of celebrity culture upon the way people interpret and react to catastrophe. I begin by considering the antecedents of celebrity culture in the rise of mass media coverage of disasters. I then enquire into the influence of celebrities on public reactions to catastrophe, upon the politics of disaster and upon the management of humanitarian activities. Examples are taken from Africa, because this continent has been a focus of interest for celebrities in rich countries. Next, I analyse the special case of the victim as celebrity, which is an artefact of the mass media culture in which we live. Finally, I endeavour to sum up the role and impact of celebrity culture in disaster risk reduction.

The origins of modern celebrity culture in the context of disasters

Over the period 1967–70 famine occurred sporadically in the Nigerian state of Biafra, where people starved to death because they were denied the ability to produce or procure food. There are some grounds for describing this as 'the prototype modern disaster'. It was the first in which television films of the effects of the famine were shown within hours of being taken – i.e., in near real-time. Viewers in rich Western countries were led to believe that Biafrans were starving because there was no food, or in other words that the causes of the disaster were natural. In reality, the Nigerian Government and Federal troops blockaded Biafra and attacked hospitals and feeding centres. Moreover, the embryonic Biafran Government behaved in no less a Draconian manner towards its own citizens (Ekwe-Ekwe 1990). In my opinion, the argument that this was the first modern disaster rests on the following evidence: first, it was a prototype of the sort of media spectacle that would later become the rule, rather than the exception. Second, attempts by NGOs, notably the French Red Cross, to maintain neutrality did more harm than good (and in fact dissatisfaction led to the founding, in 1971, of Médecins sans Frontières). Third, it was a complex disaster based on a proxy war, and interventionism of all kinds frequently did more harm than good.

Although there were strong elements of neo-colonialism in the Biafra war (Nkrumah 1966), it was a media-based participatory affair. In a very extreme example, the journalist Auberon Waugh (1939–2001), eldest son of the novelist Evelyn Waugh, named his son Nathaniel Thomas Biafra (Biafra Waugh, b. 1968). The seeds of celebrity culture were present. Meanwhile the use and glorification of foreign mercenaries by both sides kept alive the celebrity status of fighters, which had been so prevalent in 1939–45.

In the modern world, it is difficult – though not impossible – for disasters to go unpublicised by the mass media. Studies of the means by which this occurs are not copious but are accumulating slowly into a coherent body of work (Fearn-Banks 2011). In general terms it is concluded that the process of communication between the mass media and the general public is powerful but imprecise (Goltz 1984). In all its portentousness, this statement can be considered as either vague or ambiguous. However, it means either that people's opinions are deeply influenced by the mass media or, conversely, that the media have a great ability to mould opinions. The fusing of disaster with celebrity is driven by the public's insatiable appetite for personalities, human interest stories and individuals to which ordinary people can attach their fantasies (Morey *et al.* 2011). Copiousness and incessant novelty, much of it artificially manufactured, are the response of the media. Studies of press coverage of disasters reveal that at least half of what is published or broadcast consists of human interest stories (Scanlon and Alldred 1982). The fusion of these with celebrity is a natural extension that helps dramatise events and endows them with a lustre which increases their attraction to the consumers of news.

Defining celebrity and celebrity culture

A celebrity is by definition someone who is publicly celebrated because of his or her achievements. Clearly, celebrities have existed throughout human experience: in Western civilisation they were lauded in Greco-Roman epics, Scandinavian sagas, Medieval ballads, Victorian novels and many other places. Parallel hagiographies exist in other cultures: for example, the *Shahnameh*, or Book of Kings, by Abulqasim Ferdowsi (940–1020) is the history of the reigns of forty-seven Persian kings and three queens, all of whom were, to varying degrees, celebrities in Iran and neighbouring countries. There is much emphasis in this book on *farr*, the quality of majesty, regality and 'kingliness'. However it was defined, it represented a virtue that, along with riches and power, helped to perpetuate the mystique of the monarchy. Alongside the Persian monarchs were the astrologers, who for long periods of history were the celebrities as a result of their perceived ability to see into the future.

In Western culture, Michel de Nostredame ('Nostradamus', 1503–66) has been one of the most enduring figures in this respect (Nostradamus 2009).

In the modern world, power relations are constantly changing. One of the principal drivers of this process is the role of mass communication. To some extent, and for the purposes of this chapter, it can be reduced to the impact of mass entertainment (O'Neill and Harcup 2009). One of its greatest mutations concerns who, and what, are considered authoritative. Most people seem to feel a deep need to look up to someone. That person may be seen as a role model or a saint (an idealised personification of cardinal virtues), or simply as someone who is enviable or worthy of respect (Littler 2008). A leader is a person who offers both guidance and protection. This implies that such a person, if he or she is successful, is endowed with strength, intelligence, wisdom and foresight. However, that may be an idealistic vision of the qualities of leadership. A client-based relationship would merely require the leader to have power over the led. This may come from patronage and serfdom, and it survives very well in those environments in which power relations are formally structured in an exploitative manner.

A leader is not necessarily a hero, and vice versa (Toncar *et al.* 2007). That is one of the reasons why elected representatives (leaders) and the protagonists of the entertainment industry (heroes) can coexist so easily. As the latter are frequently much more enduringly popular, the former may look to them for inspiration when ratings are flagging. Celebrity culture exists when people, especially fans, look up to the celebrity because he or she is perceived to be authoritative as a result of his or her talent and fame. Hence, politicians defer to celebrities because they are revered by the public.

However, talent is not everything. In the age of mass communication, fame (like Chomsky's consent) is manufactured by barrages of publicity and hype (Herman and Chomsky 1995, Cottle 2006). Hence, there are people who are 'famous for being famous'. In many respects, this unproductive positive feedback situation underlines the power of celebrity in a world in which 'the medium [itself] is the message' (McLuhan 1964). Marshal McLuhan was one of the great theorists of media studies. Mention of his name prompts a brief examination of how valuable social theory may be to the understanding of celebrity and disaster.

A brief theoretical perspective

Whereas 'celebrity culture and disaster' is a theme that would fit – somehow – into a variety of social theories, from post-modernism to neopragmatism, the question remains outstanding as to how much benefit would be derived from forcing it into any such mould. Thus, Goffman's conception of 'identity stripping' (Goffman 1961) or Giddens's (1991) and Chomsky's (2002) ideas on the control of knowledge and power could be extended to cover the present case, but there is a serious risk of overextending them – i.e., reading more into the theory than it can offer in terms of explanatory power. At this point, one is mindful of Couch's (2000) struggles to characterise popular culture in relation to disasters, and what a heterogeneous mass of phenomena he was endeavouring to pin down.

Nevertheless, Giddens (1991) wrote of the 'sequestration of experience' (Smith 2002), meaning the ways in which those of us who are fortunate enough to live in wealthy societies have become relatively insulated from birth, death, physical injury, insanity and the sheer nastiness of life as it is lived in many less fortunate parts of the world. In my view this is an oversimplification: the experiences in question have been made more symbolic than real for those who do not experience them in first person, but they eventually catch up with us all, and then the test is whether we know how to cope with them. Nevertheless, Giddens (1991) was right to argue that in the modern age there is an unwillingness to confront major moral and existential issues. Furthermore, Williamson (1998, p. 26) suggested that we live much of our emotional lives by proxy. Blondheim and Liebes (2002) went even further in describing what they saw as the 'subversive potential' of live television, which provides the questions, but not the answers, to existential dilemmas.

By and large, we have moved from astrology and the interpretation of portents to the modern soothsaying of the 'talking heads' who forecast and comment via our mass media. Martin Amis lamented that 'what everyone has in them, these days, is not a novel, but a memoir. We live in the age of mass loquacity' (Amis 2000, p. 6).

Renn et al. (1992, p. 139) argued that 'hazards interact with psychological, social, institutional, and cultural processes in ways that can heighten or attenuate individual and social perceptions of risk and shape risk behaviour'. Much of the development of the social amplification of risk theory is based on the assumption that the process happens naturally, spontaneously or autonomously. What if it is deliberately managed? Pidgeon et al. (2003, p. 22) wrote as follows:

> Although the dramatisation of risks and risk events in the media has received much attention, the circularity and tight interrelations between the media and other components of social amplification processes (e.g., contextual effects, historical settings, interest group activity, public beliefs) render it difficult to determine the specific effects of the volume and content of media coverage.

Whereas the sheer copiousness and heterogeneity of media outputs make that observation generally true, it is less so when dealing with the more direct relationship between celebrities, the causes they espouse and the people who follow them.

After reviewing the cults and spectacles of history, Penfold (2004, p. 300) argued that the cultural reception of celebrities has developed beyond anything previously experienced in centuries past. This is perhaps an exaggeration, given, for example, the veneration of emperors, but there is no denying the change in values that has occurred.

The great rallying call that has echoed down to us from the Edwardian era is E. M. Forster's 'only connect'. Writing in his novel *Howard's End*, Forster meant one should connect across class barriers and their associated, rather contrived, moralities. Forster had to be discrete about his own homosexuality, which was then a crime in the eyes of the law. A Cambridge academic, he was very much the anti-celebrity. He would probably be amazed at how his rallying call has been taken up in the modern world, but also how it has lost its original meaning and come to symbolise the fear of being left out, not a loathing for artificial social boundaries.

The following investigation will, I hope, show that the cult and status of celebrity has a strong bearing on social attitudes to disaster, but that these remain fickle. One thing is certain: disaster is not better understood by the general public as a result of the intervention, or the manufacture, of celebrities. People's reactions to it are, however, easily moulded, even if only temporarily.

The power of celebrity

At the time of writing there is a major scandal in the United Kingdom over the activities of the late Jimmy Savile (1926–2011), entertainer and philanthropist. Amid a wave of public revulsion and official alarm, Savile has been posthumously accused of being a sexual predator and abuser of children. Although some accusations were made during his lifetime, his reputation was relatively unscathed throughout half a century of activities in the public view. In some measure, this is because attitudes towards sexual abuse were different fifty years ago and only changed slowly towards something more fair and accommodating from the perspective of the victims. In part it reflects the power of celebrity, which has grown and grown in recent decades. Savile apparently used his status as a public figure to cloak his many and monstrous predations, and in large measure he succeeded. In part this is because mass media celebrities have acquired power in the public arena, and in part it is because some of them are able to use their status, and the functions of the media, to control the image of them that is presented to the public.

Of course, fame and notoriety have always existed. They have frequently conferred power and influence upon those who acquire them. I do not wish to conflate royalty, leadership and celebrity more than is reasonable, but there is no doubt that great kings, queens and leaders have been celebrated throughout history, and in a significant number of cases merely because of the positions they held. Good examples of the powerful and celebrated exist in the Roman emperors Caligula, Hadrian and Augustus, or in Alexander the Great or Genghis Khan. However, the rise of the electronic mass media has not only greatly increased the sphere of recognition of the famous or notorious, but it has also altered the balance of who is a celebrity and what one needs to do to become one. It is a system that can easily be manipulated by whoever has the microphone and the public's attention, providing they can master the art of using these assets. Thus stars such as Paris Hilton and Kim Kardashian are photogenic, fashionable and 'famous for being famous'.

The 'Geldof/Bono factor'

My use of the term 'Geldof/Bono factor' refers to the charitable involvement of Robert Frederick Zenon (Bob) Geldof (b. 1951) and Paul David Hewson (alias 'Bono', b. 1960), both of whom are singers in rock bands. Geldof has a second career in venture capitalism that has reputedly made him a billionaire. Both individuals have become enormously influential on the basis of their efforts to mobilise the entertainment industry in favour of charitable work in Africa. Geldof's efforts began with the hit song 'Do They Know It's Christmas?' (possibly not, as many of the intended beneficiaries are not Christians!). He went on to organise the Band Aid (1984), Live Aid (1985) and Live 8 (2005) rock concerts. Bono has concentrated on disease reduction and debt cancellation in Africa.

Since the days of the industrialists Andrew Carnegie (1835–1919) and John D. Rockefeller (1839–1937), philanthropy has benefited the philanthropists as well as the recipients of their largesse. It appears to render the accumulation of immense wealth less unethical. It keeps the donors in the public eye (if that is where they wish to remain). It endows them with apparent moral authority, increases their connections with powerful decision makers and gives them influence over world affairs (Furedi 2010). High profile charity work by entertainment stars increases their celebrity and, not inconceivably, their earnings as well. Geldof in particular embodies the spirit of the age and the energy of its youth culture through the iconoclasm inherent in his use of offensive language and facile judgements. I am aware that the latter statement is Draconian, but it is backed up by the analyses of authors such as Müller, who penetratingly deconstructed the rock-concert-based approach to charity and in so doing observed (2013, p. 68) that 'The Band Aid representation of famine in Ethiopia has emerged more generally as a potent symbol of African collapse and the crisis of the post-independence project'. Yet Africa has not collapsed: indeed, it is growing vigorously in many ways, as one would expect of a continent with immense social, economic and cultural diversity.

On 15 December 2005, the writer Paul Theroux published an op-ed article in *The New York Times* called 'The Rock Star's Burden' (the title is derived from Rudyard Kipling's poem 'Take up the White Man's burden –/The savage wars of peace –/Fill full the mouth of famine/And bid the sickness cease'). Theroux (2005) criticised stars such as Bono, Brad Pitt and Angelina Jolie, labelling them as 'mythomaniacs, people who wish to convince the world of their worth'. Theroux, who lived in Africa as a Peace Corps Volunteer, added that 'the impression that Africa is fatally troubled and can be saved only by outside help – not to mention celebrities and charity concerts – is a destructive and misleading conceit'. Elsewhere, Bono has been criticised, along with other celebrities, for '[ignoring] the legitimate voices of Africa and [turning] a global movement for justice into a grand orgy of narcissistic philanthropy'. In a cartoon published in the British satirical magazine *Private Eye*, two emaciated Africans say 'We're holding a famine in favour of fading rock-stars'.

According to Theroux (2005) and Müller (2013), the main criticisms of the 'celebrity philanthropy' approach are as follows:

- Interventions are made on the basis of snap judgements about economic and aid problems that are not backed by adequate research. Many of the problems are too subtle and sophisticated to be encapsulated in a slogan or remedied by a simple action.
- Most celebrity philanthropists have ignored the political causes of hunger and instability in Africa, yet tackling these may be fundamental to the solution.
- There have been suggestions that aid was misappropriated, for example, by being given to insurgents, but this has been vigorously contested by the organisers of the various charity initiatives.
- Celebrity philanthropy tends to adopt a paternalistic rather than inclusive approach. It is a well-known axiom of human and economic development that it must help and encourage people to take control of their own destinies, as passive receipt of aid can be debilitating rather than helpful.

In the final analysis, the impact of celebrity philanthropy tends to be superficial rather than substantial.

One of the most controversial acts of celebrity philanthropy was the adoption, finalised over the period 2006–2009, of two Malawian children, David Banda Mwale and Chifundo James, by the American singer Madonna Louise Ciccone (alias 'Madonna', b. 1958). The adoptions, and Madonna's other charitable initiatives, were challenged by a Malawian NGO, the Centre for Human Rights and Rehabilitation. The attitude of the Malawian Government and Judiciary has been inconsistent as they have been torn between the benefits of charitable acts (Malawi has two million orphan children) and the need to safeguard minors against potentially unscrupulous behaviour. In this respect, Madonna's first adopted child is not an orphan, and there are suggestions that neither is the second. Hence, part of the basis of the challenge has been that children are being removed, unfairly, from their cultural and family milieu rather than given the chance to thrive in it. At the very least, Madonna's approach could be regarded as heavy handed. Her 'on–off' approach to the funding of schools in Malawi has been criticised by those whose interest in education and development in Sub-Saharan Africa is more enduring.

The difference between Madonna's strategy and that of Geldof and Bono lies in the role of popular support. Madonna acted unilaterally – at least, according to a spokesperson for the Malawian Government (Bronfen 2010, p. 180). Geldof and Bono rely upon mobilising the economic power of (mostly young) people who identify with the popular culture that they spearhead. As with all popular culture, the strategy relies upon being able to simplify issues.

Hitherto, the discussion has dealt exclusively with people who approach disasters from the starting point of already being celebrities. However, there is another class of person for whom a disaster is the opportunity to become a celebrity – those who, in the Napoleonic fashion, 'have greatness thrust upon them'. Among these there are people who accept the challenge reluctantly or with genuine altruism and those who bask in fame or notoriety for its own sake.

Celebrity victimhood

Fame, or at least notability, can be achieved through being a victim, but in the active, not the passive sense. Active victims typically fight for one of the following: justice, a safer future, recognition of a cause or compensation for losses or reprisals against those who are perceived to have instigated the harm. Hence, the following is the array of agendas that can transform active victims into celebrities:

- *Justice.* The disaster is presumed to have occurred as a result of an injustice. For example, people are killed when houses collapse in a disaster because they have not been built to conform to the prevailing building codes. The builder or designer may therefore be considered culpable. However, there is a substantial 'grey area' of responsibility in which vulnerability to disaster comes from ignorance of the consequences which may, at least in part, be excusable, and from divided responsibilities.
- *A safer future.* The victim wishes to fight to ensure that risks are reduced and in the future a disaster with the same characteristics of impacts does not occur again.
- *Recognition of a cause.* Public initiatives connected with, for example, reducing risks of increasing safety may be susceptible to failure through lack of official recognition or lack of funding and support. The victim feels that he or she has a personal stake in ensuring that the initiative succeeds.
- *Compensation.* A victim may become well-known for fighting for compensation, possibly on a grand scale, as a result of the alleged culpability of parties considered responsible for the disaster or its effects.
- *Reprisals.* Compensation suits are one of the main ways in which victims gain reprisals over those people or organisations that they deem are responsible for their misfortunes in disaster. Some of these actions can become high profile cases.

There is, of course, no inherent reason why pursuing any of these aims should convert the supplicant into a celebrity. However, the portrayal of victimhood by the modern mass media has transformed the whole process of advocacy by victims, presenting such people with the choice of whether or not to exploit the glare of publicity by projecting their personalities in certain ways. Victims may be part of 'disaster subcultures' (Granot 1996), in which their lives gain shape and substance from advocacy and association. The celebrities in disaster subcultures are usually those that exert a leadership function and keep the subculture alive.

Let us now examine how the remarkable transformation of victimhood has come about over the last 75 years or so.

The Oxford English Dictionary defines a victim as a person who has been killed or injured as a result of an event or circumstance. A survivor is defined as a person who remains alive after experiencing danger, an accident or a disaster, or continues to exist in spite of such a contingency. The following working definition will be adopted in this paper: a victim or survivor will be considered here to be a person who has been fully involved in a major incident, disaster or catastrophe and who has in some way suffered, but who pulls through and is able to recount or discuss the experience afterwards.

A century ago, celebrity and victimhood were poles apart. That is no longer true. The significance of victimhood has changed over the last century. In modern society, people who have suffered deep travails achieve a special status (Lifton 1980). Whereas in the past victims were often regarded merely as people who had suffered disgrace (with or without culpability, according to circumstance or credence), now they are listened to by investigators, politicians and the mass media with a special respect, sometimes almost with reverence. The modern survivor has captured the moral high ground. Ordinary citizens can feel thankful that they have not been put through the same mill as the victims who they see or hear interviewed on news broadcasts. But like those officials who conduct enquiries into disasters, viewers and listeners can appreciate the sense of moral outrage that nowadays accompanies victimhood. It relates very well to Horlick-Jones's (1995) model of disaster as a betrayal of trust by the authorities who, by means of procedure and regulation, were expected to keep people safe.

At 1400 hours on 25 January 2003, a London Underground train derailed in a tunnel just outside Chancery Lane station. The train left the tracks and scraped along the tunnel wall. Thirty-two people were slightly injured and many more suffered a brief entrapment in smoke and darkness. They emerged onto the platform covered in soot and dust. Accounts of their experiences were published by the BBC the following day. The interviews reveal a sense of self-importance and a desire to dramatise the incident to the level of a veritable 'brush with death'. The moral outrage may come from the uneasy contemporary relationship between citizens and civil authorities, but the real status of victimhood is conferred by the mass media (BBC News 2003). In cases such as this, exaggeration and dramatisation are the means by which it is achieved.

When unusual adverse events such as disasters occur, people struggle to endow them with meaning and explanation. At the physical level of natural hazards, this has led to, for example, confusion between the meaning of 'weather' and 'climate' and lack of understanding of the established fact that global warming can lead to more general extremes of weather, including excessive or prolonged cold spells. Natural *hazards* at least have the advantage of being morally neutral phenomena – *Natura enim non nisi parendo vincitur* (Nature, to be commanded, must be obeyed). The problem with *vulnerability* to hazards (whether natural or anthropogenic) is to find a focus for the outrage. In the words of Horlick-Jones (1995, p. 305),

'Disasters in modern society contain strong elements of a release of repressed existential anxiety, triggered by a perceived betrayal of trust by contemporary institutions'. Horlick-Jones (1996) went on to analyse the concept of blame as part of this release mechanism. As Bucher (1957, p. 467) put it, 'blaming for disasters arises out of seeking a satisfactory explanation for something which cannot be accounted for conventionally'. That, of course, is not enough to apportion blame, which requires some assumption of culpability, or that the responsible agencies will not take action to prevent a recurrence. At its worst, blame can be an attempt to deflect or transfer responsibility to other people.

At the root of the fusion of victimhood, celebrity and outrage is the modern culture of *public intimacy*: grieving, rage and sharing intimate details with strangers. It is coupled with *disinhibition*, the very public demonstration of intimate emotions (Hjorth and Kim 2011). The corollary in times of peace is the daily televised diet of emotion and money, which engenders the game show mentality through constant displays of gambling for money and celebrity. Hence, the public disinhibition of victims is coupled with a degree of public *voyeurism*. An alternative view is provided by West (2004, p. 14), who argued that 'these public displays of emotion [. . .] have a cathartic function, and serve as a means to "(in)articulate our own unhappiness"'.

In the televising of emotion, grieving and outrage, there are strong temptations to overplay the scene, which leads to a culture of exaggerated offendedness – exemplified by the common declaration 'I was devastated'. The demonstration of moral outrage is exploited as a weapon of self-aggrandisement. Because there is a consensus that something is wrong, very rarely are those who show their manufactured rage, grief or offence in front of the cameras questioned or cut down to size. They are instead taken seriously.

At this point, at the risk of being accused of abdicating my responsibilities as a researcher, I will leave it to the reader to decide how much substance there is in the world of victims and celebrities when they are confronted by the painful realities of death and destruction in disasters. Ever since the times of B. T. Barnum and his circus, showbiz has wanted to demonstrate that it has a heart. This is its way of trying to compensate for its endless orgy of self-absorption. Some of its charitable work has been absolutely laudable, other aspects have been debatable – as in the case of the Geldof concerts – and other elements have masked a much less respectable agenda, as in the case of the entertainer Jimmy Savile.

Conclusions

In conclusion, celebrity culture appears to have gone about as far as it can in influencing world affairs. Major celebrities have beat a path to the high table of international politics, for example at the Davos World Economic Forum. They sing and perform at major state events such as political inaugurations and funerals. Their opinions are heeded by politicians and officials. In some cases they are even given a semi-official role by being appointed as 'cultural ambassadors', a role that has been brilliantly satirised by the comedian Barry Humphries, who portrayed such a character – perhaps unfairly – as the lecherous, boorish and ignorant Australian 'cultural attaché' Sir Les Patterson, master of the vulgar witticism (Wikipedia 2013).

Celebrities vary in the extent to which they become knowledgeable about the complex humanitarian and environmental issues in which they have decided to involve themselves. Some have become quite authoritative, and revealed a talent for dealing with the issues, while others have miscalculated through ignorance or, above all, a tendency to oversimplify. In neither case has this had a negative influence on their standing with their followers, in whose eyes the advocacy of causes tends to be a sideshow that accompanies the business of entertainment.

In the past, monarchs could direct or govern the response to disasters by perceived divine right, the exhibition of power, their ability to command or, latterly, their constitutional role as leaders. In two decades of reconstruction after the 1755 Lisbon earthquake, the Marquis of Pombal was only able to employ Draconian measures and create innovative recovery strategies by expropriating, as Viceroy, some of the authority of a weak and ineffectual king (França 1983). In order to maintain support and status, contemporary monarchies, bound by parliaments and constitutions, increasingly have to appear – and perhaps behave – in the same way as media celebrities such as the stars of film or popular music. A convinced republican might say that they are equivalent to those stars who are simply 'famous for being famous'. Bronfen (2001, 2010) saw this in the death of Princess Diana. Although she was known for no particular exploit, her image became a universal referent for a host of feelings and aspirations (Thomas 2008). In the end, one has to hope that celebrities of any kind who involve themselves in humanitarian causes are sufficiently influenced by technically competent advisors.

Contributing money to a cause is a voluntary action that should be the result of a free decision about whether or not to give charity. However, it has long been known that the mass media can turn on or turn off the flux of donation according to the publicity given to an event or a cause (Olsen *et al.* 2003). However, modern means of communication – and donation – have 'ramped up' this phenomenon to unprecedented levels. Despite this, every day we are bombarded with information, and this endless and indiscriminate process has increased the incidence of 'attention span deficit' and 'donor fatigue'. It has also much reduced the ability of adverse events to shock people.

Regarding disasters, the magnitude of donations is closely related to the sense of involvement experienced by the donors. The Indian Ocean tsunami of 26 December 2004 led to public donations of US$4.5 billion (matched by even larger official ones). At the same time, a United Nations appeal for $30 million to fund crisis response in Darfur failed to reach its target, even though this was only two thirds of a percentage point of what was contributed to the survivors of the tsunami and earthquake. For the donors, Darfur appeared remote and perhaps incomprehensible, but on the beaches of Thailand, Bali and India, the casualties included many Westerners and this brought the catastrophe home to them. Indeed, the loss of 543 lives was the worst disaster mortality for Sweden for almost 300 years (Olofsson 2011). If familiarity can stimulate monetary involvement in a disaster aftermath to this extent (under the tacit assumption, 'it could have happened to me on my last tropical holiday'), one can imagine what more identification with celebrities can do. In fact, the case of Bob Geldof and hunger and disease in Ethiopia illustrates how swearing at potential donors and abruptly facing them with starkly simplified moral dilemmas is capable of eliciting a donation rate of £300 (€350, $475) per *second*.

Despite such dramatic statistics, it is as well to remember that celebrity is no guarantee of influence – or longevity of reputation. Toncar *et al.* (2007, pp. 272–3) found that celebrity endorsement did not necessarily render a public service emergency message legitimate. Many commentators have sought to treat celebrity involvement in humanitarianism and other causes as evidence of a desperate desire to hang on to celebrity status when contracts for musicianship or acting have ceased to be awarded. Only in a very few cases does celebrity victimhood endure and then not necessarily for happy motives, but more often as a result of continued suffering. In the United Kingdom, for example, Doreen Lawrence remains a celebrity figure (and a much respected one) not merely because her son Stephen was the victim of a high profile racist murder in 1993, but because since then her model of dignified advocacy for racial justice has been constantly in demand, and her family have never been free of racial persecution. Meanwhile, there are journalists who feel profoundly uneasy about the role of the mass media in creating and sustaining celebrity (cf. Ponce de Leon 2002, Snyder 2003).

Lastly, it would be interesting to explore how similar processes have acted to stimulate support for religious fanaticism, so-called 'fundamentalism' and the celebrity status of people such as Osama Bin-Laden. However, that would require another chapter.

References

Amis, M., 2000. *Experience*. London: Vintage.
BBC News, 2003. Tube crash: Eyewitness accounts. *BBC News*, 25 January 2003 [online]. Available from: http://news.bbc.co.uk/1/hi/england/2694503.stm [accessed 12 October 2012].
Bennett, R. and Kottasz, R., 2000. Emergency fund-raising for disaster relief. *Disaster Prevention and Management*, 9 (5), 352–9.
Blondheim, M. and Liebes, T., 2002. Live television's disaster marathon of September 11 and its subversive potential. *Prometheus*, 20 (3), 271–6.
Bronfen, E., 2001. Fault lines: Catastrophe and celebrity culture. *European Studies*, 16, 117–39.
Bronfen, E., 2010. Celebrating catastrophe. *Angelaki: Journal of the Theoretical Humanities*, 7 (2), 175–86.
Bucher, R., 1957. Blame and hostility in disaster. *American Journal of Sociology*, 62, 467–75.
Chomsky, N., 2002. *Media Control: The Spectacular Achievements of Propaganda*. New York: Seven Stories Press.
Cottle, S., 2006. Mediatized rituals: Beyond manufacturing consent. *Media, Culture and Society*, 28 (3), 411–32.

Couch, S. R., 2000. The cultural scene of disasters: Conceptualizing the field of disasters and popular culture. *International Journal of Mass Emergencies and Disasters*, 18 (1), 21–38.

Ekwe-Ekwe, H., 1990. *The Biafra War: Nigeria and the Aftermath*. Lewiston, NY: Edwin Mellen Press.

Fearn-Banks, K., 2011. *Crisis Communications: A Casebook Approach*. Boca Raton, FL: CRC Press.

França, J-A., 1983. *Lisboa pombalina e o illuminismo*. Lisbon: Bertrand.

Furedi, F., 2010. Celebrity culture. *Society*. 47, 493–7.

Giddens, A., 1991. *Modernity and Self-Identity: Self and Society in the Late Modern Age*. Stanford, CA: Stanford University Press.

Goffman, I., 1961. *Asylums: Essays on the Social Situation of Mental Patients and Other Inmates*. New York: Doubleday.

Goltz, J. D., 1984. Are the news media responsible for the disaster myths? A content analysis of emergency response imagery. *International Journal of Mass Emergencies and Disasters*, 2 (3), 345–68.

Granot, H., 1996. Disaster subcultures. *Disaster Prevention and Management*. 5 (4), 36–40.

Herman, E. S. and Chomsky, N., 1995. *Manufacturing Consent: The Political Economy of the Mass Media*. New York: Vintage.

Hjorth, L. and Kim, K-H. Y., 2011. The mourning after: A case study of social media in the 3.11 earthquake disaster in Japan. *Television and New Media*, 12 (6), 552–9.

Horlick-Jones, T., 1995. Modern disasters as outrage and betrayal. *International Journal of Mass Emergencies and Disasters*, 13 (3), 305–15.

Horlick-Jones, T., 1996. The problem of blame. In: C. Hood and D. Jones, eds, *Accident and Design*. London: UCL Press, 61–70.

Lifton R. J., 1980. The concept of the survivor. In: J. Dimsdale, ed., *Survivors, Victims and Perpetrators: Essays on the Nazi Holocaust*. New York: Hemisphere.

Littler, J., 2008. 'I feel your pain': Cosmopolitan charity and the public fashioning of the celebrity soul. *Social Semiotics*. 18 (2), 237–51.

McLuhan, H. M., 1964. *Understanding Media: The Extensions of Man*. New York: Mentor.

Morey, Y., Eagle, L., Kemp, G., Jones, S. and Verne, J., 2011. Celebrities and celebrity culture: Role models for high-risk behaviour or sources of credibility? *Second World Non-Profit and Social Marketing Conference*, 11–12 April 2011, Dublin [unpublished].

Müller, T. R., 2013. 'The Ethiopian famine' revisited: Band Aid and the antipolitics of celebrity humanitarian action. *Disasters*, 37 (1), 61–79.

Nkrumah, K., 1966. *Neo-Colonialism: The Last Stage of Imperialism*. New York: International Publishers.

Nostradamus, 2009. *The Prophecies by Nostradamus*. Oxford: Acheron Press.

Olofsson, A., 2011. The Indian Ocean tsunami in Swedish newspapers: Nationalism after catastrophe. *Disaster Prevention and Management*, 20 (5), 557–69.

Olsen, G. R., Carstensen, N. and Høyen, K., 2003. Humanitarian crises: What determines the level of emergency assistance? Media coverage, donor interests and the aid business. *Disasters*, 27 (2), 109–26.

O'Neill, D. and Harcup, T., 2009. News values and selectivity. In: K. Wahl-Jorgensen and T. Hanitzsch, eds, *Handbook of Journalism Studies*. New York: Routledge, 161–74.

Penfold, R., 2004. The star's image, victimization and celebrity culture. *Punishment and Society*, 6 (3), 289–302.

Pidgeon, N., Kasperson, R. E. and Slovic, P., eds, 2003. *The Social Amplification of Risk*. Cambridge: Cambridge University Press.

Ponce de Leon, C. L., 2002. *Self-Exposure: Human-Interest Journalism and the Emergence of Celebrity in America, 1890–1940*. Chapel Hill, NC: University of North Carolina Press.

Renn, O., Burns, W. J., Kasperson, J. X., Kasperson, R. E. and Slovic, P., 1992. The social amplification of risk: Theoretical foundations and empirical applications. *Journal of Social Issues*, 48 (4), 137–60.

Scanlon, T. J. and Alldred, S., 1982. Media coverage of disasters: The same old story. *Emergency Planning Digest*, 9, 13–19.

Smith, C., 2002. The sequestration of experience: Rights talk and moral thinking in 'late modernity'. *Sociology*, 36 (1), 43–66.

Snyder, R. W., 2003. American journalism and the culture of celebrity. *Reviews in American History*, 31 (3), 440–8.

Theroux, P., 2005. The rock star's burden. *The New York Times*, 15 December [online]. Available from: www.nytimes.com/2005/12/19/opinion/19ihtedtheroux.html?page wanted=all&_r=0 [accessed 31 January 2013].

Thomas, J., 2008. From people power to mass hysteria: Media and popular reactions to the death of Princess Diana. *International Journal of Cultural Studies*, 11, 362–76.

Toncar, M., Reid, J. S. and Anderson, C. E., 2007. Effective spokespersons in a public service announcement: National celebrities, local celebrities and victims. *Journal of Communication Management*, 11 (3), 258–75.

West, P., 2004. *Conspicuous Compassion; Why Sometimes it Really is Cruel to be Kind*. London: Civitas (Institute for the Study of Civil Society).

Wikipedia, 2013. *Sir Les Patterson* [online]. Available from: http://en.wikipedia.org/wiki/Sir_Les_Patterson [accessed 31 January 2013].

Williamson, J., 1998. A glimpse of the void. In: M. Merck, ed., *After Diana: Irreverent Elegies*. London: Verso, 25–8.

11

DISASTER MANAGEMENT CULTURE IN BANGLADESH

The enrolment of local knowledge by decision makers

Brian R. Cook

Introduction

The question of how disaster managers operationalise local knowledge is central to understanding the role of culture within disaster risk reduction (DRR). This is particularly true for anyone who views culture as a way of 'localising' or 'demo-cratising' risk management (Kearnes *et al.* 2012, Wilcock 2013). Disaster managers are key governance actors, with significant influence on DRR. But they also have less obvious influence on the knowledge claims that are admitted into management discourse. In this way 'disaster management culture' can be analysed according to its treatment of 'local culture'. By this, I mean that there exists a culture among disaster managers – one that has to date been neglected by risk researchers – and that how those managers interpret, value and deploy local knowledge is central to understanding the role of culture within DRR (Hewitt 2012). Local culture is not inserted into a vacuum when drawn into DRR. Rather, it comes into an existing, powerful and often resistant context in which management culture determines its place, purpose and influence. To date, much discussion of culture and DRR has been located 'outside' of decision making, with critiques aimed at exposing or disrupting existing hierarchies of power (cf. Epstein 1995, Irwin 1995, Epstein 1996, Wynne 1996). By exploring 1) what decision makers think of local knowledge and 2) the degree to which local knowledge is incorporated into discussions of disaster management, we can assess the impact(s) of efforts to promote the integration of 'culture', what I refer to as the 'cultural critique'. Exploring how disaster managers perceive local knowledge can help us consider the wider successes or failures of cultural critiques, particularly of the hierarchies of knowledge-power that have traditionally privileged experts and science over locals and lay forms of knowledge.

Some of the most persuasive examples of cultural critiques of risk management are situated in Bangladesh (cf. Sillitoe 1998, Yunus 1999, Sillitoe 2000, Sillitoe and Marzano 2009), with many exploring the issue of flood management (cf. Haque 1988, Rasid 1993, Rasid and Mallik 1993, Rasid and Mallik 1995, Duvail and Hamerlynck 2007, Cook and Lane 2010, Paul and Routray 2010, Dewan 2013). Perhaps most radical are those directed at flood management rooted in the Bangladeshi language (Paul 1984, Alam 1990), which distinguishes between *barsha* (beneficial floods or aspects of floods) and *bonna* (detrimental floods or aspects of floods). Reviews of flood management in Bangladesh are given elsewhere (cf. Paul 1997, Brammer 2004, Höfer and Messerli 2006, Cook 2010), allowing this analysis to focus on the state of efforts to legitimise local knowledge and participatory decision making, what I believe to be key tenets of the cultural critique. To do this, I analyse the ways in which decision makers interpret, value and deploy local and indigenous knowledge when discussing flood management in Bangladesh.

I argue that the cultural critique of expert-led, scientific flood management (cf. Cook 2010 for a review of this dominant form of flood management) appears to have been fully embraced. Decision makers accept, value and incorporate a wide range of non-scientific knowledge claims into their understandings and justifications of specific flood management measures. In line with critiques of dominant forms of governance, local and indigenous knowledge has been embedded in scientific and expert rationalisations of flood management. Despite this openness towards non-scientific knowledge, unfortunately it appears that the idealised power sharing and participation that is implicit within cultural critiques and advocacy for the inclusion of local knowledge has failed. Rather than democratise flood management, the decision makers interviewed for this project appear free to enrol almost any knowledge in support of their views. This includes knowledge that would have previously been excluded as untrustworthy or biased. More strikingly, the examples of local and indigenous knowledge drawn into discussions are often personal recollections, hearsay and anecdote apparently deployed to maintain the hierarchy between governors and those governed. In this case, rather than empower locals, it appears that the legitimisation of local knowledge has helped widen the gap between those who decide and those who are affected. Overall, decision makers appear to have appropriated the cultural critique of expert-led scientific flood management, further entrenching their positions of privilege and power.

This chapter begins with a brief description of water and flood management in Bangladesh. The technological and technocratic identity of flood management, and the resulting cultural critiques, are summarised to show the controversy that has arisen. This raises questions on the impact of culture-based critiques of flood risk management, and about efforts to incorporate concepts of 'culture' into Bangladeshi flood management more specifically. The conclusions represent a warning to those attempting to integrate 'culture' into DRR. Efforts to include a wider range of knowledge into governance, and the related (but more difficult) efforts to ensure democratic representation within decision making, remain constrained by the sturdy boundaries of the disaster management culture. This culture, much like that associated with local and indigenous knowledge, turns out to be a powerful and adaptive force that seems impervious to logic or moral argument. Instead, those seeking to insert local culture into DRR – and in so doing reshape disaster management culture – must be alive to the ways in which dominant cultures are able to respond and reshape themselves in order to maintain hierarchies of power.

The technological basis of flood management

Flood management in Bangladesh is a technological and technocratic issue dominated by engineers and a scientific framing. This is for good reasons. The Ganges–Brahmaputra–Meghna basins are large, dynamic and densely populated. Each of these rivers flows into the Bay of Bengal through Bangladesh. The often-quoted figure is that Bangladesh controls only 7% of the basin, thereby limiting its ability to manage floods (Brammer 2004, Höfer and Messerli 2006, Chellaney 2011). The climate in the region is dominated by the monsoon, leading to claims that most of the water passing through Bangladesh occurs in just a few months (Höfer and Messerli 2006). In addition, it is the confluences of these rivers and the timings of their peak flows that tend to determine the severity of floods (Brammer 2004). The geography, then, was central to the emergence of a scientifically based and technologically dependent approach to flood management.

Given the population, poverty and urbanisation within the wider basins, but particularly concentrated in Bangladesh, damaging floods have long been a characteristic of this part of South Asia. In particular, the agricultural system and dependence on rice as the primary staple mean that flood management is synonymous with food management. Until the 1970s, when small scale pumps replaced large scale irrigation, flood control was needed to protect farmland in order to ensure sufficient production and to avert widespread famine (Sen 1981, Clay 1985, Yunus 1999). The agricultural needs and the potential for famines drove approaches to flood management that necessitated control, and which involved large scale irrigation. While this approach is no longer acclaimed, it remains central to water and flood management in Bangladesh, as do the associated science based and technological approaches.

Together, the geography and the socioeconomic context – as well as mimicry of the developed world – have contributed to a flood management regime that has historically depended on engineers and engineering works for protection. In response to this concentration of power, and the resulting mismanagement, critics attacked decision making for being too removed from local contexts, values and knowledge (Paul 1997, Cook 2010). The cultural critique, which is best exhibited by the 'living with floods' movement (Shaw 1989, Zaman 1993, Ahmed 1999), sought to legitimise local knowledge with the aim of integrating local individuals into decision making. This 'alternate culture of flood management' was extremely influential and, in tandem with worldwide disenchantment with technical flood management, the cultural critique helped to prevent a massive World Bank supported project (the Bangladesh Flood Action Plan) to construct polders[1] for the entire nation (Boyce 1990, Paul 1995, Cook and Wisner 2010). What is less clear is the impacts of these relatively successful criticisms in the longer term, which is the basis of this chapter.

The analysis of decision maker perceptions is inhibited by the sample. Flood management is a diffuse and contingent process involving countless individuals, things and ideas. How can researchers assure themselves that their collection of individuals can be representative of the amorphous collection of people who contribute to something as cross disciplinary and cross sectional as flooding? For this project, I avoided this challenge by choosing, instead, to take individuals and organisations involved in negotiating past flood management as representative of the assemblage of 'power holders'. By retrospectively identifying who had influence, this project avoids the need to predetermine who shapes flood management by noting who has shaped flood management. A sample of fifty-four individuals and organisations were interviewed who contributed to the 'Options for flood risk and damage reduction in Bangladesh' report following the catastrophic floods of 2004 (Siddique and Hossain 2006). These participants, from NGOs, government, the academy, donors, consultancy firms and international organisations (e.g., ADB, IMF, World Bank), were interviewed for between thirty and ninety minutes during November 2007 and February 2008. A semi-structured interview template was used that focused on floods and flood management and the data transcribed, coded and analysed by the author.

How decision makers interpret, view and value local knowledge

Discussions with decision makers in Bangladesh and those who contribute to flood management show that local people, local knowledge and participation have become central to the flood management discourse. Throughout the interviews, experts and authorities wove tapestries of science, expertise, local knowledge and participatory governance to inform specific views concerning floods and flood management. As a group, though, the discussions have within them a more duplicitous theme: acceptance and supposedly valuing local knowledge still enables locals to be excluded from meaningful participation. This enrolment of local knowledge and people is key to this analysis because it shows the entrenchment of an unequal relationship between experts and locals, as portrayed by experts (Davies 2008).

The findings are divided into three groups, drawn from discussions in which decision makers spoke about locals, local knowledge and/or participation: 1) examples of foolish decision making; 2) personal knowledge and recollections; and 3) complementary knowledge. Each group, in decreasingly obvious, overt or direct ways, (re)asserts the traditional power relationship between experts and locals and between science and local knowledge. This claim is revisited in the ensuing discussion.

Foolish knowledge and decision making

During discussions of flood management in Bangladesh, the decision makers recount countless examples of decisions that were made for spurious reasons. Their disappointment is directed primarily at the government and at the ways flood management is done, including the knowledge on which it is based. These opinions tend to contrast corruption, short-termism and self-serving efforts against more rational and evidence-based (i.e., scientific) positions:

> We can travel anywhere in the country by road, but unfortunately the roads were built without proper knowledge. So, whatever vulnerability we have now, it is because of actions made during both the Pakistani and Bangladeshi periods [. . .]. [Those in power] make the decision on the basis of personality. For example, in my area, if there is an influential minister, so what he thinks they will do. That will be how decisions are made, without study. Even worse when studies are made with the intention of reflecting his will.
>
> (Academic, 8 December 2007)

This respondent, who is an academic who also consults for government, expands on this claim by re-emphasising the poor basis on which flood management is grounded:

> Our politicians, the tradition has come in such that political decisionmaking is not knowledge based. The political parties do not have think tanks of their own. They have some intellectual groups; they use them to support their initiatives. So the political decision making is their own, it is whimsical.
>
> (Academic, 8 December 2007)

For many of the flood managers, having worked in large scale, technical projects, the political economy of flood management was blamed. These critiques tended to emphasise the international community. Confirming the cultural critiques that resulted from mismanagement, an engineer argued that local knowledge and participation are needed to balance international cronyism and government complicity:

> I will blame the Dutch and American government. They did not know the environment of the country: they don't know the people of the country; their plans explain the problems in their countries. At the time our capacity – our political institutions – were not able to deal with it. In any case, just simple transfer of development technology is not always helpful unless you look at the local knowledge or the local history.
>
> (Academic Engineer, 15 December 2007)

Explaining a similar view, the head of an NGO that specialises in water and disaster recovery adds:

> What I see, is that [donors] do not consider the long term impacts of their investment. For example, building a bridge may create a disaster in fifteen or twenty years, and construction of roads in many places my cause water logging or increase flood risk. Those sort of long term perspectives are not being given consideration by donors. That is on one side, on the other side the politicians of the country think 'I am here for five years'. So he thinks that he has to implement something in that time to show the people that he is working hard. This kind of political commitment has happened with all of the governments that we have had. They only think about their power, their existence. Therefore, things are not very integrated, not very well planned-ahead.
>
> (NGO, 30 January 2008)

Overall, the respondents present a strikingly uniform view of government in the context of flood management: that it is driven by political and economic forces in ways that do not align with 'best practices'. In most cases, the tone of the discussions was frustration and exasperation, as the managers struggled to implement measures that would protect Bangladesh from catastrophic floods. Returning to local knowledge, this characterisation of government and the processes of governance delineated clearly between the experts and politicians, and between science and political economy. These views were, more forgivingly, also attributed to local people and to local knowledge.

In most cases, criticism of local knowledge emphasised the scale of flood management in Bangladesh. Local knowledge was not wrong so much as wrong for the task. A foreign consultant long involved in Bangladeshi flood management provided a representative view, explaining that flood management requires knowledge that is outside the reach of locals:

> In this current project we are involved in, flood protection and some of the fixes are a little bit of tough love. And they involve flood walls that are in some cases three metres high. In some cases the older people have access problems and they've lost their view of the river. And in these cases the flood may not come for five or ten years. And you have to maintain the walls and make sure that the pumps work. People have short memories, in a couple years there are sections missing. People get frustrated and break them down.
>
> (Consultant, 2 February 2008)

According to the decision makers who inform this analysis, local people, like the politicians, are influenced by concerns that inhibit effective flood management. Access and aesthetics, in this example, are used to explain that locals – who had requested flood protection to begin with – could not deal with the ramifications of that protection. Expressing a similar view, a modeller explains how locals are presumed to inform their decisions:

People make decisions based on their memories or past experience 'now I have to do this' but they don't actually link it with available data. It is an extremely complex river network. You cannot actually design or decide about a plan by looking at one particular river. You cannot be sure that the water isn't coming from another river. To know all of these things you must have a clear picture of the locality in terms of the national river system.

(Engineer/Modeller, 26 January 2008)

As a group, the decision makers commonly combine different types of knowledge, which appears to be a way of showing that they are aware of the local context and local knowledge. This association with 'local' and 'local knowledge' appears to be a way of communicating that the decision makers 'get' or 'are' local, possibly burnishing their status as knowledgeable experts. This continues within the second group of interview extracts (personal knowledge and recollections), in which these individuals responsible for shaping flood management – which has been criticised for its detachment from local contexts – personalise flooding. Unlike the direct criticisms of government and local knowledge, though, these personalisations of flooding in Bangladesh less overtly maintain the hierarchy between expert and lay knowledge.

Personal knowledge and recollections

A social scientist who advises government and NGOs on development and flood management provides a detailed example of how many of the respondents drew on their personal histories and experiences to personalise floods and local knowledge. In these instances, the decision makers used their memories to inform particular claims; interestingly, the memories were also locally situated, often with reference to 'home villages' outside of urban centres:

We have small and big floods, and the small is ok, and actually this year we had a medium flood, but the big ones almost change Bangladesh into an ocean. But you still see people laughing and joking and I said 'my gosh' because I was mainly stationed in Dhaka [. . .]. I saw the flood and thought to myself that it was unthinkable. And it changes so fast, the way people live. Even they start cooking on the roof, and they are playing cards. It is unthinkable. They worry, I should not minimise that, but their resilience is far more superior. And then when the water leaves they go straight out into the field and start working and when you return in six months the rice is back and you don't recognise the place as the one you saw flooded.

(Academic, 2 January 2008)

In this case, an experience from the individuals' early career showed that conceptions of vulnerability and resilience needed to be situated and informed by local adaptive capacities. Interestingly, this example was made as part of a discussion in which the decision maker was advocating 'living with floods', which has itself become emblematic of alternate flood management and culture based critiques. 'Living with floods' prioritises human adaptation rather than river management and, in the Bangladeshi context, is based on recognition that the scale of the rivers is often beyond human control (Shaw 1989, Zaman 1993, Ahmed 1999). Importantly, though, the decision maker was advocating living with floods only for rural locations which, because of his experience, he categorised as 'not terribly affected' by floods. This view, like that in the following extract, shows how experience with local contexts – and the resulting local knowledge – is drawn into descriptions of flooding in Bangladesh. More importantly, though, this localised knowledge is used to buttress the expertise held exclusively by these decision makers. This deployment of local knowledge, albeit personalised, reinforces their authority by displaying their awareness of local concerns and experiences; moreover, they are able to integrate the local and the expert. An engineer repeats this process as part of a discussion in which sedimentation is blamed for failed river engineering, rather than the engineering itself:

> Drainage and also the flooding with waterlogging creates a lot of problems and lost lands. Let me give you an example: in Mohakhali this sort of flooding does not take place from the major river, but from water logging because the water cannot find its way to the sea. I come from that area. In my student life, we never had any problems with water logging; over the years, sedimentation has occurred in the outflow channel. And now they need to dredge the outflow channels. Since dredging is very costly, the government has abandoned the practice and now every year there is flooding, not because of the river, but because of the sedimentation, even when there is no flooding anywhere else in the country.
>
> (Engineer/Modeller, 26 January 2008)

This combination of personal and expert knowledge to argue a particular position relating to flood management is repeated by a government advisor, who draws from childhood experiences as a way to justify efforts to empolder Bangladesh's coastal region, and to defend those now criticised efforts (Rasid and Mallik 1993, Warner 2010):

> IECO [an international engineering firm] was also appointed to look into the problems in the coastal areas. Today, the people of the coastal areas have forgotten what it was like to be so vulnerable. In the 1950s and 1960s, my father was posted in Mohakhali and I remember every new and full moon the whole area would go underwater. I would have to wade through knee-deep water and go to the railway line, which would be above water, and walk the five kilometres to my school. I was ten but this is vivid in my memory. This was flooding that resulted from primarily from tidal action. So what was suggested was polders. So they built polders.
>
> (Government Advisor/Lead Consultant, 15 January 2008)

Like the preceding excerpts that criticise government and local knowledge, the personal knowledge and recollections again, though less overtly, reassert a hierarchy between experts and lay individuals. With accounts of their localised experiences, the decision makers respond to cultural critiques of scientific and expert-led governance. Moreover, they are drawing local knowledge into their rationalisations and discussions of flood management, showing that they value such knowledge. This personal knowledge is not challenged nor out of place, which shows how these types of knowledge have become essential to discussions of effective flood management. That it makes the expert even more central to governance is, likewise, not considered. The result, though, is clearly an empowerment of the expert: it is their knowledge, their experiences, their histories and their interpretations of local knowledge that are legitimised within the discourse. These excerpts maintain, and possibly entrench, the authority of the decision maker, but these personalised knowledge claims are – despite their contributing to a power hierarchy between experts and locals – obvious. More disconcerting are discussions that assert the value of local knowledge and participations, but which simultaneously and less overtly reassert the power and privilege of experts.

Complementary knowledge

Nearly all of the respondents who inform this research genuinely value local knowledge. Moreover, they recognise that their expert knowledge provides only a partial account of a locality, leading many to recognise that local knowledge is essential to 'good' flood management. A leader from a local NGO provides an extremely clear account for the need for local knowledge:

> People are very conscious. They do not know the science of climate change, but they know the issues of climate change. They know that they are being affected. They know that in the local markets they cannot store their goods or get good prices. So, if you go to them, learn their problems through their eyes. From the eyes of the vulnerable. Then you can match it into your policies. You must respect and believe in your people. All groups. We are educated, we have masters and PhDs, and we think that we do not need to ask people; but it should not be like that. People know what they want and they know their priorities.
>
> (NGO, 3 January 2008)

This valuing of local knowledge is not exclusive to NGOs. For example, the head of an engineering consultancy firm explains that local knowledge is essential for refining models and projections. At his organisation, engagement with locals is how they ensure that their models are superior to 'imported' models, which cannot explain local events. Through regular engagement with locals and their knowledge, this organisation is able to raise the standard of their flood models:

There are three stages to our projects. To understand the people and their views about floods. Because local knowledge is very important. The scientists and the modellers, they depend on the computers with the theories in their minds, but when you go and speak to the people, that is very important information as well.

(Consultant, 10 January 2008)

This view is also found within government. For example, the head of water resources within a government ministry explains that expert knowledge, due to government policies concerning transfers, is too general to apply to specific locations. This requires direct engagement with locals and an appreciation for local knowledge:

The trouble is often with management. I work here for now, but the maximum time I can manage this office is three years and then I will be transferred. So I may not know all the details of the local circumstances. So local people have better understanding of the system of the area and this helps a lot.

(Government Official, 15 January 2008)

Finally, an academic engineer, following similar advocacy for local knowledge and engagement, concludes that science and expertise cannot always determine solutions to flood problems in specific places. In such circumstances, locals represent a source of alternate interpretation that can up-end expectations or recast the challenge in ways that the experts did not expect:

If it were an embankment and it didn't really help the people, or if it causes harm to some people, then you cannot go back to the people. I think, you know, to share our knowledge with them at the beginning, we did the analysis and this will probably not help you. Now we need other ideas.

(Academic Engineer, 15 December 2007)

Unlike 'foolish decision making' or 'personal recollections', the 'complementary knowledge' excerpts show sincere valuing of local knowledge and a willingness to collaborate with locals. These examples are representative across the interviews conducted for this project, showing an empowered group of experts who recognise gaps within their expertise and accept that locals hold specific and concrete knowledge that, when drawn into management, improves governance. In these cases, local knowledge is something outside the reach of experts, and their decision making is improved for having opened itself. Like the preceding groupings, though, the hierarchy between expert and lay person is maintained, though less overtly. In general, the decision makers describe instances where the boundary between expert and lay knowledge is 'lowered' in order to admit knowledge that would traditionally be excluded. This has at first the appearance of participation and collaborative decision making, but the reality is that the experts maintain tight control over what knowledge is admitted and the basis on which that knowledge is heard. In terms of the cultural critique of expertise and local knowledge, the first two groups make for easy targets, as they overtly maintain the divide between expert and local and between scientific and lay knowledge. The third group, though, gives the appearance of being more receptive to local knowledge without disrupting the hierarchy.

Knowledge-power (participation)

Among decision makers, attitudes towards local knowledge appear to show that criticisms have been heard, leading to the acceptance and valuing of local knowledge. Despite this generally positive outcome, it also appears that the associated acceptance of participation and power sharing has been less successful. The power of the decision makers remains central, unchallenged and possibly solidified, meaning that these decision makers appear empowered because they act as gatekeepers over the admission of both scientific and local knowledge. Alternatively, and more problematically, by valuing local knowledge and advocating participation – all the while maintaining the dominant science based hierarchy – the decision makers might appear to be extremely progressive and thereby temper criticisms of expert authority and governance. In discussing their views, the decision makers are free to draw on local knowledge, anecdotes and personal memories to justify their positions. This is itself an important point, as we cannot simply infer that a willingness to utilise these knowledge claims when discussing floods with a foreigner will be repeated in other arenas where such knowledge might be less accepted. What is more important, these decision makers are developing their worldviews by drawing local knowledge into their expert opinions. This knowledge can then be expunged (though presumably not absent) when they either engage with tradition governance or it has become more widely accepted. Either way, local knowledge has become a key actor within the flood management discourse, which deserves further discussion.

Conceptually, local knowledge and participation are inseparable. If local knowledge is legitimate, then there is no basis for exclusion of local people from governance. In practice, though, it appears that acceptance of local knowledge is easier for decision makers than is meaningful participation (i.e., power sharing). This is a particularly important finding, as much of the academic debate surrounding participation emphasises 'why' it should be done and the 'benefits' that will result (cf. Hanchett 1997, Callon 1999, Mustafa 2002, Jasanoff 2004, Kesby 2005, Kindon *et al.* 2007, Cook and Lane 2010), but with less consideration of the decision making systems into which participation is being inserted. While it is inadvisable to let the practices of governance dictate critiques of governance, there is need to question the situations into which local knowledge is introduced. The result is that disaster management culture appears able to divide knowledge from practice. This in turn allows a partial acceptance of the cultural critique of expert authority and a response in which local knowledge is embraced, but also where power remains concentrated among a small segment of the population.

Conclusion

The cultural critiques of flood management emphasise power through analysis of knowledge. This research project has shown a response among decision makers that has been to emphasise knowledge but to disregard power. While disappointing, this should not be surprising. Disaster management represents a concentration of power, prestige and privilege that few people would renounce. Furthermore, relative to their past or more traditional colleagues, these decision makers are extremely progressive concerning the value of local knowledge and the role of locals. What makes this analysis extremely complicated is this openness towards culture, in the context of their less overt and possibly unconscious treatment of locals within flood management. Despite indications that the power hierarchies surrounding flood management remain stable and tilted towards experts, these individuals view themselves as part of progressive efforts to break from top–down and decontextualised governance. As one engineer explains: 'What you need to know is that the reason that the engineers today are better educated is because of these criticisms' (International Org., 29 January 2008).

That the distribution of decision making power ultimately remains unchanged is cause for concern, particularly when outside appearances might suggest that culture based criticisms have been successful. The end result of this research appears to be that management culture, by way of its authority and boundaries, is able to subsume local culture. While not terribly original, showing the processes and logics through which expert, science based authority is maintained is a valuable contribution. Unfortunately, showing that this is in many ways a negative outcome of a genuinely progressive effort is disappointing; there are no villains in this maintenance of power, but rather people seeking to do their best within the existing system.

It is the appearance of collaborative decision making that is most daunting as it may lead those seeking more democratic decision making to temper their calls for coproduction (Callon 1999). In such contexts, the genuineness of the decision makers cannot distract from the system they inhabit or the concentration of power therein. Simultaneously, analyses of these systems of power cannot assume the worst. For all of the respondents who contributed to this study, rather than conniving or being disingenuous, I believe they honestly viewed their efforts as contributing to a more democratic and sensitive form of governance. They wanted more effective flood management and believed that local knowledge was invaluable. For a small number of the respondents, the addition of local knowledge to their expertise represents a deep and conscious concession. For most, though, further steps along this effort to democratise decision making (Arnstein 1969, Cook et al. 2013) appears to be completely outside consideration. It is interesting that the decision makers genuinely want this power sharing but, through self-discipline, are unable to implement it as it appears outside their worldview (Kahan et al. 2006, Kahan et al. 2007).

These accounts and anecdotes may not be 'true'. They are, possibly, fabrications or collections of hearsay and innuendo amassed by these experts in order to justify their positions. This is not a criticism, though it is critical to recognise that this discussion is more accurate about how individuals develop understandings. Worldviews are dynamic tapestries woven by individuals based on numerous factors, culture being central (Douglas and Wildavsky 1982). Again, this is not written to disparage experts but to show that the divide between expert knowledge and lay knowledge is unjustifiably strong and that overstating this divide can result in the misapplication of science (Wynne 1996). The worldviews presented by the experts appear to be much more like 'local knowledge' than like idealised 'expertise'. This is an absolutely critical point, as it shows the divide between locals and experts to be much less prominent, which ought to make the case for participation and power sharing easier.

Finally, the case for participation requires comment. In this discussion, I have aimed to show that the range of knowledge drawn into governance by decision makers has expanded to include local knowledge. I maintain that this has occurred because of culture-based criticisms of expert knowledge-power. By exploring the types of local knowledge used, and the ways that local knowledge is enrolled and deployed by these individuals, it can be seen that local knowledge has become central to flood management, but that the power sharing many would expect to be part of that democratisation has not occurred. Decision makers remain in (almost) complete control of the knowledge that informs governance, including the examples of local knowledge that shape the discourse. If meaningful participation is to be realised, it must be more explicitly linked to local knowledge within cultural critiques of governance. If not, the case of flood management in Bangladesh may act as a warning in which a 'half-victory' may have resulted in a complete loss.

Note

1   Polders are large areas of land encircled by embankments. Beginning in the 1960s, they were used to fortify the coastal region against cyclones and tidal surges; they were also aimed at reclaiming land from the sea in an effort to expand agricultural development.

References

Ahmed, I., 1999. *Living with Floods: An Exercise in Alternatives*. Dhaka: The University Press Limited.

Alam, N., 1990. Perceptions of flood among Bangladeshi villagers. *Disasters*, 14 (4), 354–7.

Arnstein, S., 1969. A ladder of citizen participation. *Journal of the American Institute of Planners*, 35 (4), 216–24.

Boyce, J. K., 1990. Birth of a megaproject – Political-economy of flood-control in Bangladesh. *Environmental Management*, 14 (4), 419–28.

Brammer, H., 2004. *Can Bangladesh be Protected from Floods?* Dhaka: The University Press Limited.

Callon, M., 1999. The role of lay people in the production and dissemination of scientific knowledge. *Science, Technology and Society*, 4 (1), 81–94.

Chellaney, B., 2011. *Water: Asia's New Battleground*. Washington DC: Georgetown University Press.

Clay, E., 1985. The 1974 and 1984 floods in Bangladesh: From famine to food crisis management. *Food Policy*, 10 (3), 202–6.

Cook, B. R., 2010. Flood knowledge and management in Bangladesh: Increasing diversity, complexity and uncertainty. *Geography Compass*, 4 (7), 750–67.

Cook, B. R. and Lane, S. N., 2010. Communities of knowledge: Science and flood management in Bangladesh. *Environmental Hazards*, 9 (1), 8–25.

Cook, B. R. and Wisner, B., 2010. Water, risk and vulnerability in Bangladesh: Twenty years since the FAP. *Environmental Hazards*, 9 (1), 3–7.

Cook, B. R., Kesby, M., Fazey, I. and Spray, C., 2013. The persistence of 'normal' catchment management despite the participatory turn: Coproduction and consultation framings. *Social Studies of Science*, 43 (5), 754–79.

Davies, S. R., 2008. Constructing communication: Talking to scientists about talking to the public. *Science Communication*, 29 (4), 413–34.

Dewan, A. M., 2013. Vulnerability and risk assessment. In: A. M. Dewan, ed., *Floods in a Megacity: Geospatial Techniques in Assessing Hazards, Risk and Vulnerability*. Heidelberg, New York, London: Springer, 139–77.

Douglas, M. and Wildavsky, A. B., 1982. *Risk and Culture: An Essay on the Selection of Technical and Environmental Dangers*. Berkeley, CA: University of California Press.

Duvail, S. and Hamerlynck, O., 2007. The Rufiji river flood: Plague or blessing? *International Journal of Biometeorology*, 52 (1), 33–42.

Epstein, S., 1995. The construction of lay expertise: AIDS activism and the forging of credibility in the reform of clinical trials. *Science, Technology & Human Values*, 20 (4), 408–37.

Epstein, S., 1996. *Impure Science: AIDS, Activism, and the Politics of Knowledge*. Berkeley, CA: University of California Press.

Hanchett, S., 1997. Participation and policy development: The case of the Bangladesh flood action plan. *Development Policy Review*, 15, 277–95.

Haque, C. E., 1988. Human adjustments to river bank erosion hazard in the Jamuna floodplain, Bangladesh. *Human Ecology*, 16 (4), 421–37.

Hewitt, K., 2012. Culture, hazard and disaster. In: B. Wisner, JC Gaillard and I. Kelman, eds., *Handbook of Hazards and Disaster Risk Reduction*. Abingdon: Routledge, 85–96.

Höfer, T. and Messerli, B., 2006. *Floods in Bangladesh: History, Dynamics and Rethinking the Role of the Himalayas*. Tokyo, New York, Paris: United Nations University Press.

Irwin, A., 1995. *Citizen Science: A Study of People, Expertise and Sustainable Development*. Routledge: London.

Jasanoff, S., 2004. *States of Knowledge: The Co-production of Science and the Social Order*. New York: Routledge.

Kahan, D. M., Slovic, P., Braman, D. and Gastil, J., 2006. Fear of democracy: A cultural evaluation of sunstein on risk. *Harvard Law Review* 119 [online]. Available from: http://ssrn.com/abstract=801964 [accessed 26 March 2014].

Kahan, D. M., Braman, D., Gastil, J., Slovic, P. and Mertz, C., 2007. Culture and identity-protective cognition: Explaining the white-male effect in risk perception. *Journal of Empirical Legal Studies*, 4 (3), 465–505.

Kearnes, M. B., Klauser, F. and Lane, S. N., eds, 2012. *Critical Risk Research: Practices, Politics and Ethics*. John Wiley & Sons: Sussex.

Kesby, M., 2005. Re-theorizing Empowerment-through-participation as a performance in space: Beyond tyranny to transformation. *Signs*, 30 (4), 2037–65.

Kindon, S., Pain, R. and Kesby, M. eds, 2007. *Participatory Action Research Origins, Approaches and Methods. Connecting people, Participation and Place*. Abingdon: Routledge.

Mustafa, D., 2002. To each according to his power? Participation, access, and vulnerability in irrigation and flood management in Pakistan. *Environment and Planning D: Society & Space*, 20 (6), 737–52.

Paul, B. K., 1984. Perception of and agricultural adjustment to floods in Jamuna floodplain, Bangladesh. *Human Ecology*, 12 (1), 3–19.

Paul, B. K., 1995. Farmers responses to the Flood Action Plan (FAP) of Bangladesh – an empirical study. *World Development*, 23 (2), 299–309.

Paul, B. K., 1997. Flood research in Bangladesh in retrospect and prospect: A review. *Geoforum*, 28 (2), 121–31.

Paul, S. K. and Routray, J. K., 2010. Flood proneness and coping strategies: The experiences of two villages in Bangladesh. *Disasters*, 34 (2), 489–508.

Rasid, H., 1993. Preventing flooding or regulating flood levels? Case studies on perception of flood alleviation in Bangladesh. *Natural Hazards*, 8 (1), 39–57.

Rasid, H. and Mallik, A., 1993. Poldering vs. compartmentalization – The choice of flood-control techniques in Bangladesh. *Environmental Management*, 17 (1), 59–71.

Rasid, H. and Mallik, A., 1995. Flood adaptations in Bangladesh – Is the compartmentalization scheme compatible with indigenous adjustments of rice cropping to flood regimes. *Applied Geography*, 15 (1), 3–17.

Sen, A., 1981. *Poverty and Famine: An Essay on Entitlement and Deprivation*. Oxford: Clarendon.

Shaw, R., 1989. Living with floods in Bangladesh. *Anthropology Today*, 5 (1), 11–13.

Siddique, K. U. and Hossain, A. N. eds, 2006. *Options for Flood Risk and Damage Reduction in Bangladesh*. Dhaka: The University Press.

Sillitoe, P., 1998. The development of indigenous knowledge: A new applied anthropology. *Current Anthropology*, 39 (2), 223–53.

Sillitoe, P., 2000. *Indigenous Knowledge Development in Bangladesh: Present and Future*. London: Intermediate Technology Publications.

Sillitoe, P. and Marzano, M., 2009. Future of indigenous knowledge research in development. *Futures*, 41 (1), 13–24.

Warner, J., 2010. Integration through compartmentalization? Pitfalls of 'poldering' in Bangladesh. *Nature and Culture*, 5 (1), 65–83.

Wilcock, D. A., 2013. From blank spaces to flows of life: Transforming community engagement in environmental decision-making and its implications for localism. *Policy Studies*, 34, 455–73.

Wynne, B., 1996. May the sheep safely graze? A reflexive view of the expert-lay knowledge divide. In: S. Lash, B. Szerszynski and B. Wynne, eds, *Risk, Environment & Modernity. Towards a New Ecology*. London: Sage, 44–83.

Yunus, M., 1999. *Banker to the Poor: The Story of the Grameen Bank*. London: Aurum Press.

Zaman, M. Q., 1993. Rivers of life – Living with floods in Bangladesh. *Asian Survey*, 33 (10), 985–96.

12

CULTURE'S ROLE IN DISASTER RISK REDUCTION

Combining knowledge systems on small island developing states (SIDS)

Ilan Kelman, JC Gaillard, Jessica Mercer, Kate Crowley, Sarah Marsh and Julie Morin

Introduction

This chapter examines some approaches for understanding the role of culture in disaster risk reduction by focusing on how different knowledge systems could be combined. The case studies used are Small Island Developing States (SIDS). SIDS are a United Nations designated group covering fifty-two small countries and territories (for example Bahrain, Mauritius and Vanuatu) that are mainly, but not exclusively, islands. SIDS experience similarities in livelihoods, development and sustainability challenges along with parallel approaches for dealing with those challenges, especially in terms of dealing with disasters and disaster risk reduction (UN 1994, 2005).

SIDS have differences too, as exemplified by the acronym itself. 'S' means 'small', although the SIDS of Papua New Guinea (PNG) is larger in population and land area than New Zealand (which is not a SIDS). 'I' stands for 'island', although Guyana, Belize and Guinea-Bissau are SIDS but are not islands. 'D' represents 'developing', but comparatively developed countries such as Bahamas and Singapore are part of the SIDS group. The final 'S' means 'state', even though several non-sovereign territories are SIDS including Montserrat and Guam.

Certainly, the SIDS are a construct, aimed at and recognised by UN processes (e.g. UN 1994, 2005) but led by the SIDS governments themselves who see significant value in the grouping, particularly for climate change negotiations but within wider sustainability challenges. While it is legitimate to question the constructedness and utility of the SIDS grouping, it is also important to recognise the reasons that the SIDS came together, to share stories and ideas; to recognise the similarities of the sustainability challenges faced; and to give themselves a common and powerful voice that is otherwise ignored by larger countries and powers.

That does not bury the differences and variety among the SIDS. The diversity of SIDS is particularly represented in their various cultures, from predominantly Muslim Maldives to Fiji's ethnic mix to the slave and British traditions of Barbados. The case studies used in this chapter will demonstrate such diversity for combining knowledge systems that are internal and external to SIDS in order to effect disaster risk reduction, but will also demonstrate the commonalities among SIDS that helps to bring them together and legitimise the constructed grouping.

Local leadership is key for disaster risk reduction (e.g. Lewis 1999, Twigg 1999–2000), with SIDS geographies epitomising the contribution of local approaches for dealing with disaster. Use of communities' internal knowledge for disaster risk reduction is especially important in SIDS as many SIDS communities are isolated and marginalised, meaning that external assistance is not always forthcoming rapidly, so the SIDS peoples have within their culture and history the understanding of self-help (Gaillard 2007, Mercer *et al.* 2007). The importance of island communities that are particularly small and isolated, as opposed to other spatial and social settings (see also McCall 1996, Baldacchino 2007), is further important due to the compression of spatial scales, in that the local level is frequently the same as the national level. For disaster response and disaster risk reduction, there is limited option to scale up in the short-term, furthering the necessity of taking care of one's own disaster needs with minimal external help, instead relying on the islanders' own experience and knowledge.

An example of a programme to work with SIDS on such terms, to boost the advantages of their own knowledge while overcoming any limitations by supplementing it with external knowledge, is Many Strong Voices or MSV.[1] MSV brings together Arctic and SIDS peoples to deal with climate change in the context of other disaster and development concerns. Over the years, the programme has been shifting from focusing purely on climate change to enveloping a wider disaster risk reduction framing within the context of development. While the Arctic and SIDS have clear environmental and cultural differences, many of the characteristics articulated for SIDS apply to Arctic peoples and communities. The coastal nature of many of the communities leads to strong parallels in addition to their community based natural resource dependence. These communities particularly enjoy sharing stories with each other, in which their own knowledge is embedded in order to learn how to use each other's knowledge for disaster risk reduction, alongside published scientific papers and wider policy discussions.

This knowledge sharing applies directly to culture and disasters. Culture here is a broad framework of individual and group action, incorporating peoples' knowledge, values, views and beliefs across various aspects of societies. Knowledge, and how different knowledge forms interact within a community, are part of the culture of dealing with disasters (e.g. Mercer *et al.* 2012). Based on several English dictionaries, a succinct definition of 'knowledge' can be 'information, skills, and awareness acquired through education or experience'. The scientific literature (Wisner 1995,

Shaw *et al.* 2009) has numerous ways of dividing local and non-local knowledge. Examples are indigenous knowledge and scientific or Western knowledge; internal and external; traditional and contemporary; traditional and expert; and local and external. This particular chapter focuses upon the role and development of internal and external knowledge in the context of disaster risk reduction in SIDS. The scientific literature further provides different ways in which these knowledge forms are combined for disaster risk reduction. Wisner *et al.* (1977) wrote about 'peoples' science', referring to people having a deep understanding of their own situations and of adjusting to environmental hazards within those situations. Glantz (2003) highlights 'useable science' meaning that research findings are targeted directly towards helping people on their own terms. That knowledge and the adjustment processes based on them can often be distorted by external influences – which can include the domination of modern technology and Western science, with advantages and disadvantages.

Gaillard (2007), Mercer *et al.* (2007) and Gaillard and Mercer (2012) summarise much work on combining traditional and non-traditional knowledge forms for disaster risk reduction. They, and the work on which they are based, demonstrate how no single knowledge system can provide everything needed for dealing with disaster, before, during or after. Instead, they show how all knowledge has limits but, when combined properly, the strengths of each knowledge form can be used to overcome the gaps.

From this discussion, and to a large degree irrespective of specific vocabularies used, two principal cultures of knowledge emerge for disaster risk reduction. These cultures of knowledge have been directly shaped and formed by the cultural framework in which they continue to develop. Internal or local knowledge is a body of knowledge existing within or acquired by people in a specific location over a period of time through accumulating experiences, community practices and institutions. It is usually unique and contextual to a community, culture or society. External knowledge is more globally known or accepted, especially among the literate population. It frequently includes elements developed through attempts to de-contextualise and objectify the knowledge through cross-cultural consensus with the Western scientific process being an example. Combining internal and external knowledge forms for disaster risk reduction is now explored on a case study basis for three SIDS case studies from different SIDS regions.

Haiti: social hierarchies during emergency response and recovery

Haiti is a SIDS even though it shares its island of Hispaniola with another SIDS, the Dominican Republic. Most disasters in Haiti are caused not by the specific hazards, but by the insidious creation and failure to deal with vulnerability (Nicolas *et al.* 2009). That vulnerability was created and perpetuated by successive waves of exploitative, corrupt and incompetent leaders supported by ex-colonial powers, mainly France and the US, who further prevented Haiti from accessing its own resources in order to govern itself (Schuller and Morales 2012). In fact, Haiti's abject poverty and horrendous underdevelopment is not a given for the country, since it was once known as the Pearl of the Antilles, being the first Caribbean country to produce sugar and to mine precious gems (Nicolas *et al.* 2009).

Due to the current vulnerabilities, damage during hurricanes – from rainfall-related flooding, landslides and coastal surges – is exacerbated by the rapid deforestation, partly to serve timber interests and partly through Haitians doing their best to survive day-to-day with few livelihood options. Recently, Hurricane Jeanne in 2004 killed over 3,000 people. Although hurricanes have struck regularly, little government infrastructure has been in place to respond, with the main national response prior to 2010 coming from the Haitian Red Cross (Nicolas *et al.* 2009).

The 12 January 2010 brought an immense change when a magnitude (Mw) 7.0 earthquake at 13 km depth struck 25 km from the capital, Port-au-Prince. Given the poor infrastructure, the resulting destruction was vast. The death toll is disputed, ranging from under 100,000 to over 300,000 (Schuller and Morales 2012), representing 1–3% of the country's population at the time. Perhaps one quarter of the country's population was displaced. The response, focused on Port-au-Prince, was profoundly shaped by the local urban culture based in the country's history (Clermont *et al.* 2011). That includes the characteristics and behaviours of urban residents – specifically their hierarchies, communities of inclusion and exclusion and their hopes and aspirations (Jenks 2004).

Urban areas attract a broad range of people who are often actively seeking 'a better life', which is expected to be achieved through urban employment. That better life often does not appear, leading to poverty-stricken slums alongside high rates of crime and, often, the appearance of social structures and hierarchies, which are different from traditional social settings in rural areas, such as gangs. Where employment is available, it leads to a cash economy, with collaborative and community efforts, including local knowledge and previous experience, diminishing in comparison to their importance in rural locations (Clermont *et al.* 2011). The entrepreneurial nature of a cash-based economy became prominent in Haiti when certain community committees in post-disaster settlements were encouraged to start their own small businesses, for example, selling water. That has advantages for developing a self-sustaining system of local livelihoods, but it should also be important to maintain and support local knowledge and skills developed over generations as part of that post-disaster development.

As part of that approach to combining urban and traditional knowledge forms, Build Change[2] in collaboration with CAFOD and Caritas Port-au-Prince further developed the knowledge of five teams of twenty-five local masons in earthquake resistant housing construction. In addition, they trained three Caritas engineers in supervisory skills so that the local civil society organisations could continue the work independently from external support. This approach enables Haitians with existing construction skills to learn and apply advanced techniques integrating knowledge that, through employment creation, contributes to overcoming the contemporary urban issues (e.g., increased crime and gangs) while supporting the local economy. It further enabled the construction of permanent homes rather than short-term transitional shelters.

Although the shattered social environment in the context of increased crime, gangs and urban slums was a challenge for building knowledge to deal with post-disaster activities, some agencies attempted to use the existing strong social networks as a basis for camp committees and community action. Patrick (2011, p. 3) noted that 'even the most devastated communities and governments retain capacities. Even if the physical/material infrastructure is destroyed, the communities still have strong relationships, personal skills, organisational abilities, important norms and values, effective leaders and the ability to make decisions'.

According to Nicolas *et al.* (2009), Haitians have a cultural strength shaped by their history and reinforced by the social support systems, particularly strong kinship ties, including the extended family and bonds to the church. This strong cultural system, although fractured to some degree by the decades of poor governance and the growth of Port-au-Prince's urban culture, has previously acted as a source of childcare, jobs and livelihoods, shared transport and household equipment, informal borrowing and lending associations, and retaining community knowledge in times of stress.

After the 2010 earthquake, the survivors initially aimed to maintain their 'neighbourhoods' and community ties in order to effect self-help. As humanitarian resources became available within the Internally Displaced Peoples camps, many people abandoned their homes, leading to an increase in camp populations between March and June 2010. This increase could indicate that traditional coping mechanisms based on local knowledge were being overwhelmed or were being subverted by the camp-focused relief effort. Most likely, it was a combination of the two.

Where camps did appear to reflect the pre-earthquake situation, that was more focused on the relatively new and usually difficult complex urban social hierarchies rather than the more traditional social structures. For example, before the earthquake, the Solino neighbourhood of Port-au-Prince was considered to be a poor, marginalised and overcrowded area of the city. The resulting social characteristics and behaviours were transferred into the 'Solino' camp culminating in a continuation of that urban gang-based influence on everyday life in the camp, which relief efforts at post-disaster shelters tried to overcome or in some cases work with (Hirano 2012).

Yet humanitarian agencies found that certain areas in Port-au-Prince were too insecure to carry out their full programme of work. Like Solino, these areas were already considered to be dangerous prior to the earthquake, in particular being notorious for gangs and high crime. Hirano (2012) describes how gang members sometimes saw humanitarian agencies as a threat and instead created 'camp committees' themselves, which perpetuated the pre-earthquake urban 'culture' supporting their power structures. Intimidation, violence and threats were the usual mode of letting the committees' interests be known, with those controlling the camp committees and camp neighbourhoods often seeking to create tension with the humanitarian agencies. Sometimes, it took more than a year of humanitarian staff having a presence in a location before the communities trusted the external interventions.

The Haiti case study indicates different dimensions of everyday life cultures in a complex and rapidly changing urban environment. A balance was needed among building on the beneficial components of local knowledge; overcoming increased violence, crime and gangs; and external interventions bringing the external relief and reconstruction to a disaster-devastated location.

Comoros: internal knowledge, cultural obligations and local resources

Internal as well as external knowledge must be contextualised for disaster risk reduction, as people always confront their knowledge of potential threats with other topics and constraints affecting their daily life. Their behaviour in facing vulnerabilities to hazards stems from a balanced perspective of possible positive outcomes and potential drawbacks of any actions.

The way some local communities in Comoros make sense of their knowledge of vulnerabilities and hazards provides a vibrant example of the complexity of interactions among knowledge forms. Comoros is an archipelagic SIDS lying between Madagascar and Mozambique. It is perceived by most international organisations such as the UN (Programme des Nations Unies pour le Développement 2010) as being one of the poorest countries in the world. Many Comorians struggle to sustain their daily needs associated with shelter, food and schooling while trying to fulfil their cultural obligations and aspirations. The latter pertains mainly to the so-called '*grands mariages*' (costly social ceremonies), which permit people to rise up in the social hierarchy and grant them the privilege to speak out in public (Rouveyran and Djabiri 1968, Walker 2002).

The archipelago is further afflicted by political crises (coups d'état and island secession attempts) as well as health (notably malaria and cholera) and other hazards (mainly cyclones, floods and volcanic eruptions). Mt. Karthala volcano constitutes one of the most prominent and lingering threats to the communities of the largest island of Ngazidja. As one of the largest active volcanoes in the world with about thirty-three eruptions since the nineteenth century, local communities have built up a historical repository of knowledge dealing with volcanic hazards. In the aftermath of the latest major eruption of Mt. Karthala in April 2005, lahars (volcanic mudflows) buried several villages at the foot of the mountain. Over the course of this eruption and associated protracted hazards, the locals have developed particular knowledge of lahars (Morin *et al.* 2009). This and other historical knowledge is the only available information they have with regard to dealing with living next to the volcano, since they tend to lack access to scientific knowledge. In fact, the government-driven information campaign about the potential dangers associated with further volcanic activity proved culturally inappropriate and failed to trickle down to the grassroots level to the villagers who needed the information (Morin and Lavigne 2009).

The local culture of facing Karthala's volcanic hazards, particularly lahars, reflects the way people interpret their internal knowledge in the context of everyday livelihood and cultural obligations. In the small village of Vouvouni, which was badly affected in April 2005, many people realised that the vast amount of sand brought by successive lahars could easily be exploited as a valuable resource (Morin and Gaillard 2012). Indeed, the *grands mariages* require celebrants and their relatives to build large cement houses, which, in turn, result in a huge demand for construction materials. The lahars were thus timely, as they provided a large quantity of sand just a few weeks ahead of the start of the *grands mariages* season in June. To meet cultural obligations, local people used internal construction knowledge and creativity to turn a hazard into a resource.

In fact, the owners of the lots that acquired the largest amount of lahar materials initially prevented access to their property to secure control over the sand. In response, other villagers diverted the local river towards their own property to claim their share of the lahar resource, although they were perfectly conscious that this may endanger their lives as large lahars sweep down the mountain. As expected, the river diversion led subsequent lahars to inundate the entire village, affecting hundreds. Many families were forced to abandon their houses while others lost valuable crops buried by lahar deposits.

However, sand was now abundant and freely accessible in many private and public spaces. Furthermore, extracting sand requires few skills and tools, so that people whose crops were affected were able to earn comfortable and relatively easy income. Meanwhile, many students earned money as labourers to extract sand during the school holidays. In many cases, the extracted sand was sold to wealthier local traders who then retailed construction materials, earning cash. In late 2005, the entire community of Vouvouni had used their internal knowledge to generate livelihoods based on quarrying sand for the construction sector, which supplied what was needed for local cultural rituals.

Eventually, the increase in sand supply saturated the market. Simultaneously, persistent lahars progressively filled up pipes in toilets, bathrooms and kitchens. The mud floods spread rubbish in food and water supplies, contributing to gastroenteritis outbreaks. The economic and social costs of diverting lahars to extract the sand thus progressively turned into a serious burden for the people of Vouvouni. Internal knowledge utilised to diversify livelihoods based on quarrying sand had now contributed to increased vulnerability. The number of households relying on sand extraction as a major means of livelihood decreased from 50 in 2006 to only 20 in 2009 (Morin and Gaillard 2012).

In 2008 and 2009, community members, realising the problems, collaborated in building communal levees to protect their village – this built upon internal knowledge aiming for disaster risk reduction while responding to an immediate need to protect life and livelihood. But external knowledge should have been combined with the internal knowledge to complete an adequate job, because these levees proved to be inadequate and failed, so that lahars seriously affected Vouvouni in 2009. The next approach using local capacity was small-scale protection measures such as sandbagging and constructing small rock walls in front of their houses. Most of the households also placed their belongings in higher locations within their house in the event of flooding. These measures, while effective, were comparatively haphazard, required continual vigilance and were implemented in response to an immediate threat. Again, combining internal and external knowledge forms on the people's own terms could have improved these disaster risk reduction approaches.

Nevertheless, this example shows that through their own experiences, the people of Vouvouni gained an understanding of the positive and negative outcomes of lahars – and of how they choose to manage the lahars as a hazard and as a resource. Their individual and collective behaviour evolved over time, based on the development of their internal knowledge. In that sense, internal knowledge consists of both an individual and collective set of beliefs and skills as well as how people individually and collectively make sense of them. That is particularly poignant in the Comorian village setting due to the isolation and marginalisation of the affected villages from the country's power brokers and resource controllers, further situated in the context of Comoros' comparative isolation and marginalisation from external knowledge sources. As Wisner *et al.* (2014) suggest regarding operational disaster risk reduction, 'there can be no knowledge that completely separates from what others know and have known in the past'.

Such knowledge is enmeshed in the reality of people's daily lives, dealing not just with lahars and their vulnerabilities to lahars, but also to the everyday risks and opportunities that arise in going to school, tending crops and seeking clean drinking water. The internal knowledge that the villagers applied to the lahars, with successes and failures, cannot be dissociated from their larger cultural values and norms, the social and economic hardships and opportunities they experience every day, as well as the power relations within and beyond their communities. External knowledge can help to situate these challenges in wider contexts, as long as it does not attempt to dominate or override internal knowledge, as with the government's attempt to 'educate' the locals about lahars.

Internal knowledge is not perfect, as shown by the troubles created by trying to manage the lahars. But internal knowledge is intrinsically and contextually dynamic, not a static ideal with the 'noble savage' living in harmony with their environment. That makes internal knowledge non-explicit and hard-to-access for many outsiders, meaning that external interventions must place external knowledge within the context of internal knowledge, values and norms to ensure that knowledge forms are combined on the people's terms to address disaster risk reduction.

Timor-Leste: knowledge integration for climate change adaptation

The final case study is Timor-Leste (e.g. Mercer *et al.* 2014), for which the focus here is the cultural processes at work through national level knowledge in order to build a programme of climate change adaptation which is one dimension of disaster risk reduction. As a relatively new independent state, born in May 2002 (e.g. Steele 2002), Timor-Leste is still building its policies and institutional structures for climate change, part of which involves applying different forms of knowledge within the national government. Within Timor-Leste's Government, gaps in resources, capacities and data, as well as in knowledge exchange and development mechanisms, are impeding the potential for disaster-related challenges to be incorporated into national and sub-national development planning (Norton and Waterman 2008). Timor-Leste provides a useful case study for understanding how knowledge forms can be combined for a new SIDS grappling with independence as well as other development challenges such as environmental hazards, including climate change.

The climate change impacts that Timor-Leste will experience are not fully known (Wasson 2001, Kirono 2010). Historical climate and weather data are not available; monitoring systems providing a consistent baseline do not exist; models for downscaled projections are not available; and the baseline data needed, such as bathymetry and topography for sea level rise, need to be collected.

Recognising the gaps, national governance structures are being created to address them, part of which is incorporating different knowledge forms into these structures. The National Directorate of Meteorology and Geophysics (NDMG) has placed a high priority on expanding climate data collection, monitoring and analysis – a focus on external, scientifically collected knowledge. Similarly for external knowledge, NDMG is collaborating with the Pacific Climate Change Science Program (PCCSP) to re-establish a meteorological network, to undergo professional education for the topic and to recover, digitise and analyse climate data collected under the Portuguese and Indonesian administration periods (da Silva and Moniz 2010).

Meanwhile, responsibility for climate change adaptation in Timor-Leste currently resides in the Secretariat for Environment within the Ministry of Economy and Development. Initiatives for climate change adaptation tend to be focused on external processes. Timor-Leste signed the United Nations Framework Convention on Climate Change (UNFCCC) on 8 January 2007 meaning that the country must develop and submit a National Communication Document. The process for developing the National Communication Document is led by the United Nations Development Programme (UNDP). Additionally, Timor-Leste is a Least Developed Country (LDC) within the UNFCCC, so it has written a National Adaptation Programme for Action (NAPA), which involved the National Disaster Management Directorate (NDMD) and the Ministry for Social Solidarity. Yet few of the staff assigned to this work had much experience of such processes or in the topics, which meant that they focused on learning by doing, effectively acquiring as much knowledge of any form as feasible within the limited time and other resources available.

The focus tended to be on external knowledge, without fully realising the complementary contribution that internal knowledge could and should make for Timor-Leste's climate change adaptation work. The NAPA, in fact, was produced without much consultation at sub-national governance levels, in that only five out of the thirteen districts were consulted during the process and there was limited consultation at the local level. The NAPA also did not include an in-depth analysis of the country's climate historically (e.g. Lape and Chin-Yung 2008) or at the present. As such, the NAPA process was comparatively top-down, being national-led and using only limited amounts of external knowledge without combining this with the extensive internal knowledge available (Oxfam 2011). Such knowledge includes historical and present-day information on changing weather patterns and longer term trends – so linked to climate change – as well as on local adaptation techniques such as drainage canals, reforestation, implementation of *tara-bandu* (local laws governing use of environment), agro-forestry, inter-cropping and erosion control methods, one example of which is gabions.

Overall, climate change is a new (even if growing) concern in Timor-Leste, but more knowledge of all forms is required for the processes of developing laws, regulations, policies and plans – followed by monitoring, enforcing and updating of the laws, regulations, policies and plans. That includes an inclusive development process of ensuring that Timorese are involved through full and fair consultation; that different knowledge forms are used and combined; and that actions are based on the best available knowledge of all forms combined in such a way that no knowledge form dominates. In particular, to achieve appropriate disaster risk reduction with respect to climate change, both mitigation and adaptation need to be addressed in tandem, rather than keeping them as separate activities. Those actions further need to be enfolded within disaster risk reduction and wider development processes.

As an example, forests are important for both climate change mitigation and adaptation (see Godinho *et al.* 2003 for further details on climate change and forests in Timor-Leste) while contributing to other disaster risk reduction processes, such as reducing flood and landslide risk (although see cautions in FAO 2005). Deforestation is noted as perhaps the most prominent environmental concern in Timor-Leste (Sandlund *et al.* 2001, Kirono 2010). Large scale logging was banned in 2000 followed by small-holder logging in 2008. Some small-scale logging continues, wood is used for fuel and harmful slash-and-burn agriculture is still practised. Local communities and Timor-Leste's Government are starting to recognise the impacts of these activities upon land degradation yet, without alternative livelihoods, the people have little choice. Bringing in external knowledge regarding the consequences of deforestation on the people's terms while providing alternative livelihoods based on their internal knowledge could contribute towards enforcing the logging ban.

Much external knowledge is still lacking. Data regarding forest cover and degradation exhibit major discrepancies. The country is currently developing a forest inventory system. Monitoring regimes are sparse, if existing at all. The availability of internal knowledge with respect to forests has not been properly identified. As forestry management progresses in Timor-Leste, for climate change adaptation along with other disaster risk reduction approaches, trying to combine internal and external knowledge from the beginning would be a significant step forward.

Naturally, that principle applies beyond forestry management. Combining knowledge forms to better understand all hazards and vulnerabilities, and linking disaster risk reduction endeavours to day-to-day livelihood activities, would contribute towards successful disaster risk reduction that does not interfere with day-to-day development. There is no need to separate disaster risk reduction from wider development while climate change adaptation sits perfectly within wider disaster risk reduction, rather than being a separate sector (e.g., Shaw *et al.* 2010a, 2010b). Combining knowledge forms and utilising historical internal knowledge that may address identified knowledge gaps could address all these topics simultaneously to ensure that addressing one problem does not create or exacerbate others.

Conclusions

Combining different knowledge forms for disaster risk reduction involves recognising and respecting the differences among knowledge forms. No amount of oral knowledge passed down through the generations can provide analyses over geological time scales, which are needed for understanding the high-end extremes of potential hazards. If a SIDS community in the past was wiped out by a hazard, internal knowledge will not be available. Many SIDS cultures, even when indigenous, are young compared to the geological time frame of their islands (e.g., Dickinson 2009). If a volcano erupts for the first time since a location was settled, internal knowledge can provide little regarding the possible eruption safety zones. No amount of internal knowledge can provide projections of the future under climate change or extremes beyond what humans have experienced.

Instead, from external knowledge, climate change studies (e.g., IPCC 2007) are helping to suggest scenarios that the different SIDS cultures might need to deal with in the future. External knowledge within and outside SIDS can also look back into the past at multiple time scales, giving information on potential hazards that internal community knowledge cannot have. External knowledge has limitations as well (e.g., Martin 1979). The modern scientific method is comparatively young. If an unusual phenomena occurs only once a century, that might be difficult for scientific fields to accept, because one tenet of many disciplines is repeatability.

Overall, any knowledge form, internal or external, has advantages and limitations; it must, by definition, be bounded. Respecting and bringing together different knowledge forms builds a much deeper and broader understanding of disaster and disaster risk reduction challenges faced, along with the options for addressing the challenges. Where knowledge forms seem to contradict each other, that does not mean that one is right and one is wrong. Instead, it means understanding why the differences arise and determining how different perspectives can be reconciled for action. SIDS case studies highlight these issues due to the need to frequently deal with the challenges they face with limited external support.

All forms of knowledge build on the past. That past history and culture should be integrated into all learning without becoming mired in only the past. Human history has built up approximately 10,000 years of knowledge that can be applied to disaster risk reduction. Yet the planet is entering environmental and social regimes new for humanity, such as under climate change (demonstrated for Timor-Leste), a huge global population often choosing or being forced to live in hazardous locations (demonstrated by Comoros), and megacities with large concentrations and densities of people (demonstrated by Haiti). By combining all knowledge forms as part of the melding of cultures witnessed on SIDS (and around the world), the sum of those knowledges can lead to a greater wisdom to achieve sustainable disaster risk reduction, including dealing with the hazard of climate change, without exacerbating or creating other social or environmental problems.

Notes

1   www.manystrongvoices.org
2   www.buildchange.org

References

Baldacchino, G., ed., 2007. *A World of Islands: An Island Studies Reader.* Charlottetown, Luqa: Institute of Island Studies and Agenda Academic.

Clermont, C., Sanderson, D., Sharma, A. and Spraos, H., 2011. *Urban Disasters: Lessons from Haiti: Study of Member Agencies' Responses to the Earthquake in Port au Prince, Haiti, January 2010.* London: Disasters Emergency Committee.

da Silva, S. and Moniz, T. F., 2010. *Climate, Climate Variability and Change of Timor-Leste.* Dili: Pacific Climate Change Science Program.

Dickinson, W. R., 2009. Pacific Atoll Living: How Long Already and Until When? *GSA Today,* 19, 4–10.

FAO, 2005. Forests and floods drowning in fiction or thriving on facts? RAP Publication 2005/03, Forest Perspectives 2. *Food and Agriculture Organisation of the United Nations (FAO).* Bangkok, Bogor Barat: Center for International Forestry Research (CIFOR).

Gaillard, JC, 2007. Resilience of traditional societies in facing natural hazards. *Disaster Prevention and Management,* 16, 522–44.

Gaillard, JC and Mercer, J., 2012. From knowledge to action: Bridging gaps in disaster risk reduction. *Progress in Human Geography,* 37 (1), 93–114.

Glantz, M. H., 2003. *Climate Affairs: A Primer.* Washington, DC: Island Press.

Godinho, L., Nacuray, E., Cardinoza, M. M. and Lasco, R. D., 2003. Climate change mitigation through carbon sequestration: The forest ecosystems of Timor-Leste. *Proceedings from the 1st National Workshop on Climate Change,* 19 November 2003, Dili.

Hirano, S., 2012. *Learning From the Urban Transitional Shelter Response in Haiti.* Baltimore, MD: Catholic Relief Services.

IPCC (Intergovernmental Panel on Climate Change), 2007. *IPCC Fourth Assessment Report.* Geneva: Intergovernmental Panel on Climate Change (IPCC).

Jenks, C., 2004. *Urban Culture: Critical Concepts in Literacy and Cultural Studies.* Abingdon: Routledge.

Kirono, D., 2010. *Climate Change in Timor-Leste. A Brief Overview on Future Climate Projections.* CSIRO (Australia's Commonwealth Scientific and Industrial Research Organisation), Australia.

Lape, P. V. and Chin-Yung, C., 2008. Fortification as a human response to late holocene climate change in East Timor. *Archaeology in Oceania*, 43 (1), 11–21.

Lewis, J., 1999. *Development in Disaster-Prone Places: Studies of Vulnerability.* London: Intermediate Technology Publications.

Martin, B., 1979. *The Bias of Science.* Canberra: Society for Social Responsibility in Science.

McCall, G., 1996. Clearing confusion in a disembedded world: The case for nissology. *Geographische Zeitschrift*, 84 (2), 74–85.

Mercer, J., Dominey-Howes, D., Kelman, I. and Lloyd, K., 2007. The potential for combining Indigenous and Western knowledge in reducing vulnerability to environmental hazards in small island developing states. *Environmental Hazards*, 7 (4), 245–56.

Mercer, J., Gaillard, JC, Donovan, K., Shannon, R., Alexander, B., Day, S. and Becker, J., 2012. Cultural awareness in disaster risk reduction: Lessons and opportunities. *Environmental Hazards*, 11 (2), 74–95.

Mercer, J., Kelman, I., do Rosario, F., de Deus de Jesus Lima, A., da Silva, A., Beloff, A. M. and McClean, A., 2014. Nation-building policies in Timor-Leste: Disaster risk reduction and climate change adaptation. *Disasters*, 38 (4), 690–718.

Morin J. and Gaillard JC, 2012. Lahar hazard and livelihood strategies on the foot slopes of Mt. Karthala Volcano, Comoros. In: B. Wisner, JC Gaillard and I. Kelman, eds, *Handbook of Hazards and Disaster Risk Reduction.* Abingdon: Routledge, 705–6.

Morin, J. and Lavigne, J., 2009. Institutional and social responses to hazards related to Karthala Volcano, Comoros. Part II: The deep-seated root causes of Comorian vulnerabilities. *Shima: The International Journal of Research into Island Culture*, 3 (1), 54–71.

Morin, J., Lavigne, J., Bachelery, P., Finizola, A. and Villeneuve, N., 2009. Institutional and social responses to hazards related to Karthala Volcano, Comoros. Part I: Analysis of the May 2006 eruptive crisis. *Shima: The International Journal of Research into Island Culture*, 3 (1), 33–53.

Nicolas, G., Schwartz, B. and Pierre, E., 2009. Weathering the storm like Bamboo: The strengths of Haitians in coping with natural disasters. In: A. Kalayjian, D. Eugene and G. Reyes, eds, *International Handbook of Emotional Healing: Ritual and Practices for Resilience after Mass Trauma.* Westport, CT: Greenwood Publishing Group, 93–106.

Norton, J. and Waterman, P., 2008. *Reducing the Risk of Disasters and Climate Variability in the Pacific Islands: Timor-Leste Country Assessment.* Washington, DC: World Bank.

Oxfam, 2011. *Climate Change Impacts upon Communities in Timor-Leste.* Dili: Oxfam.

Patrick, J., 2011. Haiti earthquake response: Emerging evaluation lessons. *Evaluation Insights*, June (1), 1–11.

Programme des Nations Unies pour le Développement, 2010. *Rapport National sur le Développement Humain: Cohésion Sociale et Développement Humain en Union des Comores.* Moroni: Programme des Nations Unies pour le Développement.

Rouveyran J. C. and Djabiri, A., 1968. Le 'dola n'kou' ou grand mariage comorien. *Revue Tiers Monde*, 9 (33), 95–127.

Sandlund, O. T., Bryceson, I., de Carvalho, D., Rio, N., da Silva, J. and Silva, M. I., 2001. *Assessing Environmental Needs and Priorities in East Timor: Issues and Priorities.* Dili: UNDP.

Schuller, M. and Morales, P. eds, 2012. *Tectonic Shifts: Haiti since the Earthquake.* Sterling, VA: Kumarian Press.

Shaw, R., Sharma, A. and Takeuchi, Y. eds, 2009. *Indigenous Knowledge and Disaster Risk Reduction: From Practice to Policy.* Hauppauge: Nova Publishers.

Shaw, R., Pulhin, J. M. and Pereira, J. J., ed., 2010a. *Climate Change Adaptation and Disaster Risk Reduction: An Asian Perspective. Community, Environment and Disaster Risk Management.* Bingley: Emerald.

Shaw, R., Pulhin, J. M. and Pereira, J. J., eds, 2010b. Climate change adaptation and disaster risk reduction: Issues and challenges. *Community, Environment and Disaster Risk Management*. Bingley: Emerald.

Steele, J., 2002. Nation building in East Timor. *World Policy Journal*, 19 (2), 76–87.

Twigg, J., 1999–2000. The age of accountability? Future community involvement in disaster reduction. *Australian Journal of Emergency Management*, 14 (4), 51–8.

UN, 1994. *Report of the Global Conference on the Sustainable Development of Small Island Developing States*. Document A/CONF.167/9. New York: United Nations.

UN, 2005. *Draft Mauritius Strategy for the Further Implementation of the Programme of Action for the Sustainable Development of Small Island Developing States*. Document A/CONF.207/CRP.7. New York: United Nations.

Walker, I., 2002. Les aspects économiques du grand mariage de Ngazidja (Comores). *Autrepart*, 23, 157–71.

Wasson, M., 2001. East Timor and climate change: Security and sustainable development. In: R. Anderson and C. Deutsch, eds, *Sustainable Development and the Environment in East Timor*. Melbourne: Cleveland Press, 38–41.

Wisner, B., 1995. Bridging 'expert' and 'local' knowledge for counter-disaster planning in urban South Africa. *GeoJournal*, 37 (3), 335–48.

Wisner, B., O'Keefe, P. and Westgate, K., 1977. Global systems and local disasters: The untapped power of peoples' science. *Disasters*, 1 (1), 47–57.

Wisner, B., Gaillard, JC and Kelman I., 2014. Hazard, vulnerability, capacity, risk and participation. In: A. Lopez-Carresi, M. Fordham, B. Wisner, I. Kelman and JC Gaillard, eds, *Disaster Management: International Lessons in Risk Reduction, Response and Recovery*. London: Earthscan.

13

CULTURE, GENDER AND DISASTER

From vulnerability to capacities

JC Gaillard, Maureen Fordham and Kristinne Sanz

Introduction

People's ability to face hazards and disasters is shaped by an array of social, cultural, economic and political factors that are deeply enmeshed (Wisner *et al.* 2004, 2012). Those factors particularly influence people's vulnerability. Vulnerability reflects people's inability to access resources and means of protection that are available to those with more power. It is thus rooted in structural issues, which lie beyond the direct reach of those who are vulnerable (Gaillard 2010). Capacities refer to the set of knowledge, skills and resources to which people resort in dealing with hazards and disasters. Capacities are not the opposite end of vulnerability on a single spectrum, because highly vulnerable communities may also display a large set of capacities (Cadag and Gaillard 2013).

Culture has been identified as a key driver of people's vulnerability (cf. Hewitt 2009, Mercer *et al.* 2012) but also encapsulates the varied sources of capacities that people tap in times of disasters. Religious beliefs and attachment to place are some examples of essential components of culture that matter greatly in explaining why people and communities decide to consciously live in hazardous areas. Ethnicity and caste affiliation also prove critical in shaping people's access to resources and means of protection (Gaillard 2012).

Gender in disaster research has been framed by various theoretical approaches. However, researchers have primarily drawn either from liberal feminist or gender and development theory (Enarson *et al.* 2007). Those theorising have put primary emphasis on political and economic structures to the detriment of cultural issues (Fordham 2012). Curiously, the interaction between culture and gender has not yet gained significant attention in disaster literature.

Combining a cultural and gender lens in disaster research and practice could broaden our understanding of the plurality of identities, histories, experiences and knowledge that are often silenced and homogenised in dominant disaster discourses. This nuanced understanding could challenge stereotypical notions about how gender is portrayed in disaster literature. There remains the tendency to equate gender with women and view gender as a stable system of relations that operate between binary sex categories making invisible other marginalised social groups, such as persons who do not identify as either male or female. These tendencies are reflected in disaster policy and action that perceive these diverse social groups as passive victims and as coherent subgroups, having similar needs and concerns.

The chapter places emphasis on the need to recognise these diverse identities and to challenge the implications arising from the intersection of gender with other social markers. At the same time the chapter explores how gender is socially and culturally constructed in order to gain understanding of both the root causes of inequalities and discrimination as well as people's capacities in facing hazards and disasters.

Interlinking gender and culture: some theoretical considerations

Gender and culture, as an interrelated set of constructs, has been central in the analyses of scholars in the 1970s and 1980s who were trying to understand and explain what appeared to be a universal subordination of women (Ortner 1972). The formulation nature/culture – female/male has gained prominence as it relates 'sexual ideologies and stereotypes to the wider system of cultural symbols and to social roles and experiences' and its appeal lies in helping explain how the symbolic association of women with nature makes women's subordinate status appear 'natural' (Moore 1994, pp. 15–16). This approach explains that a woman's body and its functions are involved more than men's in 'species life' – childbirth and child-rearing – while men, lacking this natural ability to procreate, engage in activities outside the domain of the family; for instance, involvement in activities such as art and public affairs, which 'make culture' (Mascia-Lees and Black 2000, p. 75). This approach has been widely challenged, however. First, the categories 'nature' and 'culture', like 'man' and 'woman', are cultural constructs themselves, mediated by a particular intellectual tradition and history specific to the West (Moore 1994, p. 20). In addition, it remains fixated on a binary that excludes other ways of being and becoming.

Culture, as used in these instances, refers to the 'meanings embodied in symbols [. . .]', a vehicle through which individuals 'communicate, perpetuate, and develop their knowledge about and attitudes toward life' (Geertz 1973, p. 89). Culture is a complex repertoire (Swidler 2001) of resources, (e.g., traditions, rituals, behaviour, values, beliefs, artefacts) that social actors draw from in order to construct meanings about the world and define their strategies for actions.

In disaster contexts, culture is seen as both a source of vulnerability and capacity for social groups. As examples, cultural norms and practices that disallow women to go outside the house unaccompanied by male relatives or the requirement for women to wear clothing to cover most of their bodies (even including faces), impede women's ability to save themselves during disasters and resulted in the large mortality gap between women and men in the aftermath of the Indian Ocean tsunami (International Federation of Red Cross and Red Crescent Societies 2010). On the other hand, cultural values such as strong kinship ties or solidarity among neighbours form invaluable support networks in times of disasters.

As an 'institutionalized cultural and social status' (Lorber and Farrell 1991, p. 8), gender categories influence the kinds of opportunities, if any, and the access to such by individuals and groups. It plays a significant role in determining access to education and work, authority within the family, choices and decisions about sexuality, the production of culture and knowledge itself, and participation in political life (Lorber and Farrell 1991, p. 2). While we cannot deny the deep entanglement of gender systems in our daily lives, gender is neither natural nor irrefutable and should be seen as a site of both oppression and resistance. Rich ethnographic accounts of how other cultures perceive gender show that gender is 'relatively fluid and open in nature' (Lamb 1993, p. 231) and not 'a constant determined by dichotomous and fixed physical differences between women and men' (Lamb 1993, p. 231).

The engagement of disaster scholarship with gender has brought to light important analyses about the specific vulnerabilities and capacities of women in disasters. This focus has lasted long enough for gender to become almost synonymous with women. It is only recently that the focus of gender analysis in disaster research has broadened to include other identities, in particular, people who identify outside the male–female binary such as lesbians, gays, *aravanis* (biologically male who dress as women but who claim neither a male nor female identity in India) and *warias* (biologically male with distinct feminine features and identity in Indonesia), among others (cf. Pincha 2008, Balgos *et al.* 2012). These emerging narratives, combined with what we already know about the discrimination faced by marginalised women and girls in disasters, serve as powerful reminders of how gender as a social structure, embedded in the individual, interactional and institutional dimensions of society (Risman 2004), intersects with other markers such as age, ethnicity, race and class to organise and maintain social order and hierarchy in many societies. For many, the dominant system of gender norms is experienced as a daily form of 'gender tyranny' (Doan 2010, p. 635).

The need to hear the stories and voices of marginalised social groups in disasters is vital as Hill Collins (1999, p. 3) notes: 'suppressing the knowledge produced by any oppressed group makes it easier for dominant groups to rule because the seeming absence of dissent suggests that subordinate groups willingly collaborate in their own victimisation'.

The emphases on 'openness, fragmentation and diversity' (Mac an Ghaill and Haywood, 2007, p. 37) premised by queer theory could benefit disaster scholarship and practice as it renders inadequate existing sexual categories vis-à-vis the nuances of people's lived experiences and 'destabilize socially given identities, categories and subjectivities around the common-sense distinctions between homosexuality and heterosexuality, men and women, and sex and gender' (Mac an Ghaill and Haywood, 2007, p. 37).

Theoretical approaches in understanding and explaining gender and gender systems are varied and sometimes even contradictory. It is not the intention of the chapter to provide a comprehensive account of these, rather we sought to explore current and emerging theories that could serve as take-off points for further debates to nuance disaster policy and research and reflect multiple identities and voices in disasters.

The following section highlights the experiences of social groups in disasters – the Garifuna women from Honduras and the *bakla* from the Philippines – who are placed in disadvantaged positions by virtue of their socially constructed, historically specific and culturally located genders – yet adapt, negotiate and resist the implications of these power relations in order to ensure their survival and contribute positively to their societies.

Gender and disaster in context: a case from Honduras

It might be expected that a case study of Garifuna women and disaster in Honduras would paint a picture of vulnerability, powerlessness and a lack of capacities. Clearly, structural inequalities based on gender, race/ethnicity, income and the colonial legacy mean the Garifuna are a marginalised sub-group within a wider context of social disadvantage as emphasised in both the Human Development Index and Gender Inequality Index of the United Nations Development Programme (UNDP 2013). However, this will be a case study of capacities and action rather than vulnerability and passivity.

The Garifuna are a distinct ethnic group descended from African slaves and indigenous Carib and Arawak Indians. Following forced exile by their colonial oppressors, the Garifuna settled in a number of coastal locations in Central America including Honduras, Nicaragua and Guatemala. The largest Garifuna population is in Honduras. The Garifuna people have a distinct cultural heritage recognised by the 2003 United Nations Educational, Scientific and Cultural Organisation (UNESCO) declaration referring to their 'intangible cultural heritage' (Brondo and Woods 2007, p. 13). Nevertheless, despite this recognition, they are the inheritors of enduring colonial and postcolonial racial ideologies and stereotypes (Mollett 2006, pp. 77–8). However, partly due to migratory processes, the Garifuna are also a matrilocal culture and thus women occupy a position of relative advantage compared to some other non-Garifuna women in the region (Anderson 2007).

When Hurricane Mitch struck Honduras in 1998, external aid was slow to arrive in Garifuna communities. Women took the lead in organising response and in the subsequent recovery process. Grassroots women organised boats to rescue people and became leaders of the *Comité de Emergencia Garífuna de Honduras* (Garifuna Emergency Committee of Honduras). In line with their subsistence farming tradition and expertise, they led collective farming and fishing initiatives, and developed tool banks to share farming tools and community seed banks to ensure post-disaster food security in the longer term. As so many physical resources had been destroyed by Hurricane Mitch, they led collective, hurricane-safe, housing construction (benefiting from a learning exchange with a Jamaican Women's Construction Collective) and bought land for relocation to safer areas.

The Garifuna women also engaged at a political level: participating in a planning meeting coordinated by the Urban Planning Ministry; training, at the invitation of local Mayors, local government representatives in vulnerability reduction; and creating a coordinated plan for disaster response (Fordham and Gupta 2011).

In these disaster risk reduction (DRR) efforts they were supported by a network of grassroots women's groups: GROOTS (Grassroots Organisations Operating Together in Sisterhood).[1] GROOTS helped set up peer learning exchanges and other opportunities for grassroots women to learn from each other how best to strategise for community and household benefit. Beyond the local grassroots groups such as the Garifuna Emergency Committee of Honduras they have reached out to the national, regional and global level through their attendance at, and engagement with, global fora of the United Nations (as one example). In this way, Garifuna women became recognised as expert practitioners and partners rather than beneficiaries; as women speaking for themselves and their constituencies, rather than having others speak on their behalf (Schilen, Box 7.1 in Vielajus and Haeringer 2011).

This case study has used a cultural lens to reveal a complex pattern of both vulnerability and actions among grassroots women ordinarily represented as subjugated. Here they can be seen as active agents of change, helped partly by their own cultural history and partly by processes of networking to aid in accessing those with power (Vielajus and Haeringer 2011). Such transnational networks facilitate collective working and scaling up of local initiatives. The Garifuna women and their communities are vulnerable in so many ways and yet they display a large array of capacities in facing disasters based firmly on development activities that seek to strengthen a range of resources:

- human basic requirements (ensuring food security and sustainable livelihoods; engaging in knowledge exchange/transfer and building of self-esteem);
- social and political (organising communities and networks; networking and creating partnerships for DRR; facilitating good governance);
- financial (influencing governments to allocate community funds and budgets; forming savings and credit groups, cooperatives and federations; developing housing and community infrastructure finance options);
- natural (conserving natural resources and protecting biodiversity; practising sustainable agriculture); and
- physical (securing housing and shelter; accessing community tools and equipment; building community resource centres, and women and child-friendly spaces).

(Fordham and Gupta 2011)

This is an example of local capacities, embedded in particular cultural practices, aided by external networking at a range of scales to shift the balance from considerable vulnerability to a resistance and action to reduce the risk of disaster.

Beyond male and female: a case from the Philippines

The Southeast Asian and Pacific world settled by Austronesian speakers hosts significant minorities sharing similar gender identities beyond the male–female binary, e.g., the *whakawahine* of New Zealand-Aotearoa, the *fa'afafine* of Samoa, the *mahu* of Hawaii and the *waria* of Indonesia. Most are biological males who perform cultural roles in their societies. In that sense, sex matters less than gender in defining their identity (cf. Schmidt 2003, Lomax 2007). Such a minority exists too in the Philippines where they are traditionally called *bakla*.

The identity of the *bakla* does not refer only to a particular sexual behaviour. It rather expresses specific roles within the household and society or an ability to swing from male to female tasks and responsibilities (Garcia 2008). It is, therefore, more a question of gender than a simple differentiation upon a divergent sexual identity. This is evident in the term *bakla* (noun), which is actually the contraction of *babae* (woman) and *lalaki* (man). When used as an adjective, *bakla* further means uncertainty or indecisiveness (Tan 1995).

Bakla openly claim their identity and are recognised for their leadership and initiative when it comes to community activities (Tan 2001). Yet, they often suffer from mockery and discrimination when in the presence of men and women, especially in rural areas. On the larger political scene, and despite several legal requests, marriage between *bakla* has not yet been recognised. However, progress towards the recognition of *bakla*'s rights has recently been made as their congressional party list was authorised to file candidacy during the 2010 elections. Within the family, young *bakla* are frequently marginalised and tasked with demanding house chores that span across the usual responsibilities of both boys (e.g., fishing, fetching fire wood and water) and girls (e.g., cleaning the house, doing the laundry, caring for babies).

These everyday forms of discrimination against the *bakla* are reflected during disasters as are their skills and resource sets that they draw upon from their daily lives (Gaillard 2011). In Irosin, a small town located at the southern tip of Luzon, young *bakla* are often asked by their parents to do the dirty chores, for example, cleaning up the house in the aftermath of flash floods. In Masantol, located in the delta of the Pampanga River, *bakla* teenagers were asked to look after young children and do the laundry at home and also to fetch water and firewood amid deep flood water after a powerful cyclone in 2011. In Quezon City, Metro Manila, some *bakla* youth were left to eat last and least when their households were affected by two back-to-back powerful cyclones in late 2009. When evacuated in crowded churches or public buildings, the *bakla* are never identified as such. Official lists of affected people are always limited to male and female categories thus depriving *bakla* of assistance that caters for their specific needs. The *bakla* suffer from lack of privacy, being uncomfortable with either women or men. Their personal grooming needs are also objects of jokes from men in male comfort rooms where they are assigned.

Although suffering from the foregoing discrimination, *bakla* display significant capacities in facing hazards and disasters (Gaillard 2011). *Bakla* rely on endogenous resources and activities that reflect their everyday role within the Philippine culture and society. These resources particularly mirror their ability to swing from male to female tasks and responsibilities as well as their sense of initiative and leadership. In Irosin, there are those who spontaneously walk around their neighbourhood to collect relief goods. In Quezon City, young *bakla* organised larger relief operations in the aftermath of the back-to-back cyclones of 2009. For such, they went to request support not only from their neighbours, but also from the local chief executive who provided them with relief goods and logistical support. In evacuation centres all throughout the country, *bakla* are often those who spontaneously care for babies and young children. Some do the cooking and the cleaning.

Such capacities provide invaluable resources for planning DRR. In Irosin, young *bakla* were involved in a DRR project as a group and as members of the larger community (Gaillard 2011). Isolated group discussions have enabled the assessment of their particular roles and needs in the face of natural hazards. They then participated in participatory activities with the larger community. *Bakla* identified their houses on a map in order to delineate specific areas where each of them would collect relief goods in time of disaster. Their potential contribution to the life of the community while evacuated in public buildings was also discussed. These activities conducted in the presence of men and women contributed to the recognition by the larger community of the contribution of *bakla* to DRR. When a consultation was conducted in early 2011 with government officials of Irosin regarding the possibility of reproducing similar DRR initiatives in all neighbouring villages, the male leader of the community that was covered by the foregoing project instantly and spontaneously mentioned the case of the *bakla* who were able to contribute to their local activities the previous year.

Unfortunately, *bakla*'s particular needs as well as their capacities in facing hazards and disasters are not formally recognised in policy (Gaillard 2011). The otherwise progressive legal framework for DRR and disaster management in the Philippines does not mention the actual or potential contribution of the *bakla*. Neither do the dense network of NGOs and civil society organisations that focus on reducing the risk of disaster in the country. The specific needs of the *bakla* in facing natural hazards and in times of disaster are similarly ignored. This is despite recent attempts at the recognition of such rights by *bakla* and larger LGBT lobby groups at the national level.[2]

Reclaiming spaces in disasters

We place emphasis on the recursive relationship between social structure (i.e., gender) (Risman 2004) and individuals (Giddens 1984) by presenting the experiences of two social groups in disasters to show that, while social structure shapes individuals, individuals and groups simultaneously shape the social structure. DRR policy and scholarship need to recognise these diverse social groups as political actors who face hazards and disasters on a regular basis yet remain capable of effecting change within the limits of their social locations.

It is, however, important to bear in mind the different levels of constraints faced by the Garifuna women and the Philippine's *bakla*, which are apparent in terms of their participation in the public sphere. For instance, while some Garifuna women have been able to participate actively and engage with powerful political actors at the international level, identifying as a *bakla* in the Philippines subjects one to discrimination and political participation for this social group remains limited.

In narrating the experiences of the *bakla* during disasters, the social context within which the *bakla* identity operates and exists has to be made visible. The Roman Catholic Church, followed by 90% of Filipinos, is widely known for its antagonistic stance against homosexuality and remains a powerful institution that plays a significant role in shaping other social institutions and organisations as well as behaviours and consciousness of the Filipino society. With this in mind, we can then understand why identities and relationships outside of the accepted sex categories are considered a minority in the Philippines and are subjected to marginalisation and discrimination.

Studies of culture and gender in disasters have often stressed the contribution of cultural norms and practices in shaping vulnerability of various social groups in facing hazards (cf. Enarson and Morrow 1998, Phillips and Morrow 2008, Hewitt 2009, Knight *et al.* 2012). The foregoing case studies indeed highlight that, in some contexts, women as well as individuals identifying outside of the dominant sex categories are denied access to resources and means of protection that are available to men or other dominant social groups. The United Nations Development Programme's country evaluation for Honduras recognises that 'gender inequality is deeply embedded in the culture of Honduras' (UNDP 2006, p. 42). This is expressed through inequality of opportunity in relation to economic and social resources and rights and also women's generally poor representation in the political sphere. There is a lack of enforcement of legal, institutional and policy frameworks that would 'make rights and opportunities real for women' (Delaney and Shrader 2000, p. 4). In the Philippines, *bakla* suffer from everyday marginalisation in their families and from the larger society where heterosexuality remains the norm. It must be said, however, that in some settings, men prove more vulnerable than women because of the role assigned to them by cultural norms or by social practices and behaviours (Enarson 2009, Mishra 2009). For example, Klinenberg (2002) shows that more men died during the 1995 heat wave in Chicago because they tend to lose social relationships that women often retain.

The foregoing case studies yet emphasised that despite their situations the Garifuna women of Honduras and the *bakla* of the Philippines are not helpless nor simply victims in facing hazards and disasters. Both groups mobilised themselves to claim their spaces within the 'battlefield of actions' (Gaillard and Mercer

2013, p. 93). Garifuna women and *bakla* both pulled together a wide range of knowledge, skills and resources that have proved useful in overcoming the impact of various natural hazards that affected their communities. The Garifuna case also shows the importance of working closely together with existing grassroots groups to enhance the capacities they already have in order to ensure that interventions do not contribute to the further marginalisation of these groups.

Garifuna women and *bakla*'s capacities in facing disaster both reflect the strength, skills and resources of social groups at the margins in facing the consequences of gender roles and expectations influenced by their histories and cultures. These capacities intrinsic to social groups in disasters warrant particular attention and inclusion into both disaster policy and practice.

A call for genuine participation

DRR policies need to recognise the interactions between culture and gender as both sources of vulnerabilities and capacities for diverse social groups in disasters. We have shown the cases of Honduras and the Philippines to reflect on the different identities and contexts that are often homogenised in these policies. Despite both being considered as vulnerable social groups, Garifuna women and the *bakla* from the Philippines have experienced and responded to disasters differently.

The active participation of these social groups has been recognised as a viable alternative to the shortcomings of dominant DRR practices (cf. Wisner *et al.* 1977, Chambers 1984, Maskrey 1984) which remain Western-orientated, technocratic and gender blind. By active participation we mean the reclaiming of space for political engagement. In DRR this means defining their own needs and determining potential solutions, alongside other stakeholders (international agencies such as the United Nations, International NGOs, among others), which impact their lives and livelihoods. Genuine participation should aim for transformative empowerment. Unfortunately, participation is often corrupted to serve the interests of outside actors who need to justify the 'involvement' of locals in DRR activities they have designed beforehand. In fact, participation is often conceived as an outcome accountable to funding agencies, rather than a process where accountability should actually be downward to those who participate. In addition, participation is frequently devised for specific groups, e.g., women, excluding other marginalised groups.

However, DRR must be inclusive not exclusive (Gaillard and Mercer 2013). In this view, reducing the risk of disaster requires the collaboration of a large array of stakeholders, including international organisations, government agencies, scientists, NGOs, the private sector, schools, faith groups and all segments of local communities, including the most marginalised. The latter include all gender groups, i.e., women, men and those claiming an identity beyond those two categories, as in the case of the Philippine *bakla*. The collaboration between all these actors requires dialogue and trust, which, more often than not, does not exist in current DRR practices, including those that encourage a tokenistic form of participation.

Dialogue and trust are essential for all stakeholders to recognise, value and integrate others' vulnerabilities and capacities into comprehensive DRR policy and practice. Many of the marginalised groups across cultures and societies, including women and gender minorities, know what their needs and resources are in facing hazards and disasters. The issue for these groups is usually to make their vulnerability and capacities tangible and recognised by others. In that sense, it is often insufficient for DRR practitioners to only work with a particular marginalised group, for example women, in isolation from the larger community, including men. This marginalised group should interact with those with power within the community or larger society otherwise DRR initiatives remain clustered and fail to address the unequal power relationships that led disasters to occur. Such a culturally and gender sensitive integrative initiative for reducing the risk of disaster proves powerful as for the *bakla* of Irosin in the Philippines (Gaillard 2011).

Obviously, fostering dialogue across social and cultural boundaries is difficult. It requires political will to recognise and value cultural realities. These realities are unfortunately frequently ignored by contemporary dominant policies geared towards reducing the risk of disaster, framed by homogenising Western cultural references. As in the case of the Philippines, there do not seem to be any legal instruments for DRR anywhere in the world that officially recognise the particular vulnerabilities and capacities of gender minorities, although the archipelago is one of the very few countries where most recent law acknowledges the particular needs of women and other marginalised groups.

Prelude to further challenges in disaster policy and scholarship

While gender serves as a powerful organising feature in many societies, it does not exist in isolation and must be analysed in connection with other social markers such as age, ethnicity, abilities, race, class and sexual orientation. Gender 'produces, reproduces, and legitimates the choices and limits that are predicated on sex category' (West and Zimmerman 1987, p. 147). However, there are various ways in which these are subverted as shown by Garifuna women and the *bakla* from the Philippines.

Given our increasing understanding of diverse identities that exist outside of dichotomous sex categories, disaster policy and scholarship need to broaden its engagement with gender. There has to be an interrogation of the complex political, social and cultural processes that work together to produce and reproduce meanings, relationships, identities as well as knowledge of and about social groups in disasters and how these are resisted and negotiated by the same.

Reconceptualising gender requires radical change on a whole gamut of gender relations which exist at the individual, institutional and interactional levels:

> the social subordination of women, and the cultural practices which help sustain it; the politics of sexual object-choice, and particularly the oppression

of homosexual people; the sexual division of labour, the formation of character and motive, so far as they are organised as femininity and masculinity; the role of the body in social relations, especially the politics of childbirth; and the nature of strategies of sexual liberation movements.

(West and Zimmerman 1987, p. 126)

Culture is therefore a central issue for understanding the role of gender in disasters. In consequence, to be gender inclusive, DRR policies and practices must be informed by cultural norms and values, including those that both have a positive and negative impact on people's ability to face disasters. Ultimately, to tackle the challenging tasks at hand, political will and genuine commitment as well as a reflexive disaster scholarship are required.

Notes

1   www.groots.org
2   http://diversityandequality.ph and http://groups.to/lgbtipinoysforcalamityvictims

References

Anderson, M., 2007. When Afro becomes (like) Indigenous: Garifuna and Afro-indigenous politics in Honduras. *Journal of Latin American and Caribbean Anthropology*, 12 (2), 384–413.

Anderson, W. A., 1965. *Some Observations on a Disaster Subculture: The Organizational Response of Cincinnati, Ohio, to the 1964 Flood.* Research Note No. 6, Disaster Research Centre. Columbus, OH: The Ohio State University.

Balgos, B., Gaillard, JC and Sanz, K. (2012). The warias of Indonesia in disaster risk reduction: The case of the 2010 Mt Merapi eruption. *Gender and Development*, 20 (2), 337–48.

Blackwood, E., 2009. Tombois in West Sumatra: Constructing masculinity and erotic desire. In: E. Lewin, ed., *Feminist Anthropology: A Reader.* Malden, MA: Blackwell Publishing, 411–34.

Brondo, K. V. and Woods, L., 2007. Garifuna land rights and ecotourism as economic development in Honduras. Cayos Cochinos marine protected area. *Ecological and Environmental Anthropology*, 3 (1), 2–18.

Butler, J., 1990. *Gender Trouble Feminism and the Subversion of Identity.* London: Routledge.

Cadag, J. R. D. and Gaillard, JC, 2013. Integrating people's capacities in disaster risk reduction through participatory mapping. In: A. Lopez-Carresi, M. Fordham, B. Wisner, I. Kelman and JC Gaillard, eds, *Disaster Management: International Lessons in Risk Reduction, Response and Recovery.* London: Earthscan, 269–86.

Chambers, R., 1984. *Rural Development: Putting the Last First.* London: Longman.

Collins, P. H., 1999. *Black Feminist Thought: Knowledge, Consciousness, and the Politics of Empowerment.* 2nd ed. London: Routledge.

Delaney, P. L. and Shrader, E., 2000. *Gender and Post-Disaster Reconstruction: The Case of Hurricane Mitch in Honduras and Nicaragua.* Washington, DC: World Bank.

Doan, P. L., 2010. The tyranny of gendered spaces: Reflections from beyond the gender dichotomy. *Gender, Place and Culture*, 17 (5), 635–54.

Enarson, E., 2009. *Women, Gender & Disaster: Men & Masculinities.* Gender Note #3, Gender and Disaster Network, Newcastle upon Tyne.

Enarson, E. and Morrow, B. H., eds, 1998. *The Gendered Terrain of Disaster: Through Women's Eyes*. Santa Barbara, CA: Praeger.

Enarson, E., Fothergill, A. and Peek, L., 2007. Gender and disaster: Foundations and directions. In: H. Rodriguez, E. L. Quarantelli and R. R. Dynes, eds, *Handbook of Disaster Research*. New York: Springer, 130–46.

Fordham, M., 2012. Gender, sexuality and disaster. In: B. Wisner, JC Gaillard and I. Kelman, eds, *Handbook of Hazards and Disaster Risk Reduction*. Abingdon: Routledge, 424–35.

Fordham, M. and Gupta, S., 2011. *Leading Resilient Development: Grassroots Women's Priorities, Practices and Innovations*. New York: GROOTS/United Nations Development Programme.

Gaillard, JC, 2010. Vulnerability, capacity, and resilience: Perspectives for climate and development policy. *Journal of International Development*, 22 (2), 218–32.

Gaillard, JC, 2011. *People's Response to Disasters: Vulnerability, Capacities and Resilience in Philippine Context*. Angeles City: Centre for Kapampangan Studies.

Gaillard, JC, 2012. Caste, ethnicity, religious affiliation and disasters. In: B. Wisner, JC Gaillard and I. Kelman, eds, *Handbook of Hazards and Disaster Risk Reduction*, Abingdon: Routledge, 459–69.

Gaillard, JC and Mercer, J., 2013. From knowledge to action: Bridging gaps in disaster risk reduction. *Progress in Human Geography*, 37 (1), 93–114.

Garcia, J. N. C., 2008. *Philippine Gay Culture: Binabae to Bakla, Silahis to MSM*. Quezon City: University of the Philippines Press.

Geertz, C., 1973. *The Interpretation of Cultures*. New York: Basic Books.

Giddens, A., 1984. *The Construction of Society: Outline of the Theory of Structuration*. Berkeley, CA: University of California Press.

Gillison, G., 1980. Images of nature on Gimi thought. In: C. MacCormack and M. Strathern, eds, *Nature, Culture and Gender*. Cambridge: Cambridge University Press, 143–73.

Hewitt, K., 2009. *Culture and Risk: Understanding the Sociocultural Settings that Influence Risk from Natural Hazards*. Kathmandu: International Centre for Integrated Mountain Development.

International Federation of Red Cross and Red Crescent Societies, 2010. *A Practical Guide to Gender Sensitive Approaches for Disaster Management*. Geneva: International Federation of Red Cross and Red Crescent Societies.

Klinenberg, E., 2002. *Heatwave: A Social Autopsy of Disaster in Chicago*. Chicago, IL: University of Chicago Press.

Knight, K., Gaillard, JC and Sanz, K., 2012. Gendering the MDGs beyond 2015: Understanding needs and capacities of LGBTI persons in disasters and emergencies. *Global Consultation on Addressing Inequalities, UN Women and UNICEF* [online]. Available from: www.worldwewant2015.org/node/283239 [accessed 8 February 2014].

Lamb, S., 1993. The making and unmaking of persons: Gender and body in Northeast India. In: C. B. Brettell and C. F. Sargent, eds, *Gender in Cross-Cultural Perspectives*. New Jersey: Pearson Preston Hall, 230–40.

Lomax, T., 2007. Whakawahine – A given or a becoming? In: J. Hutchings and C. Aspin, eds, *Sexuality and the Stories of Indigenous People*. Wellington: Huia Publishers, 82–93.

Lorber, J. and Farrell, S. A., 1991. *The Social Construction of Gender*. Newbury Park: Sage.

Mac an Ghaill, M. and Haywood, C., 2007. *Gender, Culture and Society: Contemporary Femininities and Masculinities*. Basingstoke: Palgrave Macmillan.

MacCormack, C. and Strathern, M., 1980. *Nature, Culture and Gender*. Cambridge: Cambridge University Press.

Mascia-Lees, F. E. and Black, N. J., 2000. *Gender and Anthropology*. Prospect Heights, IL: Waveland Press.

Maskrey, A., 1984. Community based hazard mitigation. In: *Proceedings of the International Conference on Disaster Mitigation Program Implementation*. Jamaica: Ocho Rios, 12–16 November 1984, 25–39.

Mercer, J., Gaillard, JC, Donovan, K., Shannon, R., Alexander, B., Day, S. and Becker, J., 2012. Cultural awareness in disaster risk reduction: Lessons and opportunities. *Environmental Hazards*, 11 (2), 74–95.

Mishra, P., 2009. Let's share the stage: Involving men in gender equality and disaster risk reduction. In: E. Enarson and P. G. Dhar Chakrabarti, eds, *Women, Gender and Disaster: Global Issues and Initiatives*, New Delhi: Sage, 29–39.

Mollett, S., 2006. The Miskito and Garifuna struggle for Lasa Pulan. *Latin American Research Review*, 41 (1), 76–101.

Moore, H., 1994. The cultural constitution of gender. Polity Press, ed., *The Polity Reader in Gender Studies*. Cambridge: Polity Press, 14–21.

Ortner, S., 1972. Is female to male as nature is to culture? *Feminist Studies*, 1 (2), 5–31.

Phillips, B. H. and Morrow, B. D., 2008. *Women and Disasters: From Theory to Practice*. Bloomington: Xlibris.

Pincha, C., 2008. *Understanding Gender Differential Impacts of Tsunami and Gender Mainstreaming Strategies in Tsunami Response in Tamil Nadu, India*. Anawim Trust/Oxfam America, Tamil Nadu.

Risman, B., 2004. Gender as a social structure: Theory wrestling with activism. *Gender and Society*, 18 (4), 429–50.

Schmidt, J., 2003. Paradise lost? Social change and Fa'afafine in Samoa. *Current Sociology*, 51 (3/4), 417–32.

Swidler, A., 2001. *Talk of Love: How Culture Matters*. Chicago, IL: University of Chicago Press.

Tan, M. L., 1995. From Bakla to Gay: Shifting gender identities and sexual behaviours in the Philippines. In: R. G. Parker and J. H. Gagnon, eds, *Conceiving Sexuality: Approaches to Sex Research in a Post-Modern Research*. New York: Routledge, 85–96.

Tan, M. L., 2001. Survival through pluralism: Emerging gay communities in the Philippines. *Journal of Homosexuality*, 40 (3/4), 117–42.

UNDP, 2006. *Country Evaluation: Assessment of Development Results – Honduras*. New York: United Nations Development Programme.

UNDP, 2013. *Honduras – Country Profile: Human Development Indicators*. New York: United Nations Development Programme.

Vielajus, M. and Haeringer, N., 2011. Transnational networks of 'self-representation': An alternative form of struggle for global justice. In: M. Albrow and H. Seckinelgin, eds, *Global Civil Society 2011: Globality and the Absence of Justice*. London: Palgrave Macmillan, 88–101.

West, C. and Zimmerman, D., 1987. Doing gender. *Gender and Society*, 1(2), 125–51.

Wisner, B., Blaikie, P., Cannon, T. and Davis, I., 2004. *At Risk: Natural Hazards, People's Vulnerability, and Disasters*. 2nd ed. London: Routledge.

Wisner, B., Gaillard, JC and Kelman, I., eds, 2012. *Handbook of Hazards and Disaster Risk Reduction*. Abingdon: Routledge.

Wisner, B., O'Keefe, P. and Westgate, K., 1977. Global systems and local disasters: The untapped power of peoples' science. *Disasters*, 1 (1), 47–57.

14

A CULTURE OF RESILIENCE AND PREPAREDNESS

The 'last mile' case study of tsunami risk in Padang, Indonesia

Jörn Birkmann, Neysa Setiadi and Georg Fiedler

Introduction: the perception of hazards and risk

Before dealing with specific aspects of perception of the risks of a tsunami and the case study of tsunami early warning systems and anticipated response behaviour, this chapter outlines the development of perception geography and the core concepts it uses to understand the divergence between natural or objective hazard information and different risk perceptions. Additionally, an overview of important milestones in the development of perception geography in the context of natural hazards allows an improved understanding of *why* and *how* hazard and disaster risk reduction and adaptation research still needs to be enhanced when aiming to understand cultural, and especially religious, systems of meaning in the context of disasters. In addition, definitions of the terms 'culture of resilience' and 'preparedness' are provided.

The concept of a culture of disaster resilience

The idea of 'disaster resilience' and the promotion of a 'culture of disaster resilience' were specifically formulated within the 2005 World Conference on Disaster Reduction – which took place in Kobe, Japan shortly after the Indian Ocean tsunami of late December 2004, an event that resulted in more than 250,000 fatalities. These notions have been established as a key principle for risk management in the Hyogo Framework for Action (HFA) (cf. UN/ISDR 2005, pages 5, 7 and 10). The Hyogo Framework primarily links a culture of disaster resilience to improvements in public consultations, sustained public education about disaster risks, and the continued monitoring and assessment of hazards, vulnerabilities and risks. Interestingly, the term 'culture' in disaster risk management relates to various aspects of the understanding, the communication and the management of risks. However, a clear or universal definition has not yet evolved. In this regard, it is important to note that incorporation of the vision of a 'culture of disaster resilience' within the international discussion on disaster risks also represents a certain shift from a more top-down and hierarchical understanding of managing disaster risks towards a more participatory and consultative risk management framework, where a culture of disaster resilience would manifest itself in public awareness about risks, language of how risks are communicated and ways in which the public is involved in preparing for adverse events. Consequently, the notion of a 'culture of disaster resilience' and 'preparedness' within the HFA can be interpreted as addressing the need to revisit existing values and belief systems of, for example, organisations involved in disaster risk management, as well as to evaluate past and present routines and management processes in disaster risk reduction (DRR). In addition, this chapter uses the term 'culture of preparedness' in order to underscore that a culture of resilience should be primarily focused on improved preparedness strategies, not just on the fast recovery processes that are sometimes also associated with the notion of resilience.

The objective and subjective worlds

The historical development of perception geography is based on the need to differentiate between the subjective and objective worlds for organisational reasons. This means a core underlying assumption of perception research (cf. Tolman 1948, Piaget and Inhelder 1956, Moore 1976, Piaget 2007) and perception geography (cf. Lynch 1960, Hard and Scherr 1976, Downs and Stea 1977, Katz 1991, Geipel 1992, Hidajat 2002) is the fact that the real world – which is 'out there' – will be individually perceived and understood 'inside' every subject. This reflects different filtering effects of (neuro)biological, psychological and sociocultural dispositions of people. This includes, for example, consciousness, personality and the five senses, as well as the social and cultural contexts ingrained in people's minds when they deal with crises or disasters. These filtering effects influence, to different degrees, the individual's information, perception and cognition of the object world (Knox and Marston 2001). Hence, the perceived environment represents a subjective construction or imagination of the real world. This differentiation of a physical and real world classified as the object world, in contrast to a subject world represented by individuals and their perception of phenomena of the object world, often leads to a dichotomy that might hinder an integrative perspective. However, this dichotomy provides an important epistemological and methodological background for risk reduction strategies and in forming a culture of resilience and preparedness.

According to the 'dual worldview', two grand epistemological schools of thought prevail in geographical research (Reuber and Pfaffenbach 2005). The first school is related to critical rationalism in the tradition of Popper (1983). Based on a hypothetical realism (Gebhardt and Reuber 2011), the researcher in this system aims to move towards a supposed reality by testing hypotheses, using scientific approaches and methods of falsification to verify these. Currently, the second school of thought, (social) constructivism, also plays a significant role in perception research and human geography (cf. Berger and Luckmann 1966, Schütz and Luckmann 1975, Knorr-Cetina 1981, Glasersfeld 1995, Gergen and Westmeyer 2002). In contrast to critical rationalism, social constructivism focuses on understanding how human beings in their everyday situations produce reality through the (re)production of cultural and religious systems of meaning, among other things.

Since the so-called 'quantitative revolution' in the 1960s, geographical investigation of the object world has primarily dealt with quantitative methods, often taking up ideas of critical rationalism. Epistemologically, research on the subjective world requires, in the strict sense, a constructivist approach – hence qualitative methods. However, in the practice of research, combinations of quantitative and qualitative approaches are often applied to perception research.

Historical development trends in perception geography

In recent years, geographic perception research, which includes research on potential reactions to actual and future hazards, has gained a great deal of attention in the context of human adaptation to climate change. Also, in this context of the discourse around 'limits of adaptation', the field of perception research, and particularly its questioning of behavioural changes in day-to-day activities due to extreme events, offers an important avenue of study (IPCC 2012).

The first thread of geographic perception research was conceptually based on the behaviourism coined by Watson and Skinner in the early 1920s. In accordance with the stimulus response framework, perception geography was dominated by the understanding that an environmental hazard (or stressor) would directly lead to a chain reaction and response by groups of people or individuals. Based on the critique of this simple stimulus–reaction behaviour model, a so-called 'cognitive revolution' emerged (Chomsky 1959) that focused particularly on mental processes as important missing links between environmental phenomena and actual human responses. In this regard (and in contrast to radical behaviourism), the behaviourist school of thought underscored that human behaviour and decision making processes are not characterised by a simple stimulus–response mechanism (Weichhart 2008, p. 153). Rather, the behaviourist approach underpins that decision making processes, as well as human response behaviour to environmental hazards, are embedded in complex cognitive processes that involve perception, understanding, thinking, knowledge, planning and action. Various cognitive processes are placed in between the stimulus or hazard and the actual reaction of humans to it, thus giving some indication of different human reactions to facing an environmental hazard.

Behavioural perception analysis used within natural hazards and risk research has a long tradition; e.g., Burton, Kates, and White (1993) have examined the role of risk perception in the context of disaster risk reduction or so-called 'hazard research'. However, the focus of their research is primarily on how perceptions are influenced by natural phenomena or disaster events and how human responses to hazards are influenced by risk perceptions. Little work has been done so far on how risk perceptions and cultural symbols can be actively used and integrated in risk preparedness strategies and risk reduction measures.

Another important milestone in the discussion on how people react to natural hazards, crises and disasters was developed by Tobin and Montz. They argue that human response to natural hazards and risks, as well as the actual behaviour of people in the face of adversity, is determined by a) situational factors and b) cognitive factors (Tobin and Montz 1997, p. 135). Situational factors determine the context in which people are embedded, and hence deal with characteristics of the physical and the socioeconomic environment. Factors of the physical environment encompass, for example, the frequency and magnitude of natural hazards or the duration of environmental phenomena such as the duration of droughts or flooding. In contrast, factors that comprise the socioeconomic environment, according to Tobin and Montz, contain issues such as culture, education, income and poverty, age, gender and household size, to name just a few (Tobin and Montz 1997, p. 135; see also Figure 6). In contrast to factors that determine a given situation, cognitive factors include psychological variables and characteristics of individuals that influence their acceptance or avoidance of risk taking (Tobin and Montz 1997, p. 135). People who instead view environmental phenomena and their impact as outside of the influence of human behaviour may pay less attention to risk preparedness compared to those who view environmental hazards and their consequences as something that can be modified by human responses. In this regard, Tobin and Montz also underscore that cognitive and situational factors can work individually, in combination, or even in sequence, to influence human responses (Figure 6).

Furthermore, regarding early warnings, it is increasingly recognised that knowledge and information provided by experts might not be equal to the knowledge and perception of local communities and those at risk. In various studies (Lachman *et al.* 1961, Baker 1991, Riad and Norris 1998, Tierney *et al.* 1999, Lindell *et al.* 2005, among others), early warning in advance of a disaster did not necessarily trigger an appropriate behaviour from the affected people and many individual factors and social interactions, including perceptional factors, play a role in response behaviour. In addition, attitude-behaviour models have been developed and applied in protective actions and emergency response, for example the theory of planned behaviour (Ajzen 1991), the theory of protection motivation (Rogers 1983, Martin *et al.* 2007) and the protective action decision model (Lindell and Perry 1992, 2012). Most of these theories and models argue that the intention behind a certain protective behaviour depends on the attitude towards this behaviour and on social norms, which are influenced by various perceptions. In this regard, embedding disaster preparedness into the daily life and priorities of local people, as well as

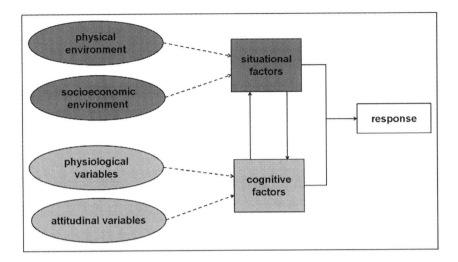

FIGURE 6 Factors that influence human behaviour and reactions to crises and hazards
Source: figure based on Tobin and Montz 1997, p. 136.

developing mechanisms suitable for the community (such as early warning message dissemination through various local channels) is a crucial aspect in building a 'culture of disaster resilience and preparedness' as defined by the HFA. It is important to understand risk perceptions and integrate disaster preparedness within an existing cultural context. Here, a broader understanding of culture, one which encompasses identity, lived culture, language issues and education and awareness levels, as well as other aspects, seems to be different from the concept of culture used by Tobin and Montz (1997). Despite these definitional challenges, a major task of building a culture of resilience and preparedness also lies in linking different knowledge types, perceptions and norms in order to be able to develop effective and integrative strategies for disaster risk reduction. Core concepts of perception geography and features of human-environmental relations are shown in Figure 7.

Figure 7, based on Haggett (1991), underscores, among other issues, that the knowledge of experts and institutions responsible for disaster risk management can significantly differ from the knowledge that individuals at risk have based on their own experience. Furthermore, Haggett shows that risk management has to deal with the challenge that some hazards might be unknown to the local people and organisations involved in disaster risk management. For instance, organisations in Sri Lanka had limited knowledge of DRR regarding tsunami risk before the Indian Ocean tsunami of 2004. Consequently, some human-environmental relations remain hidden and unknown.

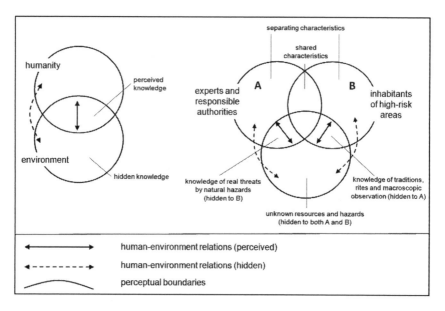

FIGURE 7 Human–environmental relations

Source: figure based on Haggett 1991, p. 311; translated from German and slightly modified by the authors.

The Indian Ocean tsunami: a catalyst for a culture of resilience and preparedness

The Indian Ocean tsunami in December 2004 was one of the most deadly disasters of the last decades. According to current estimates, it killed more than 220,000 people, made more than one million people homeless and destroyed the livelihoods of many thousands of people (cf. ADPC 2006, among others). In this regard, the Indian Ocean tsunami was a game changer for many countries exposed to potential tsunami risk. The people and coastal communities directly affected, and those not directly affected but informed about tsunami risk via modern communications media, fundamentally changed their perception about coastal risks, particularly tsunami risk. People started to develop structural and non-structural measures for tsunami preparedness and protection. With the support of the international community, many countries in the Indian Ocean, particularly Indonesia and Sri Lanka, established national and regional tsunami early warning systems as a response to the event in December 2004 (IOC–UNESCO 2013).

While major emphasis was given in the first years to the development and implementation of technical early warning systems, a second phase also emerged that underscored the necessity to better prepare people at risk within the so-called 'last mile' of disaster relief and prevention operations. It was recognised that even the best hazard detection and warning system technology fails if people at the last mile are not able to respond to warnings.

In order to strengthen potential response at the last mile, awareness–raising activities became key issues to promote, alongside technical systems and fostering a culture of resilience and preparedness. One and a half years after the Indian Ocean tsunami, the Indonesian Minister of Research and Technology, Dr. Kadiman, stated that Indonesia needs 'to build structure and culture to respond to tsunamis' (*Jakarta Post* 2006).

Hence, after the Indian Ocean tsunami, scientists and policy makers (such as Kadiman) underscored that in order to build the response capacity of people to face future tsunami risks, one would need structural measures (technology for early warning systems) in addition to strategies to empower communities at risk; for example, regular public education campaigns and evacuation drills, to build a 'culture' of resilience and preparedness.

This chapter uses a notion of 'culture' that stresses processes of identity and self-actualisation, including both spiritual and religious frameworks. However, the core focus of the chapter is not on how culture modifies risk perceptions and response behaviour; instead, it is on how disaster risk reduction strategies can be embedded within an existing culture and how these risk reduction strategies can actively use cultural symbols to promote disaster resilience and preparedness. The following case study of the city of Padang provides a good illustration of how religious beliefs and symbols play an important role in building a culture of preparedness. The example may also provide a first interpretation of what a culture of preparedness might mean in real terms on the ground.

Padang case study: risk perception and cultural symbols

Introduction

After the Indian Ocean tsunami of 2004, the city of Padang, located on the western coast of the island of Sumatra, was identified as one of the most likely sites for the next serious earthquake and potentially major tsunami (Borrero *et al.* 2006). Due to the city's topography, almost half of the population of Padang lives in low-lying coastal areas; the city and its population are thus highly exposed to coastal hazards. The city received significant attention after the 2004 tsunami, and specific approaches to confront disaster risk were developed by official governmental institutions and by entities of civil society, such as NGOs. The city and province of Padang were affected by the 9.1 magnitude earthquake that caused the major tsunami in Nanggroe Aceh Darussalam in December 2004 as well as the subsequent 8.6 magnitude Sumatra earthquake of March 2005. Against this background, the United Nations University Institute for Environment and Human Security (UNU-EHS) was involved in an interdisciplinary study that dealt with improving the preparedness of people at the 'last mile' of an early warning system (cf. Shah 2006, Birkmann *et al.* 2012). The study in the city of Padang consisted of a 2008 household survey of 933 selected households representing various socioeconomic groups and levels of tsunami exposure in the city, complemented by twenty non-structured interviews (conducted mostly between June and August 2009) as well as numerous informal conversations with local actors. Beside technical infrastructures for hazard detection and warning tools in the city of Padang, various preparedness activities have also addressed building a culture of resilience of people exposed to tsunami risk (Taubenböck *et al.* 2009, 2012). This included, for example, tsunami warning response capability of the people defined and assessed through different levels of vulnerability (Birkmann *et al.* 2008, Setiadi 2014a, 2014b). In the following, selected findings of research from the 'last mile' project (Taubenböck *et al.* 2009, 2012, among others) and the GITEWS risk assessment project (GITEWS 2013) will be presented. These will examine the nexus between risk perception and response behaviour to warnings and hazard signals on the one hand and the role and importance of cultural and religious symbols for promoting rapid and effective tsunami evacuation on the other.

Responses to the potential tsunami warning and its challenges

Assessing the potential human response behaviour to specific environmental stressors or events is a major challenge, since actual behaviour in a specific event is determined by the various factors of the situation as well as factors of cognition, as outlined before and based on Tobin and Montz (1997). Moreover, in the study of Padang, analysis of the intention and willingness to evacuate also considered behavioural and cultural factors identified within theoretical approaches, such as the theory of planned behaviour (Ajzen 1991), the theory of protection motivation (Rogers 1983, Martin *et al.* 2007) and the theory regarding the protective action decision model (Lindell and Perry 1992, 2012; Setiadi, 2014b). The city of Padang served as an interesting laboratory during the study period because another major earthquake, after the Indian Ocean tsunami, occurred in September 2007, hitting the city of Padang in particular. Consequently, it was possible to compare, in a household survey, the risk perception of individuals and their subjective world with the actual behaviour of these individuals during the earthquake in September 2007. Figure 8 provides an overview of the response behaviour of people during that earthquake, which might have also triggered a tsunami, but did not. The earthquake caused about 64 fatalities, many injuries and damage to more than 30,000 houses in the province of West Sumatra (IFRC 2007). The findings of the household survey shown in Figure 8 underscore that approximately one quarter of the people interviewed in coastal locations in the city of Padang evacuated when a warning was issued, and another quarter did not receive any warning; however, most importantly, nearly half of the household interviews reported that they did not evacuate (Figure 8). This high number does not necessarily mean that those people were fatalistic; the evacuation decision had to be taken in much uncertainty under difficult circumstances (ground shaking) and in a context where precise guidance regarding evacuation procedures and information about the potential development of a tsunami was missing. Nevertheless, further investigation of the reasons for not

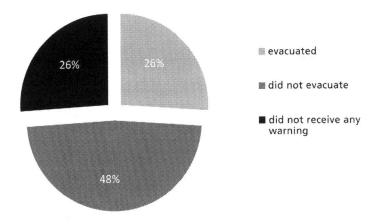

FIGURE 8 Reaction to potential tsunami warning during the September 2007 earthquake

Source: author's own analysis based on UNU–EHS Household Survey 2008.

evacuating shown in Figure 9 reveals that the main reasons for not evacuating were: a) limited trust in the warning itself, b) the assumption that the warning was only a precautionary alert without the need to act immediately, and c) the idea that it was appropriate to wait until a tsunami could be confirmed (Figure 9). However, in one expression of culture (here, in the form of a belief system), some respondents also mentioned religious aspects, such as 'trust in God' and 'following of instructions of religious and cultural leaders' as reasons why they did not follow the tsunami warning and evacuate.

The findings were also supported by the qualitative information extracted from non-structured interviews with selected households and key informants at the community level. The actual response behaviour of not evacuating was in some cases influenced by psychological factors, for example, the perceived (in)capability to evacuate. In addition, the uncertainty people faced regarding what might happen and people's assessment of their options to act (evacuation response: yes or no), which were linked both to cognitive factors (cf. Rogers 1983, Ajzen 1991, Lindell and Perry 1992, Tobin and Montz 1997, Martin *et al.* 2007, Lindell and Perry 2012) as well as to the perceived and hidden knowledge in human-environmental relations (limited knowledge of what a tsunami is) (Haggett 1991), were mentioned as a major constraint to evacuating during the event. Interviewees thus reported that they were, for example, overwhelmed by the magnitude of the earthquake they experienced and they perceived their knowledge or resources needed to quickly evacuate to a safer location as very limited. Some people even stated that evacuation would not make a significant difference in the face of major events. According to one local disaster management actor, many households made such a statement: 'If it is time to die, we would die even though we evacuate' (Interview with local actors, Padang, 2009).

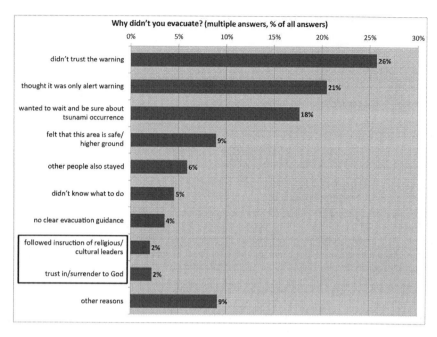

FIGURE 9 Main reasons given for not evacuating during the September 2007 earthquake

Source: authors' own analysis based on UNU-EHS Household Survey 2008.

Additionally, interviews with affected people stressed the importance of community leaders in influencing people's decisions and their response behaviour in a crisis situation or in the light of natural hazard risks, particularly from earthquakes and tsunamis.

One of the cultural values in the community is to pay respect to the community leaders. Many testimonies from the people mentioned that they evacuated if the leaders evacuated. We strongly suggest considering the capacity of community leaders and their knowledge of earthquakes and tsunamis (Transcript P 10–10:1 KO).

Consequently, it may be important to shift the focus of geographic perception research away from an analysis of the influence of cognitive processes on human responses to an environmental hazard during its occurrence and instead to focus on how cognitive aspects and cultural symbols can be pro-actively used to strengthen risk preparedness and evacuation behaviour of people in high risk zones, such as the city of Padang.

Against this background, a locally run NGO – first organised by students from a Padang university – was established in Padang after the Indian Ocean tsunami of 2004, working intensively on raising community awareness of the tsunami risk. The NGO, called KOGAMI (Komunitas Siaga Tsunami), stated that at the beginning of its awareness-raising activities it was rather difficult to find positive responses to their work. Some people even got angry and perceived the activities of the NGO to prepare for another tsunami disaster as 'inviting disasters'. On the other hand, the culture of tsunami preparedness did gradually change as more and more people perceived the activities of the NGO, as well as of the municipality and international agencies, as something that would provide them with increased capabilities in case a tsunami were to strike in the future. In this regard, the concept of culture clearly refers to knowledge and awareness about disaster risk, but also to a change of belief systems and the language used to describe such risks. Interestingly, it was particularly the involvement of community leaders that helped people to see preparedness activities in the right light (as not part of a donor programme) and allowed improvement in the effectiveness and acceptance of these activities (Interview with KOGAMI staff, July 2009). Overall, it is important to note that in Padang, the development of a culture of resilience and preparedness in the face of tsunamis was not solely steered and fostered by governmental actors but additionally, and in many cases even primarily, by non-governmental actors such as KOGAMI. Before discussing the specific challenges in using cultural symbols for disaster risk reduction and tsunami preparedness programmes, an overview of the cultural context in the city and province of Padang is provided below. The overview is not intended to be comprehensive, but it provides an important background for understanding individual risk perceptions and response behaviour.

Cultural context of the Minangkabau and the role of religion and religious leaders in disaster preparedness in Padang

First, we introduce the cultural context of our case study by using a notion of 'culture' that stresses processes of identity and self-actualisation. Often, this includes spiritual or religious frameworks as well as how risks and risk reduction issues are communicated and discussed. The majority of the population in the province of West Sumatra, including the city of Padang, belongs to the Minangkabau ethnic group. Minangkabau people have been known to define themselves as strong defenders of their culture and religion. The cultural identity of the Minangkabau is formulated as '*Adat Basandi Syarak, Syarak Basandi Kitabullah (ABS-SBK)*'. This statement was originally generated as an agreement of the Minangkabau community and leaders in the context of the civil wars at the beginning of the nineteenth century in Indonesia and meant that the culture (*adat*) of the Minangkabau is founded on the basis of the Islamic religion (*syarak*), which is based on rules and norms defined in the Quran (*Kitabullah*). This Minangkabau culture has been embedded in the various institutions and societal frameworks of West Sumatra for many decades (Abidin 2008). Most of the Minangkabau people see themselves as religious people with a strong cultural identity, and hence link their interpretations about the environment and respective environmental phenomena to the cultural and religious norms and rule systems of the Minangkabau.

This implies that the root causes and impacts of disasters are also interpreted and viewed within this cultural context (e.g., within a specific belief system). Therefore, ongoing programmes dedicated to tsunami awareness and preparedness ought to pay attention to the cultural and subjective world in which people are living. Hence, early warning systems have to be seen as complex systems that can not be reduced to the identification of processes in the objective world, such as hazard detection.

It is widely accepted that religious belief systems have a very significant influence on the risk perception of people in Padang and of Indonesia in general. Accordingly, nothing is seen to happen without God's will. One of the community stakeholders interviewed expressed the influence of religious teachings on evacuation decision making as follows: 'Some religious perceptions still influence the people's behaviour: if we do good deeds, we will be protected. If we run away, it means we are afraid or not faithful – this makes people ashamed to evacuate immediately' (Transcript P 2 – 2:15 KO).

However, religious norms and rules are not exclusively a hindrance to prepared-ness strategies. A discussion with a religious leader also indicated that religious norms can be used to promote preparedness strategies:

> Everybody [should] know, being prepared is a command from God. That it is indeed in God's hand, when disaster would happen, but we need to be alert, do our best to save ourselves. Do not commit suicide, we have to struggle.
>
> (Transcript P 4 – 4:7 AT)

The important role of religious leaders in risk communication and in actually influencing risk response behaviour and cognitive processes was clearly underscored by the results of the surveys conducted in Padang. The 2008 household survey revealed that the majority of respondents either agreed (65.4 percent) or strongly agreed (7.2 percent) with the statement that religious leaders have significant influence on the topics to be discussed by the community about tsunami risk.

Also, evacuation behaviour regarding the selection of an evacuation destination is (indirectly) influenced by cultural, and especially religious, aspects. In the 2008 household survey in Padang, 11 percent of the respondents perceived mosques as a safe place in case of potential tsunamis. In addition, in the neighbouring district, Padang Pariaman, many people would, during a strong earthquake event, seek protection at the cemetery of Syekh Burhanuddin, a respected religious leader in West Sumatra, without considering the actual safety of this location in terms of its exposure to tsunami inundation (Dewi, personal communication, 26 July 2009). This shows that cultural and religious belief systems have a strong influence on the actual evacuation and preparedness strategy of local people in Padang and Padang Pariaman. These examples underscore that organisational differentiation between the object and subject worlds can be helpful in order to understand mismatches between physical world phenomena and human responses to and interpretations of them.

One must, however, bear in mind that religious belief systems are not uniform; even in Padang, minority groups such as the Chinese (mostly Buddhist or Christian), people from Nias and North Sumatra (mostly Christian) and tribal groups may consider other factors and psychological variables that influence their attitudes to tsunami response. It is important to remember that Indonesia is a highly diverse country, home to more than 700 languages, and this also indicates a multiplicity of viewpoints and cultures. Although the research conducted did not specifically aim to examine the views and roles of minority groups, it was observed that minority groups have, in some cases, had even less access to information and disaster preparedness activities than majority groups in the city.

Cultural symbols for proactive preparedness at the community level

In the scope of community tsunami preparedness projects conducted by the KOGAMI, two model communities were identified in a village called Kelurahan Parupuk Tabing (Figure 10). The communities represented were chosen carefully, one (RW17) representing a community that would need to evacuate rapidly in the event of a tsunami, and another community (RW09) that, due to its location

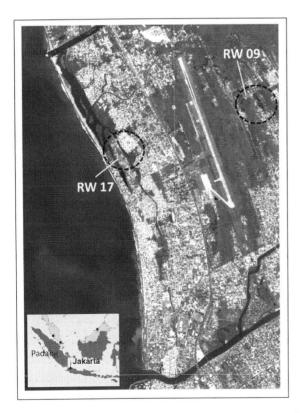

FIGURE 10 Location of two model communities in Parupuk Tabing (Padang)

Source: KOGAMI 2009.

further away from the coast, would most likely be in the position of providing shelter for the coastal community. Promoting evacuation in both the community at risk and also in the community that should serve as the host to provide shelter for other people is a challenge. Therefore, KOGAMI used religious symbols to promote acceptance of such an evacuation process (Dewi, personal communication, July 2009). In this way, the cultural context and history of the community and its symbols were taken into consideration in developing a community evacuation and preparedness strategy.

This approach for tsunami preparedness activities made use of a historical religious story about the migration of the Prophet Muhammad with his followers from Mecca to Medina. As the situation there was becoming unsafe for the Prophet, he and his followers (*Muhajirin*) left Mecca, leaving their properties and belongings behind and seeking a safe place in Medina. The people of Medina, whom they had never met before, welcomed them warmly and provided them with food and a place to stay (*Anshar*), showing their great fondness and commitment to the Prophet and also their strong sense of brotherhood with his followers. Based on these religious symbols and the norms expressed in the story, people (primarily Muslims) have been encouraged to provide help for the *Muhajirin* people (Al-Anfal verse 74) – here, interpreted as the community having to leave the unsafe place close to the coast.

Consequently, cultural symbols helped to link two different communities and, together with the religious storyline, have been an important foundation for providing arguments as to why people need to evacuate while other people in more inland communities need to help those at risk. The coastal community RW17, which in case of a tsunami needs to evacuate, has been defined by KOGAMI and local stakeholders as the *Muhajirin* community (people who travelled with the Prophet) and the community RW09, which should provide shelter for their metaphorical brothers and sisters, was defined as *Anshar* (people who helped the *Muhajirin*). The use of the religious symbols and story in cooperation with the involvement of respected religious leaders was a major tool in Padang to link both communities and to promote the acceptance of evacuation and preparedness planning. Hence it becomes evident that the broader notion of a culture of disaster resilience not only considers existing belief systems, but also critically reviews and modifies operational procedures on how response strategies for local communities are developed and articulated. Internalising individual roles in the evacuation process through religious and cultural symbols has proven to be an appropriate strategy in this case, since many people in Padang are strongly influenced by cultural, and particularly religious, symbols and norms. KOGAMI has used a religious approach in its general tsunami preparedness activities with the community and has been well accepted, since religious values provide an indubitable basis for subsequent education and preparedness activities (Dewi, personal communication, March 2012). This approach was further extended to the neighbouring district of Padang Pariaman in cooperation with Mercy Corps (PGIS 2010).

Linking risk perception, evacuation behaviour and cultural context

In a study conducted by Setiadi (2014a), several variables representing cognitive factors were tested to examine their influence on the intention of an individual to evacuate when faced with danger. Analysis revealed that the main factors influencing behaviour and perception were: basic knowledge about tsunami risk, recognition that a tsunami is a threat to oneself, recognition of personal vulnerability and the importance of preparedness, and perception of one's own capabilities and of the efficacy of evacuation. In an additional analysis, these factors were also found to influence the intention to conduct improvements in evacuation infrastructures and facilities (Setiadi, 2014b). Consideration of these factors in the development of disaster risk reduction and risk communication strategies is crucial. Doing so also provides an improved understanding of the cultural and socioeconomic contexts in which people and communities are embedded.

Based on the findings previously described, religion – understood here as one aspect of the cultural context – may have a particular influence on people's recognition of their own vulnerabilities and of the importance of preparedness (Figure 11). Religious values and the opinions of religious leaders may influence people's perceptions as to the root causes of disasters and can also determine how information about existing risks and necessary preparedness strategies is disseminated and interpreted in the community. Compared to Western societies, where religion plays a rather marginal role, the Islamic religion in Sumatra, for example in Padang and Aceh, is an overarching system of norms that structures thinking, behaviour and language. Hence, it would be false to assume that religious belief systems would represent only a minor part of culture in these societies. Overall, it is evident that religious leaders play an important role in the Muslim society of Indonesia in terms of improving risk preparedness and communication at the last mile. In return, the views of religious leaders may also influence the acceptance of and participation in necessary preparedness actions.

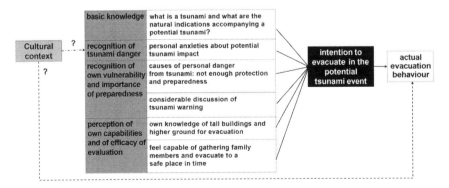

FIGURE 11 Cultural context in cognitive factors influencing intention to evacuate

Source: authors' figure, based on Setiadi 2014a.

Although belief systems that favour a rather fatalistic view on disasters can also be found in Indonesia, religious beliefs and religious leaders can provide important positive stimuli for conducting actual activities for disaster preparedness, as seen in the discussion of responses to the potential tsunami warning and its challenges above and in the example of community evacuation planning in Kelurahan Parupuk Tabing.

Summary and conclusions

This chapter has examined the role and influence of risk perceptions and cultural symbols on people's decision to prepare for tsunami risk and early warnings at the so-called 'last mile'. In contrast to studies that focus on how risk perceptions are shaped and influenced by crises or disasters, we have shifted focus by asking how cultural, and especially religious, symbols can be actively used to promote disaster resilience and lead to particularly effective evacuation during adverse conditions, such as tsunamis in Indonesia. Beside the usefulness of working with existing religious symbols of the dominant population group (Muslims) to promote a culture of resilience and preparedness in Indonesia, it is also important to note that such religious symbols might further marginalise ethnic and religious minorities, since their values and belief systems are not taken into account on a similar basis. It will be interesting to monitor the further development of these processes.

Overall, this chapter has introduced the fundamentals of geographic perception research, differentiating the real world (external) and the subjective world (internal), corresponding to the notions of constructivism. We have shown the two poles of the discussion, thereby underscoring the wide spectrum of approaches that are presently used to examine response behaviour to natural hazards and risks from a more objective and naturalistic point of view to the constructivist perspective. Against the development of geographic perception research, we argue that it is valuable to differentiate the two worlds (objective versus subjective) from an organisational point of view; however, we also underscore and show that it is even more important to focus on the interactions between these two worlds in order to understand the complexity of risk perceptions in the context of tsunamis.

Only acknowledgement of these two worlds and an understanding of how they interact in crises and disaster situations can provide a sufficient basis for developing effective preparedness strategies. Factors influencing human behaviour in such crisis situations are important elements that need to be considered in order to establish early warning systems that are truly people-centred. We showed that, particularly in the context of early warning, expert knowledge about natural hazards can be quite different from knowledge of the traditions and norms of local inhabitants that may guide their day-to-day decisions. Consequently, effective and people-centred early warning must build bridges between local, day-to-day knowledge of people at risk and the expert knowledge that is often embedded in technical approaches to hazard detection.

Based on case study research in Indonesia, the country that was hit hardest by the Indian Ocean tsunami of 2004, we have identified important gaps between the general understanding that evacuation may be needed during a tsunami and people's actual evacuation behaviour in the city of Padang in Sumatra during the region's last major earthquake in 2007. Even though an early warning was issued for this earthquake, which might have also triggered a tsunami, various inhabitants in high risk zones did not evacuate, hence underscoring the disconnect between expert knowledge and official warnings on the one hand and the local community and individual reactions to this warning and threat on the other hand. Interestingly, distrust of the official early warning and the assumption that it was only precautionary were two major reasons people mentioned for why they did not evacuate during the earthquake in 2007. In the light of these findings, we examined how religious symbols can help to promote a culture of disaster resilience, while also taking into account the specific cultural context of the Minangkabau in Padang and West Sumatra. Cultural, and particularly religious, symbols were used by a local NGO to promote both the temporary migration and evacuation of households living in the high-exposure zone close to the sea and assistance by people on higher grounds (host communities) to those displaced. The use of religious symbols and active involvement of religious leaders has been proven to be an effective approach and entry point for preparedness activities. It gives room for local people to revisit their own preparedness activities and needs within the broader religious value system (Islam), which is a dominant cultural factor in the day-to-day life of people in Sumatra. In addition, the involvement of religious leaders in raising awareness in the community to view preparedness as part of their religious deeds can be an effective way to build a 'lived' culture of resilience and preparedness. However, such an approach should be handled with care and done together with local religious leaders to ensure appropriate interpretation of religious symbols and values. In addition, the role and position of religious minorities has to be carefully considered.

Cultural and religious symbols therefore function as an important tool to communicate the necessity of temporary evacuation to people potentially affected by tsunamis and to those that may be affected by evacuations of the coastal zone. Overall, qualitative and quantitative findings underscore that religious symbols and religious beliefs still play an important role in risk perception and also in proactive preparedness strategies.

We conclude that a hybrid and interactive model concerning both the objective and subjective worlds is needed in order to understand the dynamics and interactions between these two worlds, which in the end are key in developing effective tsunami early warning and response capacities. The case study of the city of Padang on the island of Sumatra in Indonesia provided an interesting laboratory that shows that neither of the extreme poles – neither pure subjectivity nor pure objectivity – can provide sufficient information to address the so-called 'last mile'. This example underscores that preparedness strategies on the ground have to be based on a sound understanding of the physical tsunami hazard and its characteristics, as well as estimations of tsunami arrival time and inundation characteristics on the one hand, and the situational and cognitive factors that shape people's risk perception and anticipated response capacities on the other. The tsunami preparedness strategy in Padang, particularly the work of KOGAMI – a local NGO – builds a bridge between both worlds and actively links belief systems, especially religious symbols, with technical and natural science information about supposedly safe areas. Using religious symbols for promoting effective evacuation in case of a tsunami goes beyond the traditional notion of risk perception research, which often tends to focus on how people are constrained and influenced by their risk perception rather than on how to use risk perceptions and religious symbols to promote the development of a culture of resilience and preparedness.

Even though cultural symbols and norms are only one factor influencing complex decision making processes at the household level, in the case of tsunami risk in Padang it has been shown that religious belief systems are an important entry point to strengthen anticipatory activities. Since Indonesia has the largest Muslim population of any country worldwide, it is important to promote a culture of disaster resilience through religious institutions and religious leaders alongside the work of institutions and organisations officially responsible for disaster risk management. The involvement of local religious leaders is especially important to ensure appropriate interpretation and use of religious symbols in preparedness communications.

References

Abidin, M., 2008. ABS-SBK, Kesadaran Kolektif dan Norma Dasar untuk Membangun Masyarakat Beradat di Sumatera Barat [online]. Available from: http://hmasoed. wordpress.com/2008/04/08/abs-sbk-kesadaran-kolektif-dan-norma-dasar-untuk-membangun-masyarakat-beradat-di-sumatera-barat [accessed 28 March 2013].

ADPC (Asian Disaster Preparedness Centre), 2006. Regional Analysis of Socio-Economic Impacts of the December 2004 Earthquake and Indian Ocean Tsunami. Bangkok: ADPC.

Ajzen, I., 1991. The theory of planned behavior. *Organizational Behavioral and Human Decision Processes*, 50, 179–211.

Baker, E. J., 1991. Hurricane evacuation behavior. *International Journal of Mass Emergencies and Disasters*, 9, 287–310.

Berger, P. L. and Luckmann, T., 1966. *The Social Construction of Reality: A Treatise in the Sociology of Knowledge*. Garden City, NY: Doubleday.

Birkmann, J., Changseng, D. and Setiadi, N., 2012. Enhancing early warning in the light of migration and environmental shocks. *Environmental Science & Policy*, 27 (1), 76–88.

Birkmann, J., Setiadi, N. and Gebert, N., 2008. Socio-economic vulnerability assessment at the local level in context of tsunami early warning and evacuation planning in the city of Padang, West Sumatra. International Conference on Tsunami Warning, 8–11 November 2008 Nusa Dua.

Borrero, J. C., Sieh, K., Chlieh, M. and Synolkais, C. E., 2006. Tsunami inundation modeling for Western Sumatra. *Proceedings of the National Academy of Sciences of the United States of America*, 103 (52), 19673–7.

Burton, I., Kates, R. W. and White, G. F., 1993. *The Environment as Hazard*. 2nd ed. New York: Guilford Press.

Chomsky, N., 1959. A review of B. F. Skinner's verbal behavior. *Language*, 35 (1), 26–58.

Downs, R. M. and Stea, D., 1977. *Maps in Minds: Reflections of Cognitive Mapping*. New York: Harper & Row.

Gebhardt, H. and Reuber, P., 2011. Wissenschaftliches Arbeiten in der Geographie. In: H. Gebhardt, R. Glaser, U. Radtke and P. Reuber, eds, *Physische Geographie und Humangeographie*. Heidelberg: Spektrum, 81–91.

Geipel, R., 1992. *Naturrisiken. Katastrophenbewältigung im sozialen Umfeld*. Darmstadt: WBG.

Gergen, K. J. and Westmeyer, H., 2002. *Konstruierte Wirklichkeiten: Eine Hinführung zum Sozialen Konstruktionismus*. Stuttgart: Kohlhammer.

GITEWS (German Indonesian Tsunami Early Warning System), 2013 [online]. Available from: www.gitews.org [accessed 2 January 2013].

Glasersfeld, E. von, 1995. *Radical Constructivism: A Way of Knowing and Learning*. London: Falmer Press.

Haggett, P., 1991. *Geographie. Eine modern Synthese*. Stuttgart: Ulmer.

Hard, G. and Scherr, R., 1976. Mental Maps, Ortsteilimage und Wohnstandortwahl in einem Dorf an der Pellenz. *Berichte zur Deutschen Landeskunde*, 175–220.

Hidajat, R., 2002. Risikowahrnehmung und Katastrophenvorbeugung am Merapi-Vulkan (Indonesien). *Geographische Rundschau*, 54 (1), 24–9.

IFRC (International Federation of Red Cross and Red Crescent Societies), 2007. Indonesia: Bengkulu Earthquake, Information Bulletin no. 1/2007. IFRC [online]. Available from: www.ifrc.org/docs/appeals/rpts07/ideq130907.pdf [accessed 30 July 2014].

IOC-UNESCO (Intergovernmental Oceanographic Commission-United Nations Educational, Scientific and Cultural Organization), 2013, [online]. Available from: www.unesco.org/new/en/natural-sciences/ioc-oceans/single-view-oceans/news/indian_ocean_wide_tsunami_watch [accessed 2 February 2013].

IPCC (Intergovernmental Panel on Climate Change), 2012. Managing the risk of extreme events and disasters to advance climate change adaptation. Special Report of the IPCC. Cambridge: Cambridge University Press.

Jakarta Post, 2006. We Aim to Build Structure and Culture to Respond to Tsunamis. 3 August 2006.

Katz, C., 1991. Sow what you know: The struggle for social reproduction in rural Sudan. *Annals to the Association of American Geographers*, 81 (3), 488–514.

Knorr-Cetina, K., 1981. *The Manufacture of Knowledge: An Essay on the Constructivist and Contextual Nature of Science*. Oxford: Pergamon Press.

Knox, P. and Marston, S., 2001. *Context Human Geography: Places and Regions in Global Context*. New Jersey: Prentice-Hall.

Lachman, R., Tatsuoka, M. and Bonk, W. J., 1961. Human behavior during the tsunami of May 23, 1960. *Science*, 133 (3462), 1405–9.

Lindell, M. K. and Perry, R. W., 1992. *Behavioral Foundations of Community Emergency Planning*. New York: Hemisphere Publishing Company.

Lindell, M. K. and Perry, R. W., 2012. The protective action decision model: Theoretical modifications and additional evidence. *Risk Analysis*, 32, 616–32.

Lindell, M. K., Prater, C. S. and Peacock, W. J., 2005. *Organizational Communication and Decision Making in Hurricane Emergencies*. College Station, TX: Hazard Reduction & Recovery Center, Texas A&M University.

Lynch, K., 1960. *The Image of the City*. Cambridge, MA: MIT Press.

Martin, I. M., Bender, H. and Raish, C., 2007. What motivates individuals to protect themselves from risks: The case of wildland fires. *Risk Analysis*, 27 (4), 887–900.

Moore, G. T., 1976. *Environmental Knowing: Theories, Research, and Methods*. Stroudsburg, PA: Hutchinson & Ross.

PGIS (Participatory Geographic Information System), 2010. Memadukan Pengetahuan Lokal dan Kegiatan Pengurangan Risiko Bencana dalam Budaya Minang. *PGIS News*, 14 May 2010 [online]. Available from: http://pgis-sigap.blogspot.de/2010/05/pengetahuan-lokal-pengurangan-risiko.html [accessed 28 March 2013].

Piaget, J., 2007. *The Child's Conception of the World*. Lanham: Rowman & Littlefield.

Piaget, J. and Inhelder, B., 1956. *The Child's Conception of Space*. London: Routledge & Kegan Paul.

Popper, K. R., 1983. *The Logic of Scientific Discovery*. London: Hutchinson.

Reuber, P. and Pfaffenbach, C., 2005. *Methoden der Empirischen Humangeographie*. Braunschweig: Westermann.

Riad, J. K. and Norris, F. H., 1998. *Hurricane Threat and Evacuation Intentions: An Analysis of Risk Perception, Preparedness, Social Influence, and Resources*. Newark, DE: University of Delaware, Disaster Research Centre.

Rogers, R. W., 1983. Cognitive and physiological processes in fear appeals and attitude change: A revised theory of protection motivation. In: J. Cacioppo and R. Petty, eds, *Social Psychophysiology*. New York: Guilford Press.

Schütz, A. and Luckmann, T., 1975. Strukturen der Lebenswelt. Neuwied: Luchterhand.

Setiadi, N. J., 2014a. Understanding challenges at the 'last-mile' in developing an effective risk communication to reduce people's vulnerability in context of tsunami early warning and evacuation. In: Y. Kontar, V. Santiago-Fandiño and T. Takahashi, eds, *Tsunami Events and Lessons Learned (Advances in Natural and Technological Hazards Research 35)*: Springer Netherlands, 417–33.

Setiadi, N. J., 2014b. *Assessing People's Early Warning Response Capability to Inform Urban Planning Interventions to Reduce Vulnerability to Tsunamis: Case Study of Padang City, Indonesia*. Dissertation. Bonn: Rheinische Friedrich-Wilhelms-Universität zu Bonn.

Shah, H. C., 2006. The last mile: Earthquake risk mitigation assistance in developing countries. *Philosophical Transactions of the Royal Society A*, 364 (1845), 2183–9.

Taubenböck, H., Goseberg, N., Lämmel, G., Setiadi, N., Schlurmann, T., Nagel, K., Siegert, F., Birkmann, J., Traub, K.-P., Dech, S., Keuck, V., Lehmann, F., Strunz, G. and Klüpfel, H., 2012. Risk reduction at the 'last-mile': An attempt to turn science into action by the example of Padang, Indonesia. *Natural Hazards*, 65 (1), 915–45.

Taubenböck, H., Goseberg, N., Setiadi, N., Lämmel, G., Moder, F., Oczipka, M., Klüpfel, H., Wahl, R., Schlurmann, T., Strunz, G., Birkmann, J., Nagel, K., Siegert, F., Lehmann, F., Dech, S., Gress, A. and Klein, R., 2009. 'Last-mile' preparation for a potential disaster – Interdisciplinary approach towards tsunami early warning and an evacuation information system for the coastal city of Padang, Indonesia. *Natural Hazards and Earth System Sciences*, 9, 1509–28.

Tierney, K. J., Lindell, M. K. and Perry, R. W., 1999. *Facing the Unexpected. Disaster Preparedness and Response in the United States*. Washington, DC: Joseph Henry Press.

Tobin, G. A. and Montz, B. E., 1997. *Natural Hazards, Explanation and Integration*. New York: Guilford Press.

Tolman, E., 1948. Cognitive maps in rats and men. *Psychological Review*, 55 (4), 189–208.

UN/ISDR, 2005. Hyogo framework for action 2005–2015: Building the resilience of nations and communities to disasters. *UN/ISDR* [online]. Available from: www.unisdr.org/2005/wcdr/intergover/official-doc/L-docs/Hyogo-framework-for-action-english.pdf [accessed 30 July 2014].

Weichhart, P., 2008. *Entwicklungslinien der Sozialgeographie: Von Hans Bobek bis Benno Werlen*. Stuttgart: Steiner.

15

PARTICIPATIVE VULNERABILITY AND RESILIENCE ASSESSMENT AND THE EXAMPLE OF THE TAO PEOPLE (TAIWAN)

Martin Voss and Leberecht Funk

Introduction

Discussions about vulnerability and resilience within the context of disasters and climate change are based, to a significant extent, upon Western concepts. This Western discourse (Bankoff 2001) structures the practical efforts of governments, of organisations working in the field of Official Development Assistance (ODA) and of humanitarian aid agencies. From an official programmatic perspective, those affected and their needs are frequently placed at the centre of the debate. However, from a discursive angle, global, regional and local interests are seen as pitted against each other in a dispute that centres on the sovereignty of definitions and on the assertion of interests. Within this struggle, the initial differing positions of the opposing parties (affected individuals, NGOs, government organisations and global institutions, among others) result in a structural disadvantage for the vulnerable. In other words, there is a gap between official narratives and reality.

Those working in ODA today take it for granted that the participation of those affected by disasters or the effects of global environmental and climate change increases the probability for the 'sustainable' success of projects and measures. The underlying assumption asserts that the greater the success in encouraging affected persons to contribute actively to the process of defining and assessing problems, the higher the probability of achieving appropriate and robust solutions (BMZ 1999). Using the 'transdisciplinary integrative vulnerability and resilience approach (TIV)' (Voss 2008a), this chapter argues that a multidimensional approach is necessary in vulnerability and resilience research to bridge the gap between official narratives that are largely derived and connected to scientific arguments and the living realities of the people: an approach that integrates local, regional and global actors and their viewpoints in a transdisciplinary manner, i.e., an approach that is oriented towards the problems and solutions that arise from real life. At the same time, it is crucial that participation does not remain limited to the implementation of previously defined projects or measures.

Drawing upon an ethnographic example – the Tao on the Taiwanese island of Lanyu – we show that discursive and sociocultural frames define the conditions of active participation. Therefore, in order to reduce the structural inequality during the negotiation process, it is necessary for methodology and practice to reflexively take these cultural preconditions into consideration.[1]

Among the Tao people living on the Taiwanese island of Lanyu, folk models of person and emotion, of the value of life and of the conceptualisation of disaster can be found that are not in accordance with Western scientific concepts. While most disaster experts from the West would, for example, draw a clear line between the social, natural and supernatural domains on the basis of scientific concepts, the Tao perceive these worlds as one, or as overlapping each other. Consequently, the Tao's understanding of 'disasters' and their coping strategies in dealing with them are highly culture-specific.

We argue that any dialogue between 'experts' and the Tao people needs to be grounded on a basic understanding of the Tao's world view and their relationship to their environment. By simply using a top-down approach, no sustainable solutions can be found. In this chapter, we do not evaluate a concrete negotiation process that has taken place between the Tao people and governmental institutions, humanitarian aid organisations or scientists. Instead the ethnographic example of the Tao people is described as some sort of an 'ideal type',[2] which we will apply to illustrate different facets of the epistemic gap between 'scientific experts' and 'non-scientific' local people in the context of vulnerability and resilience assessments in general.

The problems we address are by no means particular to the Tao or other indigenous people worldwide; they exist in every participative process where scientific knowledge as an artificial epistemic universe converges with people's multiple and highly diverse real life perspectives, where scientific truths, e.g., of cause and effect, of a dualism of substance and accidence or of the epistemological disjunction of nature and culture, are contested. If the ambition is to assess and reduce the vulnerability of the people and to enhance their resilience (and not, what could also be an aim of governmental institutions or business organisations), the diverse perspectives such as those found among the Tao people need to be considered.

This chapter puts forward the position that a fruitful approach for assessing and weighing the vulnerability and resilience to multiple hazards has to be related to specific living conditions, social practices, and cultural belief systems; otherwise, the consequent actions will be condemned to leave the needs of the vulnerable unmet.

First, it is necessary to describe the Tao's cultural world view and their cultural models of person and disaster at length, which will be done in the ethnographic section of this chapter. A (possible) bridging of the epistemic gap may only be achieved if the reader is presented with enough material to at least imagine the Tao's dwelling perspective. We provide a brief outline of the ethnographic methods that were applied in our research. Then, we introduce the Tao and their traditional (yet still relevant) worldview, as well as their notion of souls that are not firmly anchored within the body. With the aid of two case studies, we subsequently illustrate how the Tao deal with disaster phenomena. Finally, we discuss the significance that the Tao ascribe to the spoken word, which is a matter of crucial importance for participative vulnerability and resilience research. Subsequently, we can outline the participatory approach developed by Voss (2008a) and discuss its challenges and barriers for participation in real life.

The Tao in participative processes

To make sure readers gain a general idea about the Tao's past experiences and current involvements in participatory processes, it is necessary to provide some context information about what is presently being negotiated. Participation in political decision making is a rather new experience for the Tao. During Japanese colonial times (1897–1945) and the first decades of Taiwanese rule, policies concerning the Tao and their small island were decided on and enforced by governmental officials. Japanese anthropologists were fascinated to have found an 'isolated island people' whom they treated as 'living fossils'. In order to study them and their culture, they established an anthropological reserve allowing only limited outside influences (Leach 1937). Later on, the Chinese used the island as a prison camp for criminals, forbade Tao children to talk their own language in schools and tore down the traditional semi-terrestrial houses and replaced them with low quality concrete buildings. The Tao had never been asked for their consent in any of these matters. In 1982, the Taiwanese Government built a nuclear waste disposal site on the island, telling its people that the building under construction was a 'fish cannery'. When the Tao learned about the real intended use of this site, indigenous protests arose that have continued up until the present (Fan 2006a, 2006b). The Tao's final active involvement in national politics was made possible by a democratic process that was initiated in the 1980s.

Today, the situation has changed to the better. The 'gap' between the Tao and the Taiwanese still remains, but the government is now promoting indigenous culture and participation in social affairs. There is a growing number of Western-style educated Tao who have spent many years on the main island of Taiwan to obtain secondary education and to earn money, which they need to secure a livelihood on their home island. They are familiar with Chinese culture and Western scientific concepts and, as such, fulfil the basic conditions for participating in negotiation processes with privileged 'experts'. On a national level, there are currently ongoing debates about the establishment of a 'Sea Park' on Lanyu, the culture-sensitive reorganisation of tourism (which today is the major source of income for many Tao), indigenous education programmes within school curricula, the rebuilding of the island's landscape, land rights, natural resources and sustainability (Sutej Hugu 2012), illegal intrusion on tribal fishing grounds and overfishing by Chinese fishermen, and the permanent replacement of nuclear waste from Lanyu.

On an international level, educated Tao attend indigenous rights conferences and take part in exchange programmes with Pacific Island nations with whom they share an Austronesian heritage.

Cultural models of person and disaster among the Tao (Taiwan)

Using the example of the Tao, this chapter aims to illustrate some specific problems that can arise when applying a participatory approach to a vulnerability and resilience assessment in the field. To avoid misunderstandings, it shall be repeated that what we argue here could have been exemplified with a participatory vulnerability assessment somewhere in Europe as well. Many structural problems emerging from the gap between global scientific discourse and everyday life perspectives would be comparable, although the specific contents of what is being negotiated would probably be different.

Field research and methodology

The findings presented here were based on one year of ethnographic field research among the Tao between September 2010 and August 2011. The research was conducted as part of the interdisciplinary research project 'Socialisation and Ontogeny of Emotions in Cross-Cultural Comparison', located at the scholarly intersection of social and cultural anthropology with developmental psychology. One key area of research of this project focused on cultural models of person, emotion and emotion regulation.[3]

In addition to traditional ethnographic methods such as participant observation and semi-structured interviews, the emotional vocabulary of the Tao was surveyed employing free listing of sample sentences in the Tao's native language, *ciriciring no Tao*.[4] In order to contrast and compare conceptual knowledge with actual behaviour, more than 250 observation records of emotional episodes were also drawn up. Thus, an extensive body of data was compiled, allowing both qualitative and quantitative conclusions.

The Tao and their worldview

The Tao are an ethnic group numbering 3,500 persons, whose ancestors travelled from the Philippine Batan Islands around 800 years ago and settled on the volcanic island of Lanyu, 45 nautical miles southeast of Taiwan.[5] Japanese policy during its colonial period (1897–1945) pursued non-interference and, thus, the Tao were widely able to maintain their traditional way of life (Leach 1937). However, since the 1950s, a transformation process has taken hold that has led to social inequality and to serious intergenerational differences. While the young are at least superficially integrated into Taiwanese society as a whole, thanks to their long term residence in boarding schools and exposure to national media, their grandparents are still subsistence farmers and live a life that is firmly anchored to traditional structures. Under these circumstances, it is difficult to make general statements about the Tao. Nevertheless, it is possible to determine deeper commonalities that continue to hold significance for most Tao, whether young or old.

During the first half of the twentieth century, Japanese anthropologists began to conduct lively research activities on Lanyu, though their emphasis was not on the exploration of cosmological or religious notions.[6] Early insights into these topics are found in the work of American psychiatrist Kilton Stewart, who spent several months on Lanyu in 1937 (Stewart 1947, del Re 1951). During the 1950s, Inez de Beauclair, a German anthropologist living and working in Taiwan, embarked upon several research expeditions to Lanyu, which included systematic research into cosmological and religious concepts (de Beauclair 1957, 1959, de Beauclair and Kaneko 1994). This distinctly Western first 'wave' of research was followed by surveys carried out by Taiwanese anthropologists (cf. Wei *et al.* 1972), which have been supplemented more recently by studies conducted by indigenous authors (cf. Dong 1997). In her monograph about changes in the Tao's religious practice, Sinan-Jyavizong (2009) – a native of Lanyu – presents an overview of the results of these research activities.

The studies listed above do not provide a consistent record of the name, location and function of the Tao's divine beings. Nevertheless, the authors agree upon a number of salient points, which can be summarised as follows: in the upper realms of heaven there exists a Supreme Being[7] that either directly observes the doings of the people on earth or delegates this task to Messenger Gods, who provide regular reports. When the Supreme Being hears of the misdeeds of mankind, it becomes angry (*somozi*) and administers punishments upon the people on earth (Sinan-Jyavizong 2009, pp. 70–1). Normal language usage gathers all deities under the umbrella term of *Tao do to* (literally: people up high). It is believed that the deities live in villages and lead a life there that is similar to that of the Tao (de Beauclair 1959, p. 17). In addition to the heavenly realms populated by divine beings and the human world here on earth, there is an underworld beneath these. All three strata house evil spirits, referred to as *Anito* by the Tao.

The Tao are deeply afraid of the *Anito*, invisible creatures that lurk everywhere and seek to thwart the plans of mankind. In earlier times, Tao men wore helmets and armour made of rattan as a measure of protection against these odious spirits, which they tried to drive off with the use of spears. This is not something one would observe in daily life today, though it remains common practice when a death has occurred. According to the view of the Tao, any phenomena that depart from what is considered normal can be attributed to the machinations of the *Anito*. In general, these evil spirits are regarded as cosmic forces wreaking destruction upon human life and its foundations (Sinan-Jyavizong 2009, p. 79).

After death, the vital soul of a person is said to travel to a small 'white island' that lies to the south of Lanyu, while several other of the deceased person's attributes (or lesser souls) remain in the vicinity of the village where henceforth they lead an existence as ghosts. The Tao distinguish between the spirits of the dead of their own kinship group and those belonging to other descent groups, as well as those that can no longer be assigned to a specific group. Only spirits of the latter two kinds are referred to as *Anito* and the Tao choose respectful forms of address for their own ancestors (Sinan-Jyavizong 2009, p. 75).[8] Although it is possible to appeal to ancestors in times of need, these figures also represent an element of danger, as they can quickly become envious (*manginanahet*) and – especially if their deaths occurred not long ago – yearn to be close to the living members of their group. Desperate, they try to 'tempt' the living by contaminating them with diseases, which in severe cases can lead to death. This means that even when dealing with one's own spirits, extreme caution and a certain distance is essential; these are ambivalent creatures and one must behave in a reserved and controlled manner towards them.

The term *Anito* can also be used to refer to a specific category of celestial spirits that are sent to earth by the Supreme Being in order to punish the people for their wrongdoings. This punishment is exacted, for example, by inciting the earthly *Anito* to spoil the people's crops. A plague of caterpillars that besets the sweet potato fields can therefore be seen as retribution for past misdeeds or the violation of taboos. Certain animals, such as an endemic species of owl, or a butterfly that only exists on Lanyu, are regarded as manifestations of these celestial spirits.

The Tao assume that the human and the spirit worlds exist in parallel and are closely interwoven. Though many Tao have converted to Presbyterianism or Catholicism since the 1950s, the foundations of their traditional cosmological notions have only partly shifted.

The Tao's conceptions of the soul

According to the understanding of the Tao, man is endowed with one vital soul and several additional lesser souls, the number of which varies between three and eight depending on the source quoted (Sinan-Jyavizong 2009, p. 73). The seven-soul model places the vital soul, *pahad*,[9] in the head and the other six in the shoulder, elbow and knee joints (Guo 1988, pp. 70–1; quoted in Sinan-Jyavizong 2009, p. 74). While shoulder souls supply the body with strength, elbow souls are responsible for defending against evil spirits. Knee souls, meanwhile, ensure that a person is assiduous and a diligent worker. All souls can leave a person's body temporarily and, in some cases, even permanently, prior to death. It used to be said (and some still believe this today) that when people are 'lazy' and not engaged in any activity, their elbow and knee souls have left them. Freely roaming souls are in danger of encountering an evil spirit wanting to take possession of them. It is believed that dreams are triggered by the movements of wandering souls and that the person concerned is having a nightmare when his or her soul(s) are seduced by an *Anito*. If a roaming soul manages to quickly break free from the attacking *Anito* spirit and thus cast off the spirit's evil influence, the affected person will suffer from headaches or mild illness from which he or she will usually recover swiftly. However, if the roaming soul succumbs to the force of the spirit being and is unable to break free from its influence, it is said that the *Anito* have stolen the soul; the result of this will be serious illness that may lead to death. Mental confusion and aggressive behaviour are also attributed to the influence of the *Anito*.

When the soul is stolen by the *Anito*, physical symptoms of fear such as a racing heart rate or altered breathing set in. This 'fear of the soul' (*maniahey so pahad*) can even lead to complete immobilisation of the body, during which it is not possible to move one's arms and legs. The indigenous belief is that the elbow and knee souls have left the body at this stage. Regulating such states of anxiety is thus essential for survival, according to the Tao. Because their souls are not yet firmly anchored within their bodies and are willingly distracted by the *Anito*, the souls of infants and young children are particularly vulnerable to the threat of malicious spirits. Such spirits therefore try to tempt children into dark corners or into the wild away from the village where they might easily get injured or lost and experience very intense states of fear. Only once they reach approximately 12 years of age (after completion of the six-level primary school) is a young person regarded as being mentally, emotionally and physically strong enough to resist evil spirits (Funk *et al.* 2012, p. 227, Röttger-Rössler *et al.*, forthcoming).

Cultural models of disaster

In order to establish a measure of protection against the regular summer occurrence of typhoons, the Tao used to erect semi-terrestrial houses of an ingenious design that could withstand high winds. Traditional ecological knowledge (TEK) allowed the Tao a decent livelihood on their island – at least, during normal times. However, normality was repeatedly called into question by the regular reappearance of periods of drought, which brought with them salinisation of the soil and pest infestations in the taro and sweet potato fields. At regular intervals, famines occurred and coincided with a collapse of the social order, as each lineage fought for its own survival. Japanese colonial officials who followed a policy of non-interference did not get too involved in the affairs of the Tao, at least as long as no open outbreaks of fighting occurred among them. The confusion of the Second World War led to a relative absence of administrative measures in the 1940s, so that indigenous ways of law enforcement continued up until the mid-twentieth century. Before and during this period of time, it was also not uncommon for the elderly or the feeble to die in their fields from exposure to the icy winter winds (Yu and Dong 1998, pp. 79–81). Over eighty years ago, an outbreak of cholera functioned as a disastrous event, apparently wiping out entire villages (Yu and Dong 1998, p. 49 and pp. 85–6).

Apart from devastating storms and earthquakes, floods and landslides occur occasionally on Lanyu and pose a further threat to human existence. The threat is compounded by radioactive leakage from the nuclear waste disposal site – as the Tao had long been suspecting – and which was officially confirmed in 2011. A combination of unfortunate events may lead to high contamination rates that could make the whole island uninhabitable for the Tao – a possibility that after the Fukushima incident seems even more likely today.

Typhoon

At the height of summer in 2003, construction work was being carried out in the bay below a village on the eastern coast of Lanyu. Previously, at low tide it had been tricky to pass over a shallow bank at the entrance to the bay. This did not present a major problem when using traditional canoes because, if one knew about the shallow bank, one could simply wait for a suitable wave and make passage even when the water was low. However, this approach no longer worked with modern motor boats used today. The village council therefore decided to remove parts of the underwater rocky embankment in order to allow for the safe and smooth passage of modern vessels. Engineering work on the project was done by an excavator able to reach the relevant area of the bay by using a freshly made gravel track. Just a few weeks after construction work had been completed, Typhoon Dujuan tore across Lanyu and caused severe destruction on the island. No people were hurt, but houses, boats, cars and parts of the island's infrastructure were destroyed.

The older inhabitants of the village saw a clear causal link between the construction measures and the devastating typhoon: the *Tao do to* deities had been outraged at the human arrogance that led people to change the rock formations in the bay as they saw fit. The rocks in the bay have their own toponyms and some are believed to be inhabited by spirits. Human intervention in the cosmological order had provoked the fury of the Supreme Being, which resulted in the retaliatory typhoon. Even though younger individuals nowadays doubt that these connections exist, they will generally not openly contradict their elders. Many taboos remain intact because they are taboos. At issue is not the personal attitude of people to these taboos (the critical questioning of a taboo is in itself regarded as an act of disobedience); instead, respect for the ancestors who issued these taboos is what counts most and is of central importance here.

Death

Deaths are frightening events with far-reaching consequences for the local Tao community. Particularly in cases when a death occurs outside of the home, an aggressive rebellion of the spirit world is to be expected.[10] As soon as the body of the deceased is found, necessary precautions are taken. First, all children are called home, as they are most vulnerable due to their weakly anchored souls. The *Anito* spirits smell the scent emitted by the corpse and gather at both the site of death and near the village. The relatives of the deceased cannot return the body to the village, as this is considered too dangerous. Instead, the body is immediately brought to the natural graveyard, which is located in a coastal forest in the vicinity of the village[11] and interred without delay. The men in the funeral procession wear helmets and armour made of rattan to protect their souls against the *Anito* spirits. The graveyard features an extremely high concentration of spirits of the dead and is therefore regarded as a particularly dangerous place that is not to be entered under normal circumstances. For the great risks they have taken upon themselves, the members of the funeral procession must be well-compensated, often with gold leaf or a taro field.

Using rope and wooden boards, a so-called 'spirit path' is laid to the home of the deceased. At dusk, the spirits of the dead will take this path to gather at the house of the deceased. Laying this path is an attempt to steer the spirits along fixed paths and limit the risk of contamination from them as much as possible. At dusk, armed with ritual daggers and spears, the men of the village make their way to the forest graveyard, where, shouting loudly and thrusting their spears, they advance three times against the entrance of the burial ground.

For three days after a death, public life in the village is suspended and noisy activities are forbidden. The production of food is also prohibited during the mourning period. In this way, villagers pay their respects to the family of the deceased. A death represents a liminal phase during which the presence of evil spirits in the village can easily lead to the escalation of negative interpersonal relations and socially disruptive emotions. If the boundaries between the world of man and the world of spirits become permeable, human emotions are subject to manipulation. Evil spirits attack the souls of people and make them susceptible to feelings of rage and envy. Instances of death are potentially dangerous events that interrupt daily life and feature significant levels of complexity and insecurity. There is an ever present threat that they may attract further bad events (*marahet so vazey*). Whenever a person in the village dies, a greater risk for the entire community ensues. This situation is analogous to that of a swollen river that has overflown its banks so greatly that it might spill over a dike and flood the village at any moment.

The significance of the spoken word among the Tao

The Tao believe that words have an intrinsic power and will seek their object. Those who make false accusations or who utter disproportionately severe curses are in danger of being hit by their own words and human speech always carries the latent character of blessings or curses. Young, uninformed people tend to keep silent rather than speaking out their opinion. There is a cultural claim on attuning verbal speech to social reality; from the indigenous point of view, this is a necessity as cosmic chaos may otherwise result. The following example serves as an illustration: if I address a person who is already a grandfather (*Siapen*) with the genealogical term for 'father' (*Siaman*), the consequence of my wording is that the grandchildren of the man who has been degraded from 'grandfather' to 'father' no longer exist. Effectively, I would have uttered a death curse on his grandchildren.

Consequently, referring to 'bad things' (*marahet so vazey*) by name should be avoided if at all possible, as their power increases the more they are talked about. When faced with a typhoon, it is necessary that people remain calm and an outbreak of panic must be prevented at all costs. The men of Lanyu know when to retrieve their boats from the water and carry them ashore, though today they receive relevant information about the weather from television. From within their homes, the men carefully watch to see who goes where. Without spoken announcement or invitation to meet at the landing place, a growing number of men gather there and soon everyone is assembled and the boats are brought ashore. In the narrow and manageable world of the village it is not necessary to speak; careful observation provides those who have the requisite cultural knowledge with everything they need to know. By 'not speaking about it', the impact of a negatively assessed event is not further intensified and the vulnerable area that allows the *Anito* spirits to attack is kept as small as possible.

Vulnerability and resilience assessment in social science and social ecological research on climate change and disasters

The voices of the marginalised, the hegemonic discourse and capital

Social science research on climate change and disasters investigates which people are particularly impacted by such events and asks how, why, and by which 'unwanted' processes these people are affected. For many years, researchers mainly focused on 'natural' hazards posing a significant threat to humans in specific geographical regions and how capacities for protection against these could be improved.[12] For the most part, however, the real living conditions of the affected people were not included in such considerations. Gradually, social and economic factors began to gain attention. The 'social vulnerability approach' (cf. Blaikie *et al.* 1994, Adger *et al.* 2004, McEntire 2004) states that those who have very few resources at their disposal are forced to live in relatively dangerous areas and circumstances, while those who are better off have resources that allow them to secure their living conditions. Seen from this perspective, the primary cause of unevenly distributed vulnerability is not 'dangerous nature'. Instead, vulnerability is first and foremost a consequence of social and economic exclusion (Hilhorst and Bankoff 2004). As a result, these excluded individuals have fewer opportunities to make their life circumstances more sustainable or 'robust' and are ultimately left more exposed to the 'dangerous' forces of nature.

Within the context of socio-ecological research, this view of vulnerability has been further expanded. Changes in the environment can have a positive or negative impact on human life; conversely, social practices have an effect upon the environment. In accordance with this socio-ecological view, the analysis of vulnerability must also consider interconnections that are non-human, but that are influenced by man and retroact upon man. Thus, integrative approaches are required in order to gain a better understanding of the complicated arrangement of socio-cultural, economic and ecological aspects relevant for the wellbeing of the people at risk, and to align actions more effectively.

The Transdisciplinary Integrative Vulnerability and Resilience Approach (TIV) developed by Voss (2008a) takes up these various research traditions from vulnerability and resilience research (e.g., the so-called 'natural hazard', 'social vulnerability' and 'social ecological' approaches), combines them into an integrative approach and expands on it by adding a discursive element. It is argued that social or, to be more precise, political processes of discursive exclusion have largely been neglected in the practice of research and application, although these processes exert a significant influence on vulnerability (Voss 2008a). Accordingly, one important factor with regard to vulnerability is that the vulnerable have no 'voice'. Their worldviews, needs and interests are paid no heed in the hegemonic discourse, as their chosen forms of articulation are simply neither acknowledged nor understood by those with power and influence.[13] Many Tao, for example, are not used to speaking up for themselves and instead prefer to give voice to their elders. What from a Western perspective would be judged as a 'healthy form of assertiveness' is for them 'arrogant' and 'proud' behaviour that can hardly be tolerated, since it leads to anger and envy among fellow-villagers. From an epistemological point of view, approaches that fail to embrace this discursive dimension remain entangled in a Western discourse. Inadvertently, these approaches contribute to the reproduction of existing global, regional and local inequalities and to an attendant increase in vulnerability (Voss 2008a).

Within the framework of social science research in disasters and climate change, it has long been emphasised that vulnerability cannot be equated with poverty in the monetary sense, although this represents vulnerability's most significant driver. When assessing livelihoods, researchers in this field therefore avail themselves of an expanded definition of capital that encompasses social (networks), human (education, skills and aptitudes), physical (infrastructure) and natural (land, forest, water, among others) capital, as well as its economic form (DFID 1999–2005). In sociology, a differentiated concept of capital tends to be associated with one name in particular, namely Pierre Bourdieu, who describes actors' options for action as dependent on the ownership of economic, cultural, and social capitals that are unequally distributed across social space (Bourdieu 1984). Furthermore it is important for this approach that the value of a specific capital is not objectively defined but depends on concrete sociocultural conditions. Among the Tao, maybe not surprisingly, money is not what counts most. Although the Tao are today in desperate need of cash, which is hard to earn on their island, they still stress social relationships more than sheer monetary value. In other words, they have not forgotten that 'one cannot eat money' and that livelihood depends to a high degree on the support of one's relatives and ancestral spirits. The connection to ancestral spirits, or 'supernatural capital' as Bourdieu might have called it, is usually not considered to be an important and meaningful dimension in Western discourses. Consequently, the hegemonic debate can be conceived as a discourse that is dominated by the privileged, who have at their disposal an above-average level of resources, or resources of precisely those types of capital regarded as most valuable in the social space, which in turn endows them with an above-average capacity to shape conditions as they see necessary. The social positions of the (power) elites and of the marginalised must be viewed as the result of this social process.

The TIV approach: framework and methodology for transdisciplinary integrative vulnerability and resilience assessments

Motivation behind the approach

The TIV approach aims to increase the opportunities for the vulnerable to add their own point of view regarding their vulnerability and resilience to relevant assessments and consequently to strengthen their position in such discourse. The active participation of those affected therefore becomes the centre focus and should ensure a fairer position in the negotiation process between global, regional, and local interests. However, even phrasing a question about 'vulnerability' and 'resilience' incontrovertibly stems from Western discourse. As a rule, those directly affected are not the ones who request such an assessment or define the methods needed to carry it out. Instead, based on a variety of motives, third parties representing the fields of politics, science, NGOs, or business are usually those who consider assessments of this kind as necessary, who define the most suitable methods according to their own opinion and who subsequently present these methods to those they have identified as 'affected'. In terms of an 'official programme', the vulnerable are thus placed at the focus of attention. In reality, however, they are assigned a passive position while others define what is at stake and how to go about improving their living conditions or reducing their vulnerability. This has far-reaching consequences for the whole process. Among the Tao, for example, the intervention of outsiders (*dehdeh*) in local affairs is strongly disliked, at least when the rules for interference are defined by external agents. In the past, the Tao have had very negative experiences with the Japanese colonial government and Taiwanese official policies prior to democratisation. Outsiders are not informed about ancestral lands and lack essential knowledge about the island's ecological sustainability. Hence, almost any decisions taken by outsiders have led to disastrous consequences for the Tao's eco-environment. Today Lanyu faces serious environmental problems such as soil contamination and water shortages.

The objective of the TIV approach is to mediate between global, regional and local perspectives, and thereby taking into consideration discursive or resource dependent inequalities in the initial bargaining positions of participants. Based on establishing fairer initial positions, the approach aims at achieving a more democratic structure of opportunities to express and shape assessment of problems that have a supra-regional impact, such as disasters and the consequences of climate change. This requires that stakeholders' diverse perspectives, views on a problem, and their opinions about relevant variables with regard to vulnerability and resilience, have to be brought together. Within the context of TIV, stakeholders are those presumed to be vulnerable, but are also the political decision makers, representatives of NGOs and technical experts, among others, who participate in the assessment process. It is anticipated that stakeholders have varying notions of what constitutes vulnerability and resilience, as well as which resources and means should be used in what manner in order to achieve their specific objectives. Because the aim is mediation between these different perspectives and implicitly or explicitly associated interests, it is first necessary to develop an overview of the assumedly relevant factors and their interrelations.

In the first step of the TIV approach, local affected persons should be supported by facilitators (e.g., NGOs) in their search for suitable participative methods,[14] which are then used to design an initial visualised conceptual vulnerability map.[15] Views of problems or 'hazards' that are anchored in real life should be expressed without being pre-emptively structured.[16] At this stage, any definition of vulnerability that leans towards a particular approach is explicitly avoided; instead, a very generic understanding of vulnerability as influenced by poverty, inequality, health, access to resources, and social status (Brooks 2003) is taken as a starting point for open discussions.

In the next step, the real life perspectives are compared with a visualised and scientifically sound generic list of potentially relevant factors that influence vulnerability (see Figure 12). What is sought with these two visualisations of the rather concrete real life and the rather abstract science based perspective is not yet an assessment of factors. Instead, these are an attempt to make diverse perspectives and their implications transparent. Visualisation is widely accepted as a technique to democratise decision making processes respecting different viewpoints while balancing the need for professional or expert input (Salas *et al.* 2007). Visualisation of different perspectives provides a starting point for an iterative process of rapprochement. Referring to our example of the Tao people, a striking point will be the significance of the spoken word, which cannot be reduced to a 'technical language'. The Tao's connection to the spiritual world is important and so are the words chosen when communicating about relational issues that may include other people, ancestral land and human entanglements with other than human life forces (such as certain animals and plants, as well as diverse spiritual beings).

In a joint search involving all stakeholders as equals, overlaps should now be identified and roughly sketched factors should be further substantiated. As a result, some or many of the factors present in either conceptual map may be discarded entirely, or may only require minor modification. Suitable concepts should be sought, discussion should explore how these can, in practice, be expanded upon in the local cultural context and it should be investigated how these concepts seem to interrelate. Finally, the relative significance of each identified variable should be assessed.

The components and generic indicators of the TIV framework

With regard to the scientific part of the assessment, the TIV framework proposes an analytical distinction between four components: units of reference, hazards, social and environmental conditions, and resilience.

In general, a participative vulnerability analysis using this framework begins with the selection of a unit of reference, which determines whether the analysis will focus on individuals, households, local communities, classes, organisations, entire nations, animal or plant species, or whole ecosystems, for example.

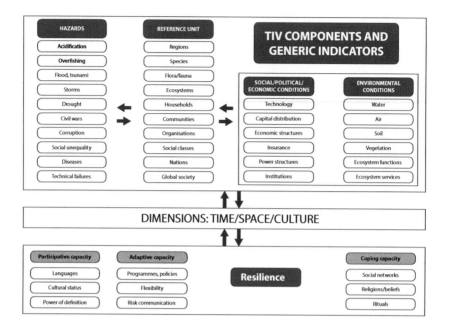

FIGURE 12 Components and generic indicators of the TIV framework

This leads to the question of which specific hazards[17] this unit is exposed to, how frequently, for how long and to what extent this exposure to hazards occurs. The term 'hazard' describes all biophysically influencing variables that have been discussed at length in disaster research: tectonic movements, heavy precipitation, spring tides and landslides, among others. However, other forms of hazard, such as financial market turbulence, food speculation, and civil war, must also be taken into account. Additionally, it is important to consider that these diverse hazards do not have distinct and isolated effects, but instead exert influence upon each other in terms of space, time, and culture. This means that effects can prove destructive in other locations and at later points in time, as the example of climate change clearly illustrates.

Hazards are the result of complex interactions between social, political, economic, and environmental conditions. Therefore, the qualities of a reference unit with regard to hazards must be assessed in relation to these conditions. As far as society is concerned, all of the following play a part in hazard analysis: capital endowment and distribution (economic, cultural, and social capital), fundamental social institutions, civil society, economic structures, perception of risks and dangers, level of insurance coverage, and the state of technological development, among others. In terms of the environment, issues could include, for example, the present state of ecosystems in comparison with the past, or natural capital (e.g., land, climate, mineral deposits) and biophysical property rights, as well as their distribution (e.g., land, climate, mineral deposits, structure and the function of ecosystems).

Resilience is integrated into the TIV approach as a fourth important component. Contemporary resilience discourse has been shaped by C. S. Holling's seminal 1973 article, in which he concluded that the relative stability of an ecosystem results from the complex and non-linear interaction of multiple variables over scales of space, time, and culture (Voss 2008a). The dynamic interplay of these variables forms a stability domain, also called a 'basin of attraction'.[18] The term 'resilience' thus indicates 'the size of stability domains, or, more meaningfully, the amount of disturbance a system [the 'basin of attraction'] can take before its control shifts ['regime shift'] to another set of variables and relationships that dominate another stability region' (Folke 2006, pp. 254–5). As such, ecosystems are complex and 'multistable' (cf. Holling 1973, Holling 1996, Gunderson and Holling 2002).

Assuming that nature and society build a more or less invariant and culture independent order, vulnerability research has to date focused on individual and relatively isolated factors (poverty, natural hazard, among others) (Voss 2006). The resilience approach instead draws attention to the complex interplay of a multitude of factors that create a dynamic type of steady state – the basin of attraction. Thus, 'stressors' cannot be assessed objectively, but only in relation to this basin of attraction, the composition of which is in turn the result of prior interactions between a wide variety of stressors. The resilience of a reference unit depends on its capacity to cope with stressors and perturbations of various kinds – to absorb or to process them so that organisational performance (structure and function) remains intact (cf. Adger 2000, p. 351, among others).[19] To be precise, resilience is not a characteristic of a unit of reference; instead, the term refers to a certain, highly demanding relationship between the unit of reference and its environment, one that ensures the structural identity of the organisational performance over time. In the Tao's basin of attraction, for instance, we can find an animist framework or spiritual dimension that is widely missing in Western societies (although not completely absent). For the Tao it is impossible to neglect the existing and established connections with the environment. Accordingly, the Flying Fish, a ritually significant animal and important staple food, has to be treated with utmost respect and may only be fished, prepared, and consumed under the observation of a multitude of taboos (*makanyo*). If the ways of the ancestors are not followed anymore the spiritual beings who call the fish swarm into the island's waters every spring will be offended and declining results from fishing will be the consequence. One could never understand what is of central importance for the structural identity of the Tao without taking this spiritual dimension into account.

This strictly relational orientation is particularly well illustrated through discussions about 'capacities'. In the literature on resilience research, authors frequently differentiate between adaptive capacities and coping capacities, though interpretation of these terms varies widely (cf. Davies 1996, Alwang *et al.* 2001, Adger *et al.* 2004, Adger 2006, among others). In the TIV approach, adaptive capacity represents the reference unit's facility for adapting to stress (i.e., through changes within political programmes or policies, or through improved forms of risk communication). These adaptations can be guided either in an active-reflexive manner, i.e., through conscious learning processes, or they can be triggered through trial and error.

What is significant is that adaptation measures cannot be assessed in isolation from a consideration of the structural genesis and development within the unique situation, i.e., separate from their environment. The unit under investigation can be considered resilient if its performance level is not substantially lowered by internal or external stressors, meaning that adaptations remain in the realm of what can be expected in a specific sociocultural and natural context, i.e., a 'normal' range. However, the meaning of the term 'normal' here can only be grasped with an understanding of the structural elements of the reference unit itself.

The so-called 'coping capacity' makes greater levels of stress bearable. The term refers to all forms of (constructive) processing of crises, disasters, and stress in general. Coping strategies allow a linkage of occurrences that are initially extraordinary with familiar, regular processes. Even those processes that appear clearly disastrous to an observer may still be imbued with meaning through coping strategies. They are extraordinarily important in vulnerability research because they force a culture-specific assessment of supposedly objective risks. The occurrence of death among the Tao is a case in example: it is the pollution of the corpse that threatens the order and integrity of the village; attracted by the sickening smell of the dead body, malicious spirits approach the village and the household of the bereaved. Extreme caution and a variety of ritualised practices are now necessary in order to cope with the dangerous situation. If things get further out of control, more deaths may occur.

The addition of participative capacity (Voss 2008a) now augments these two capacities (to adapt and to cope) with the ability to (co-)design discourse, based on the assumption that the capacity itself – or lack thereof, as briefly outlined above – significantly impacts the vulnerability of a reference unit. Participative capacity comprises two complementary elements. On the one hand, it is important to consider the sum of capitals that are relevant in a particular context and that are at the disposal of an individual. On the other hand, the value of these capitals is dependent on whether the field – or, more aptly in this case, the discourse – acknowledges the bearer of the capitals.[20] Capital can be used to acquire more recognition, provided that, for instance, ideological or racist reasons do not lead to an absolute denial of recognition, in which case any kind of capital ownership is meaningless.

In this regard, cultural capital is of particular significance, especially in its 'incorporated' form (Bourdieu 1992). It is important to both understand the discourse and to be able to move well within it – that is, according to the rules of the discourse. This requires not only language skills, but also a fundamental understanding of the 'semiotics of power' – when to address which statement to whom, whose proximity to seek out and which 'camp' to join in order to reap greater benefits in the long term. Although there is a growing number of Western-style educated Tao who may participate in national and international discourses, they too are faced with problems of how to overcome the epistemic gap between their own traditional society and that of 'experts', by describing and translating the Tao's worldview and dwelling perspective in the language of the latter. The crucial point that is not well understood by 'experts' is the spiritual nature of human engagement with the environment. Seen from this perspective, ecological sustainability on a small island is not something that can be achieved by applying natural sciences, but rather something that is embodied within people who fear and respect spiritual beings. This indeed is a very different position from that of the modern global market and consumer society; moreover, it is difficult to translate this perspective into the language of science.

Discussion

The TIV approach for the surveying and assessment of socio-ecological vulnerability and resilience assumes that the vulnerability and resilience of people are variables that cannot be objectively defined or measured. Instead, in social practice, we are dealing with political concepts and disputed terms, which in turn become weapons used by actors who have differing world views and who operate at different levels, leveraging unequally distributed resources, and pursuing a wide range of diverse strategies. Using this assumption as a starting point, the TIV approach aims for a democratisation of negotiation processes, which should include the recording and assessment of vulnerability and resilience. Participatory capacity is the focus of this approach, it being the unequally distributed opportunity to make oneself heard within negotiation processes and thus, ultimately, to wield influence. According to this constructivist understanding, vulnerability is not in the first instance a function of objectively identifiable influencing variables and of their interrelations, but is instead first and foremost the result and the producer of the discourse that assigns actors to positions in the social and physical space. However, if these influencing variables cannot simply be objectively determined, then the philosophy of the approach dictates that the aim of a vulnerability and resilience assessment must lie in a process that is 'free of domination', by which the conditions of participation are designed through shifting emphasis on participatory methods. Freedom from domination is an absolute ideal that cannot be realised in practice, as the example of the Tao exemplifies; this is certainly true for all other societies, too. It would be contrary to the intentions of the TIV approach if it did not refer to its own limitations. Using the case study of the Tao, some of these limitations have been mentioned here, showing that participation is a very demanding process with many preconditions and extends far beyond the mere inclusion of stakeholders. Some fundamental barriers will be summarised here that make participation in a globalised world such a difficult endeavour.

In the global hegemonic discourse about climate change and disasters, the dichotomy between nature and culture is fundamental. This dichotomy has no indigenous counterpart among the Tao. While Western culture views 'nature' as 'uninaturalist', it would be more appropriate to speak of 'multinaturalism' with regard to the Tao (Viveiros de Castro 1998).

Western discourse, not only among the sciences but also at an everyday linguistic level, is dominated by an informed rational view of disaster phenomena. This view is incapable of integrating the all-encompassing spiritual dimension that is so fundamental to the Tao's worldview and engagement with the environment. It only looks for causal, observable or experimentally verifiable cause and effect relationships. According to this discourse, 'societies' or 'communities' are seen as stable and objectively describable systems that are exposed to specific 'natural' hazards. Consequently, this perspective primarily seeks solution through technology. This contrasts the Tao's notion of a supernatural penetration of life, or in the jargon of the resilience approach, their constant (re)adjustment within a 'basin of attraction'. For them, human existence is made possible by the cultural knowledge transmitted by the ancestors in the form of taboo (*makanyo*) and a constant and careful observation and interpretation of 'natural' phenomena. The resilience approach in this regard opens a different view that does not simply negate modes of living that are more distant to the Western culture based concepts of a scientific mainstream.

The mere 'participation' of the Tao in the context of a vulnerability assessment or in the search for solutions to disasters and climate change would ignore the conditions that initially cause vulnerability. The example of the Tao illustrates the extent to which this highly hegemonic discourse can conflict with people's everyday life perspectives and priorities. Active participation must therefore take place at all levels, from setting the agenda to the selection and design of methods, and on to the definition and evaluation of measures perceived as necessary. Neither the local worldview nor that of the decision makers should be assigned an apolitical status that leads to one perspective rising above another. Before vulnerability and resilience are recorded and assessed in a supranational or even global context – on climate change, for instance – it is necessary to search for opportunities and ways to ensure the fair design of these assessments. The TIV approach is seen here not as the sole solution, but as a contribution to this discussion.

Notes

1 Our focus lies on research relating to the assessment and evaluation of vulnerability and resilience within the context of research into climate change and disasters. However, we believe that our arguments can also be transferred to other contexts. This applies especially, but not exclusively, to the realm of ODA and includes research into poverty or hunger, or work on watershed management.

2 The sociologist Max Weber created the 'ideal type' as a constructed ideal to accentuate most relevant elements, e.g., of a specific social problem that in reality never appears in this clarity (1949). In this sense, ideal types are simplifications. Within this article we simplify in the sense that we reduce the description of the Tao's cultural beliefs and social practices on a few selected, although important, aspects while omitting others. The denser this description becomes the less distinct and visible the generalising assumptions we made within this article would become.

3 The project was funded by the German Research Foundation and realised as part of the research cluster 'Languages of Emotion' at the Freie Universität Berlin.

4 Today this language is spoken by the elderly, while middle-aged individuals are mostly bilingual and the younger generation communicates exclusively in Mandarin.

5 For detailed ethnographic information refer to Kano and Segawa 1956, Benedek 1987 and Yu 1991. Very briefly, we would like to mention that the Tao traditionally plant taro and sweet potato and attach great economic and ritual significance to fishing. The kinship system is essentially bilateral with weakly defined patrilineages.

6 One exception is the work of the Japanese linguist Erin Asai (1936), though his survey of mythological knowledge focused on linguistically correct representation and did not ask after the precise meaning of religious ideas.

7 It is not clear if there is only one Supreme Being or many of them. Informants in the field would sometimes suggest that there is a plurality of high gods living in the upper realms of heaven. As the Supreme Being(s) are perceived as distant deities, people generally state that they do not know much about them.

8 Thus, a deceased father is referred to as 'first gone' (*pinanma tao*) and once a mother has died she becomes 'she who embraced and nurtured me' (*nizozongan*) (Sinan-Jyavizong 2009, p. 75).

9 In daily life, the remaining six souls are also called *pahad* (Mandarin: *linghun*). It is not known whether other terms also exist for them.

10 During the period of field research, there were three deaths in the village being investigated, one of which was the 'terrible death' described here.

11 Each of the six island villages has their own graveyard. The belief in *Anito* is so strong that no Christian cemeteries have been erected until now.

12 Within modern mainstream disaster research, vulnerability is still defined as the exposure of one reference unit to primarily natural risks (the so-called 'risk hazard approach' (cf. Turner *et al.* 2003), but also the 'natural hazard approach' (cf. Adger 2006). Climate research, which to a large extent follows a natural scientific orientation, limits the exposure even further to extreme weather and climate phenomena. One significant point of criticism in relation to this approach is that it neglects the characteristics of its reference unit.

13 Frequently, neither the privileged nor the excluded are conscious of these subtle mechanisms (partly semiotically anchored and affecting their patterns of perception) that they use to define who negotiates what and when, what is the subject of negotiation and what conditions apply (the '4 Ws' for setting the agenda; Voss 2008a, p. 42, 2008b).

14 Numerous other methods have been developed and tested, ranging from visualisation techniques such as 'metaplans' and 'scenario workshops', to site visits, all the way to 'participatory photography', which have all been directed towards helping individuals from different cultural and social settings and endowed with different skills and aptitudes to express themselves and introduce their views into decision and design processes. To read more about this we refer to the website of FAAN (2014), which gives an assorted overview.

15 There is no general rule how to visualise this initial conceptual map. Familiar techniques should be used, which could be paintings or photographs or similar or simple lists of supposedly relevant factors as described here in step 2 (see Figure 12).

16 Ideally, facilitators have lived and worked for long periods in the local communities concerned; otherwise, important information about the vulnerability of those who are usually excluded from the public domain might be missing. It is a false assumption that all relevant information can be solely obtained by means of interviewing. Age, gender and social status of informants are important markers that have to be kept in mind when mapping conceptual local vulnerabilities.

17 'Hazards' is used here as a general term and includes both self-caused and perceived risks, as well as dangers that are caused by others and about which little or nothing is known.

18 A 'basin of attraction' is a region in a stated space in which the system tends to remain. For systems that tend towards an equilibrium, the equilibrium state is defined as an

'attractor' and the basin of attraction constitutes all initial conditions that will tend towards that equilibrium state (Walker et. al. 2004).

19 As such the term is entirely neutral, though this is often overlooked. It does not become normative until it is, for example, linked to a specific concept of sustainability.

20 Recognition is certainly meant in the philosophical sense here for, ultimately, humans must be unconditionally respected as equals. The more this is the case the less prejudice is involved when differing arguments meet during discourse (cf. Habermas 1987). In reality, however, genuine and absolute recognition is a rare phenomenon (cf. Taylor 1994).

References

Adger, N., 2000. Social and ecological resilience: Are they related? *Progress in Human Geography*, 4 (3), 347–64.

Adger, N., 2006. Vulnerability. *Global Environmental Change*, 16, 268–81.

Adger, N., Brooks, N., Bentham, G., Agnew, M. and Eriksen, S. E. H., 2004. *New Indicators of Vulnerability and Adaptive Capacity*. Norwich: Tyndall Centre for Climate Change Research, 7.

Alwang, J., Siegel, P. B. and Jørgensen, S. L., 2001. Vulnerability: A view from different disciplines. *World Bank Social Protection Discussion Paper Series, 0115* [online]. Available from: www-wds.worldbank.org/servlet/WDSContentServer/WDSP/IB/2002/01/17/ 000094946_01120804004787/Rendered/PDF/multi0page.pdf [accessed 20 March 2014].

Asai, E., 1936. *A Study of the Yami Language – An Indonesian Language Spoken on Botel Tobago Island*. Leiden: J. Ginsberg.

Bankoff, G., 2001. Rendering the world unsafe: 'Vulnerability' as Western discourse. *Disasters*, 25 (1), 19–35.

Benedek, D., 1987. *A Comparative Study of the Bashiic Cultures of Irala, Ivatan, and Itbayat*. Thesis (PhD). Pennsylvania State University.

Blaikie, P., Cannon, T., Davis, I. and Wisner, B., 1994. *At Risk: Natural Hazards, People's Vulnerability, and Disasters*. London: Routledge.

BMZ, 1999. Übersektorales konzept. Partizipative entwicklungszusammenarbeit. *Partizipationskonzept*. BZM-Konzepte Nr. 102, September 1999.

Bourdieu, P., 1984. *Distinction. A Social Critique of the Judgment of Taste*. Cambridge, MA: Harvard University Press.

Bourdieu, P., 1992. *Language and Symbolic Power*. Cambridge, UK: Polity Press.

Brooks, N., 2003. *Vulnerability, Risk and Adaptation: A Conceptual Framework*. Norwich: Tyndall Centre for Climate Change Research, 38.

Davies, S., 1996. *Adaptable Livelihoods. Coping with Food Insecurity in the Malian Sahel*. New York: Macmillan Press.

de Beauclair, I., 1957. Field notes on Lan Yü (Botel Tobago). *Bulletin of the Institute of Ethnology of the Academia Sinica*, 3, 101–16.

de Beauclair, I., 1959. Die Religion der Yami auf Botel Tobago. *Sociologus*, 9 (1), 1–23.

de Beauclair, I. and Kaneko, E., 1994. Götter und Mythen der Yami. In: Schmalzriedt, E. and Haussig, H. W., eds, *Wörterbuch der Mythologie. Götter und Mythen Ostasiens. Bd. VI.* Stuttgart: Klett-Cotta, 372–84.

Del Re, A., 1951. *Creation Myths of the Formosan Natives*. Tokyo: Hokuseido Press.

DFID, 1999–2005. Sustainable livelihoods guidance sheets. *Department for International Development (UK), London* [online]. Available from: www.livelihoods.org/info/info_guidancesheets.html [accessed 10 February 2013].

Dong, S., 1997. *Yameizu Yuren Buluo Suishi Jiyi (Seasonal Rituals of the Yuren Tribe of the Yami).* Nantou: Taiwansheng Wenxian Weiyuanhui.

Egner, H., Schorch, M. and Voss, M., eds, 2014. *Learning and Calamities. Practices, Interpretations, Patterns.* New York/London: Routledge, forthcoming.

FAAN, 2014. Literature on Participatory Methods. *FAAN* [online]. Available from: www.faanweb.eu/bibliography/literature-participatory-methods [accessed 20 March 2014].

Fan, M., 2006a. Nuclear waste facilities on tribal land: The Yami's struggles for environmental justice. *The International Journal of Justice and Sustainability*, 11 (4), 433–44.

Fan, M., 2006b. Environmental justice and nuclear waste conflicts in Taiwan. *Environmental Politics*, 15 (3), 417–34.

Folke, C., 2006. Resilience: The emergence of a perspective for social–ecological systems analysis. *Global Environmental Change*, 16 (3), 253–67.

Funk, L., Röttger-Rössler, B. and Scheidecker, G., 2012. Fühlen(d) Lernen. Zur Sozialisation und Entwicklung von Emotionen im Kulturvergleich. *Zeitschrift für Erziehungswissenschaft* 15, 217–38.

Gunderson, L. H. and Holling, C. S., eds, 2002. *Panarchy. Understanding Transformations in Human and Natural Systems.* Washington: Island Press.

Guo, J., 1988. *Yameizu Jinji Wenhua de Xinyangguan Yanjiu (Research on the Yami's Beliefs about Taboo Culture).* Thesis (PhD). Yu-Shan Theological College and Seminary.

Habermas, J., 1987. *The Theory of Communicative Action.* Cambridge, UK: Polity Press.

Hilhorst, D. and Bankoff, G., 2004. Mapping vulnerability. In: Bankoff, G., Frerks, G. and Hilhorst, D., eds, *Mapping Vulnerability: Disaster, Development and People.* London: Earthscan: 1–9.

Holling, C. S., 1973. Resilience and stability of ecological systems. *Annual Review of Ecology and Systematics*, 4, 1–23.

Holling, C. S., 1996. Engineering within ecological constraints. In: Schulze, P. C., ed, *Engineering within Ecological Constraints.* Washington, DC: National Academy Press: 31–44.

Kano, T. and Segawa, K., 1956. *An Illustrated Ethnography of Formosan Aborigines, Vol. 1, The Yami.* Tokyo: Maruzen.

Leach, E. R., 1937. The Yami of Koto-sho – A Japanese colonial experiment. *The Geographical Magazine*, 5 (6), 417–34.

Lovejoy, T. E. and Hannah, L. J., 2005. *Climate Change and Biodiversity.* New Haven, CT: Yale University Press.

McEntire, D. A., 2004. Development, disasters and vulnerability: A discussion of divergent theories and the need for their integration. *Disaster Prevention and Management: An International Journal*, 13 (3), 193–8.

Röttger-Rössler, B., Scheidecker, G. and Funk, L., forthcoming. Learning (by) feeling: A cross-cultural comparison of the socialization and development of emotions. *Ethos*, pending.

Salas, M. A., Tillmann, H. J., McKee, N. and Shahzadi, N., 2007. *Visualisation in Participatory Programmes: How to Facilitate and Visualise Participatory Group Processes.* Dhaka: Southbound in Association with United Nations Childrens Fund (UNICEF).

Sinan-Jyavizong, 2009. *Dawuzu Zongjiao Bianqian yu Minzu Fazhan (Reform of Religion and Development of the Society of the Ethnic Group of Tao).* Taipeh: Nantian.

Stewart, K. R., 1947. *Magico-Religious Beliefs and Practices in Primitive Society – A Sociological Interpretation of their Therapeutic Aspects.* Thesis (PhD). London School of Economics and Political Science.

Sutej Hugu, 2012. *Seeking a Revival for Sustainability of an Island for the Tao People*. Paper presented at the ICCA Consortium at the 11th Conference of the Parties to the Convention on Biological Diversity, Hyderabad (India), October 2012 [online]. Available from: www.iccaconsortium.org/wp-content/uploads/images/stories/Database/issues examples/ taiwan_revival_of_sustainability_article.pdf [accessed 20 March 2014].

Taylor, C., 1994. The politics of recognition. In: Taylor, C., Anthony Appiah, K., Habermas, J., Rockefeller, S. C., Walzer, M., Wolf, S. and Gutmann, A., eds, *Multiculturalism. Examining the Politics of Recognition*. Princeton, NJ: Princeton University Press, 25–74.

Turner, B. L., Kasperson, R. E., Matson, P. A., McCarthy, J. J., Corell, R. W., Christensen, L., Eckley, N., Kasperson, J. X., Luers, A., Martello, M. L., Polsky, C., Pulsipher, A. and Schiller, A., 2003. A framework for vulnerability analysis in sustainability science. *National Academy of Sciences*, 14 (100), 8074–9.

Viveiros de Castro, E., 1998. Cosmological deixis and Amerindian perspectivism. *Journal of the Royal Anthropological Institute*, 4 (3), 469–88.

Voss, M., 2006. *Symbolische Formen. Grundlagen und Elemente einer Soziologie der Katastrophe*. Bielefeld: Transcript.

Voss, M., 2008a. The vulnerable can't speak. An integrative vulnerability approach to disaster and climate change research. *Behemoth. A Journal on Civilisation*, 3, 39–56.

Voss, M., 2008b. Globaler Umweltwandel und Lokale Resilienz am Beispiel des Klimawandels. In: Rehberg, K. S., ed., *Die Natur der Gesellschaft. Verhandlungen des 33. Kongresses der Deutschen Gesellschaft für Soziologie in Kassel 2006*. Frankfurt a.M.: Campus: 2860–76.

Walker, B., Holling, C. S., Carpenter, S. R. and Kinzig, A., 2004. Resilience, adaptability and transformability in social-ecological systems. *Ecology and Society*, 9 (2), 5 [online]. Available from: www.ecologyandsociety.org/vol9/iss2/art5/ [accessed 20 March 2014].

Weber, M., 1949. Objectivity in social science and social policy. In: Weber, M., ed., *The Methodology of Social Sciences*. Translated and edited by Edward Shils and Henry Finch. Glencoe, IL: Free Press, 49–112.

Wei, H., Yu, J. and Lin, H., 1972. *Taiwansheng tongzhigao: Yameizu pian* [Historiography of Taiwan: The Yami]. Nantou: Taiwan Provincial Committee of Historiography.

Yu, G., 1991. *Ritual, Society, and Culture among the Yami*. Thesis (PhD). University of Michigan.

Yu, G. and Dong, S., 1998. *Taiwan Yuanzhuminshi: Yameizu Shipian (The History of Formosan Aborigines: Yami)*. Nantou: Historical Research Commission of Taiwan Province.

INDEX